Ernst Kunz

Kähler Differentials

Advanced Lectures
in Mathematics

Edited by Gerd Fischer

Jochen Werner
Optimization. Theory and Applications

Manfred Denker
Asymptotic Distribution Theory
in Nonparametric Statistics

Klaus Lamotke
Regular Solids and
Isolated Singularities

Francesco Guaraldo, Patrizia Macrì
Alessandro Tancredi
Topics on Real Analytic Spaces

Ernst Kunz
Kähler Differentials

Ernst Kunz

Kähler Differentials

Springer Fachmedien Wiesbaden GmbH

CIP-Kurztitelaufnahme der Deutschen Bibliothek

Kunz, Ernst:
Kähler differentials / Ernst Kunz. — Braunschweig;
Wiesbaden: Vieweg, 1986.
(Advanced lectures mathematics)

ISBN 978-3-528-08973-3   ISBN 978-3-663-14074-0 (eBook)
DOI 10.1007/978-3-663-14074-0

AMS Subject Classification 12F, 13B, 13H, 14B 05, 14F 10

1986

Produced by W. Langelüddecke, Braunschweig

PREFACE

This book is based on a lecture course that I gave at
the University of Regensburg. The purpose of these lectures
was to explain the role of Kähler differential forms in
ring theory, to prepare the road for their application in
algebraic geometry, and to lead up to some research problems.
The text discusses almost exclusively local questions and
is therefore written in the language of commutative algebra. The translation into the language of algebraic geometry
is easy for the reader who is familiar with sheaf theory
and the theory of schemes.

The principal goals of the monograph are: To display the
information contained in the algebra of Kähler differential
forms (de Rham algebra) of a commutative algebra, to introduce and discuss "differential invariants" of algebras, and
to prove theorems about algebras with "differential methods".
The most important object we study is the module of Kähler
differentials $\Omega^1_{S/R}$ of an algebra S/R. Like the differentials
of analysis, differential modules "linearize" problems, i.e.
reduce questions about algebras (non-linear problems) to
questions of linear algebra.

We are mainly interested in algebras of finite type.
Results about arbitrary algebras are only given, if no
extra effort is necessary. However, by working with the
technic of "admissible derivations of the ground ring",
I have tried to present the results free from "separability
assumptions". Later (in § 16) traces of differential forms
are constructed in the case of locally complete intersection
algebras. These traces are strongly related to "regular

differential forms", "residues" and "duality theory", sub-
jects that are not touched upon in this book, except for
the one-dimensional case (§ 17). But I hope to come back
to the higher dimensional applications of the theory else-
where.

For standard results of commutative algebra the general
references are Matsumura [M_1] and Bourbaki [B_2]. In the early
sections of the book, in which the functorial properties of
derivation modules and differential algebras are described,
only very little is needed. Then gradually we have to rely
more and more on facts from other sources. Appendices A-G
collect, for easier reference, some material of ring theory,
mainly about complete intersections and traces. These appendices of together 96 pages do no require knowledge of the
rest of the book and can be read at the beginning. Appendix A
describes the language we use and introduces some notation.
The reader should begin by having a brief look at this
appendix.

I wish to thank the participants of my lectures and semi-
nars, and the readers of preliminary versions of these notes,
for their critical interest and valuable comments. Thanks
are also due to my present and former students S.Brüderle,
T.Grünler, R.Hübl, Dr.H.Knebl, B.Köck, Dr.J.Koch, M.Kreuzer,
G.Seibert, and Dr.R.Waldi, who have read part of the manus-
cript and have corrected many errors. Last but not least
I have to thank Frau Eva Rütz for the patience and skill
she showed while preparing various versions of this text.

Regensburg, February 1986

Ernst Kunz

CONTENTS

APPENDICES

§ 1. Derivations

In this section we shall discuss the notion of derivation. Some examples of derivations are given, and some basic rules about derivations are proved. Important invariants of an algebra are its derivation module and its module of (Kähler) differentials, which will be introduced at the end of this section.

Let R_0 be a ring, R an R_0-algebra and M an R-module. M can be regarded as an R_0-module via the structure homomorphism $R_0 \to R$.

1.1. Definition. A <u>derivation</u> of R/R_0 in M is a mapping $d : R \to M$ having the following properties:
a) d ist R_0-linear.
b) (Product formula) For all $a,b \in R$

$$d(ab) = adb + bda.$$

Such derivations will also be called R_0-derivations. In case $R_0 = \mathbb{Z}$ derivations of $R/\mathbb{Z}$ will simply be called derivations of R (absolute derivations). Since R itself is an R-module, we may consider in particular R_0-derivations $d : R \to R$.

<u>Examples.</u>
1.2. Examples from analysis.
Let $U \subset \mathbb{R}^n$ be a non-empty open set, $R := \mathcal{E}(U)$ the $\mathbb{R}$-algebra of C^∞-functions $f : U \to \mathbb{R}$. Then the partial differential operator

$$\frac{\partial}{\partial X_i} : \mathcal{E}(U) \to \mathcal{E}(U) \qquad (f \mapsto \frac{\partial f}{\partial X_i})$$

is a derivation of $\mathcal{E}(U)/\mathbb{R}$.

For $x \in U$ let $R := \mathcal{E}_x$ be the $\mathbb{R}$-algebra of germs of C^∞-functions in x. We have an induced mapping

$$\frac{\partial}{\partial X_i} : \mathcal{E}_x \to \mathcal{E}_x$$

which sends the germ $[f]$ of a C^∞-function f in x to the germ $[\frac{\partial f}{\partial X_i}]$ of its partial derivative.

$\mathbb{R}$ may be regarded as $\mathcal{E}_x$-module via the homomorphism $\mathcal{E}_x \to \mathbb{R}$, $[f] \to f(x)$. Then also

$$\Delta_i : \mathcal{E}_x \to \mathbb{R} \quad , \quad \Delta_i[f] := \frac{\partial f}{\partial X_i}(x)$$

is an $\mathbb{R}$-derivation.

Similar examples occur in complex analysis: If $R := \mathcal{O}(U)$ is the $\mathbb{C}$-algebra of holomorphic functions on a non-empty open set $U \subset \mathbb{C}$, then the derivative gives a $\mathbb{C}$-derivation of R into R.

1.3. Trivial derivations.

For any ring R and R-module M

$$d : R \to M \qquad (dr = 0 \text{ for all } r \in R)$$

is a derivation, which is called <u>trivial</u>.

1.4. Formal partial derivative in polynomial and power series algebras.

Let $R := R_o[\![X_1,\ldots,X_n]\!]$ be the R_o-algebra of formal power series in indeterminates $X_1,\ldots,X_n$. For a power series $F = \Sigma \rho_{\alpha_1\ldots\alpha_n} X_1^{\alpha_1}\ldots X_n^{\alpha_n}$ $(\rho_{\alpha_1\ldots\alpha_n} \in R_o)$ the <u>formal partial derivative</u> $\frac{\partial F}{\partial X_i}$ is defined by

$$(1) \qquad \frac{\partial F}{\partial X_i} := \Sigma \alpha_i \rho_{\alpha_1\ldots\alpha_n} X_1^{\alpha_1}\ldots X_i^{\alpha_i-1}\ldots X_n^{\alpha_n}.$$

The mapping

$$\frac{\partial}{\partial X_i} : R \to R \qquad (F \mapsto \frac{\partial F}{\partial X_i})$$

is an R_o-derivation, as is easily checked.

Let $\{dX_1, \ldots, dX_n\}$ be a set of indeterminates and let $M := RdX_1 \oplus \ldots \oplus RdX_n$ be the free R-module on the basis $(dX_1, \ldots, dX_n)$, the module of "formal differentials". Then the mapping

$$d : R \to M \qquad (F \mapsto \frac{\partial F}{\partial X_1} \cdot dX_1 + \ldots + \frac{\partial F}{\partial X_n} \cdot dX_n)$$

is an R_o-derivation of R into M. dF is called the <u>formal differential</u> of F.

If $F \in R_o[X_1, \ldots, X_n]$, the polynomial ring in $X_1, \ldots, X_n$ over R_o, then $\frac{\partial F}{\partial X_i} \in R_o[X_1, \ldots, X_n]$ and $dF \in \bigoplus_{i=1}^{n} R_o[X_1, \ldots, X_n]dX_i$. Hence $\frac{\partial}{\partial X_i}$ and d induce R_o-derivations on the polynomial ring.

In the case $n = 1$ we write for $F \in R_o[\![X]\!]$ also $F'(X)$ instead of $\frac{\partial F}{\partial X}$. Observe that <u>formal integration</u> in $R_o[\![X]\!]$ is only possible if $\mathbb{Q} \subset R_o$.

1.5. The Euler derivation of a graded ring. Let $R = \bigoplus_{n \in \mathbb{Z}} R_n$ be a graded ring. Then R_o is a subring of R. For $r \in R$ let $r = \sum_{n \in \mathbb{Z}} r_n$ be the decomposition into homogeneous elements $r_n \in R_n$. The mapping

$$d : R \to R \qquad (r \mapsto \sum_{n \in \mathbb{Z}} nr_n)$$

is an R_o-derivation, as is easily checked. It is called the <u>Euler derivation</u> of R.

1.6. Induced derivations on subalgebras.
Let R/R_o be an algebra, S/R_o a subalgebra of R/R_o such that R has a basis $\{w_\lambda\}_{\lambda \in \Lambda}$ as an S-module. Let $d : R \to R$ be

an R_0-derivation. For any $r \in R$ we write

$$dr = \sum_{\lambda \in \Lambda} d_\lambda r \cdot w_\lambda$$

where $d_\lambda r \in S$ is the (unique) coefficient of the representation of dr as a linear combination of the basis elements w_λ. Then

$$d_\lambda : S \to S \quad (s \mapsto d_\lambda s)$$

is an R_0-derivation for any $\lambda \in \Lambda$.

Clearly, d_λ is R_0-linear. Moreover, for $s_1, s_2 \in S$
$d(s_1 s_2) = \sum_{\lambda \in \Lambda} d_\lambda (s_1 s_2) w_\lambda$ and $d(s_1 s_2) = s_1 ds_2 + s_2 ds_1 = \sum_{\lambda \in \Lambda} (s_1 d_\lambda s_2 + s_2 d_\lambda s_1) w_\lambda$, hence d_λ satisfies also the product formula.

1.7. The universal derivation.

As is well-known, the kernel I of the mapping

$$\mu : R \otimes_{R_0} R \to R \quad (a \otimes b \mapsto a \cdot b)$$

is generated by the elements $r \otimes 1 - 1 \otimes r$ $(r \in R)$. $R \otimes_{R_0} R$ has two R-module structures, given by the ring homomorphisms $R \to R \otimes_{R_0} R$ $(a \mapsto a \otimes 1)$ and $R \to R \otimes R$ $(a \mapsto 1 \otimes a)$. They induce two R-module structures on I. From those, we get induced R-module structures of $M := I/I^2$, which, however, coincide: For $r \in R$, $a \otimes 1 - 1 \otimes a + I^2 \in I/I^2$ we have
$r \otimes 1 \cdot (a \otimes 1 - 1 \otimes a + I^2) = 1 \otimes r \cdot (a \otimes 1 - 1 \otimes a + I^2)$, since $(r \otimes 1 - 1 \otimes r) \cdot (a \otimes 1 - 1 \otimes a) \in I^2$.

Define a mapping

$$d : R \to M \quad (a \mapsto a \otimes 1 - 1 \otimes a + I^2).$$

Then clearly d is R_0-linear. Moreover, for $a, b \in R$

$$d(ab) = ab \otimes 1 - 1 \otimes ab + I^2 =$$
$$(a \otimes 1) \cdot (b \otimes 1 - 1 \otimes b + I^2) + (1 \otimes b) \cdot (a \otimes 1 - 1 \otimes a + I^2) =$$
$$(a \otimes 1) \cdot (b \otimes 1 - 1 \otimes b + I^2) + (b \otimes 1) \cdot (a \otimes 1 - 1 \otimes a + I^2) = adb + bda.$$

The special importance of this derivation will be discussed later (Theorem 1.19).

We are going to prove some general <u>rules about derivations</u>.

1.8. If R_O is an algebra over a ring R_O', then any R_O-algebra (any R_O-module) is also an R_O'-algebra (an R_O'-module). Clearly any R_O-derivation is also an R_O'-derivation. In particular any R_O-derivation is a $\mathbf{Z}$-derivation.

1.9. Let $d : R \to M$ be an R_O-derivation of R into an R-module M. Then:

a) $d(1) = 0$, hence $d(n \cdot 1) = 0$ for all $n \in \mathbf{Z}$.

b) $d(x_1 \cdot \ldots \cdot x_n) = \sum_{i=1}^{n} x_1 \cdot \ldots \cdot \hat{x}_i \cdot \ldots \cdot x_n dx_i$ for all $x_1, \ldots, x_n \in R$, in particular
$$dx^n = nx^{n-1}dx \qquad \text{for all } x \in R.$$

c) $P := \ker d := \{\rho \in R \mid d\rho = 0\}$ is a subring of R, and d is a derivation of R/P. The image of R_O in R is contained in P.

d) For any ideal I of R and $n > 0$ we have
$$d(I^n) \subset I^{n-1} \cdot M.$$

e) Quotient formula. If $s \in R$ is a unit, then for any $r \in R$
$$d(rs^{-1}) = (s^{-1})^2(sdr - rds)$$
in particular
$$ds^{-1} = -(s^{-1})^2 ds.$$

f) Suppose $N \subset R_O$ is multiplicatively closed and $R_O \to R$ induces a ring homomorphism $(R_O)_N \to R$. Then d is also a derivation of $R/(R_o)_N$.

g) Derivation of polynomials. Let $f \in R_o[X_1,\ldots,X_n]$ be a polynomial, $x_1,\ldots,x_n \in R$. Then

$$df(x_1,\ldots,x_n) = \sum_{i=1}^{n} \frac{\partial f}{\partial X_i}(x_1,\ldots,x_n)\,dx_i.$$

h) Derivation of determinants. Let $\Delta = \det(a_{ik})$ be the determinant of an $n{\times}n$-matrix (a_{ik}) with coefficients $a_{ik} \in R$. Then

$$d\Delta = \sum_{i,k=1}^{n} \Delta_{ik}\,da_{ik}$$

where $\Delta_{ik} := (-1)^{i+k}A_{ik}$ and A_{ik} is the determinant obtained from Δ by deleting the i-th row and k-th column.

Proof. a) We have $d(1) = d(1 \cdot 1) = 1 \cdot d(1) + 1 \cdot d(1)$, hence $d(1) = 0$. Since d is $\mathbb{Z}$-linear, we obtain $d(n \cdot 1) = n \cdot d(1) = 0$ for all $n \in \mathbb{Z}$.

b) The first formula is obtained from the product formula by induction, the second follows, if we put $x_1 = \ldots = x_n = x$.

c) Clearly, P is a subgroup of $(R,+)$. The product formula shows that P is a subring and that d is P-linear. For $r_o \in R_o$ we have $d(r_o \cdot 1) = r_o d(1) = 0$, hence P contains the image of R_o in R.

d) The elements of I^n are finite sums of product $x_1 \cdot \ldots \cdot x_n$ with $x_i \in I$ $(i=1,\ldots,n)$. The assertion follows from b).

e) From $0 = d(1) = d(s \cdot s^{-1}) = s\,ds^{-1} + s^{-1}ds$ we obtain $ds^{-1} = -(s^{-1})^2 ds$. The quotient formula follows now by another application of the product formula.

f) By e) the image of $(R_o)_N$ in R is contained in ker d. So c) can be applied.

g) Let $f = \Sigma \rho_{\alpha_1,\ldots,\alpha_n} X_1^{\alpha_1} \ldots X_n^{\alpha_n}$ $(\rho_{\alpha_1 \ldots \alpha_n} \in R_o)$.

Since d is R_o-linear, we have

$$df(x_1,\ldots,x_n) = \Sigma\rho_{\alpha_1\ldots\alpha_n} d(x_1^{\alpha_1}\ldots x_n^{\alpha_n})$$ and from b) we obtain

$$d(x_1^{\alpha_1}\ldots x_n^{\alpha_n}) = \sum_{i=1}^{n} \alpha_i x_1^{\alpha_1}\ldots x_i^{\alpha_i-1}\ldots x_n^{\alpha_n} dx_i.$$ Hence

$$df(x_1,\ldots,x_n) = \sum_{i=1}^{n}(\Sigma\alpha_i\rho_{\alpha_1\ldots\alpha_n} x_1^{\alpha_1}\ldots x_i^{\alpha_i-1}\ldots x_n^{\alpha_n})dx_i$$

$$= \sum_{i=1}^{n}\frac{\partial f}{\partial X_i}(x_1,\ldots,x_n)dx_i$$ (observe that this proof does not

work for power series).

h) If $\Delta = \det(X_{ik})$ is a determinant with indeterminates X_{ik} (i,k=1,...,n) as coefficients, then $\frac{\partial \Delta}{\partial X_{ik}} = \Delta_{ik}$, as is seen by expanding the determinant with respect to the i-th row. The formula in h) follows from g).

Examples of algebras with only trivial derivations are given in the following two propositions.

1.10. <u>Proposition.</u>   Let K/K_o be an algebraic field exten-sion and K' the subfield of all separable elements of K/K_o. Then $K' \subset \ker d$ for any K_o-derivation $d : K \to M$. In particu-lar, if K/K_o is separably algebraic, any K_o-derivation $d : K \to M$ is trivial.

Proof.   For $x \in K'$ let $f \in K_o[X]$ be the minimal polynomial of x over K_o. Since f is separable, we have an equation $f = (X-x)\cdot g$ with $g \in K'[X]$, $g(x) \neq 0$. Then $f'(x) = g(x) \neq 0$ by the product formula for derivations. By 1.9

$$0 = d(0) = d(f(x)) = f'(x)dx,$$

hence $dx = 0$ and $x \in \ker d$.

1.11. Proposition. Let R be a ring whose characteristic is a prime number p. Then $R^p \subset \ker d$ for any derivation $d : R \to M$. In particular, if K is a perfect field of characteristic $p > 0$, any derivation $d : K \to M$ is trivial.

Proof. For $r \in R$ we have $dr^p = pr^{p-1}dr = 0$. If K is a perfect field of characteristic $p > 0$, then $K = K^p$.

For an algebra R/R_o and an R-module M let $\mathrm{Der}_{R_o}(R,M)$ denote the set of all R_o-derivations $d : R \to M$. Clearly, for $d,d' \in \mathrm{Der}_{R_o}(R,M)$ and $r \in R$ also

$$d+d' \in \mathrm{Der}_{R_o}(R,M) \text{ and } rd \in \mathrm{Der}_{R_o}(R,M)$$

hence $\mathrm{Der}_{R_o}(R,M)$ is an R-module.

1.12. Example. Let X be a C^∞-manifold, $x \in X$, and let $\mathcal{E}_x$ be the $\mathbb{R}$-algebra of germs of C^∞-functions in x. Then $T_x(X) := \mathrm{Der}_{\mathbb{R}}(\mathcal{E}_x,\mathbb{R})$ is called the <u>tangent space</u> of X at x. It is shown in analysis that $T_x(X)$ is an n-dimensional $\mathbb{R}$-vectorspace, and that the derivations $d : \mathcal{E}_x \to \mathbb{R}$ are in natural one-to-one correspondence with the "tangents" defined in a more geometric way.

$\mathrm{Der}_{R_o}(R,R)$ is called <u>the module of derivations of R/R_o</u>. It is also denoted $\mathrm{Der}_{R_o}(R)$. In addition to being an R-module it also has the structure of a Lie-algebra. For $d,d' \in \mathrm{Der}_{R_o}(R)$ define

$$[d,d'] := d \circ d' - d' \circ d.$$

Clearly $[d,d']$ is R_o-linear. For $a,b \in R$

$[d,d'](ab) = d(d'(ab)) - d'(d(ab)) =$

$d(ad'b+bd'a) - d'(adb+bda) =$

add'b + dad'b + bdd'a + dbd'a - ad'db - d'adb - bd'da - d'bda

$= a[d,d'](b) + b[d,d'](a)$, hence $[d,d'] \in \mathrm{Der}_{R_o}(R)$.

One checks easily that the bracket $[,]$ is R_o-bilinear, $[d,d] = 0$ for any $d \in \mathrm{Der}_{R_o}(R)$, and that the Jacobian relation

$$[d,[d',d'']] + [d',[d'',d]] + [d'',[d,d']] = 0$$

is satisfied for $d,d',d'' \in \mathrm{Der}_{R_o}(R)$, hence $\mathrm{Der}_{R_o}(R)$ is a Lie-algebra.

For $d,d' \in \mathrm{Der}_{R_o}(R)$ in general $d \circ d' \notin \mathrm{Der}_{R_o}(R)$. Nevertheless we have the following.

1.13. Generalized Leibniz rule.

Let $d_1,\ldots,d_m \in \mathrm{Der}_{R_o}(R)$, $a_1,\ldots,a_n \in R$. For a subset $\{i_1,\ldots,i_\alpha\} \subset \{1,\ldots,m\}$ with $i_1 <\ldots< i_\alpha$ write $d_{i_1\ldots i_\alpha} := d_{i_1} \circ d_{i_2} \circ \ldots \circ d_{i_\alpha}$. For the empty subset let $d_\phi(a) := a$ for any $a \in R$. Then

$$(d_1 \circ \ldots \circ d_m)(a_1 \cdot \ldots \cdot a_n) = \Sigma d_{i_1^1 \ldots i_{\alpha_1}^1}(a_1) \cdot \ldots \cdot d_{i_1^n \ldots i_{\alpha_n}^n}(a_n)$$

where the summation is over all decompositions

$$\{1,\ldots,m\} = \{i_1^1,\ldots,i_{\alpha_1}^1\} \,\dot\cup\,\ldots\,\dot\cup\, \{i_1^n,\ldots,i_{\alpha_n}^n\} \text{ of } \{1,\ldots,m\}$$

into n disjoint (possibly empty) subsets $\{i_1^k,\ldots,i_{\alpha_k}^k\}$ with $i_1^k <\ldots< i_{\alpha_k}^k$.

For $m = 1$ this is the rule in 1.9b). The general case follows by induction on m, again making use of 1.9b).

Let N be another R-module. For each $d \in \mathrm{Der}_{R_o}(R,M)$ and each $\ell \in \mathrm{Hom}_R(M,N)$ the composition $\ell \circ d$ is an R_o-derivation from R to N, as the definition 1.1 immediately shows. Thus we obtain an R-linear map

$$\mathrm{Der}_{R_o}(R,\ell) : \mathrm{Der}_{R_o}(R,M) \to \mathrm{Der}_{R_o}(R,N) \qquad (d \mapsto \ell \circ d),$$

which makes $\mathrm{Der}_{R_o}(R,M)$ a covariant functor of M.

1.14. Proposition. $\mathrm{Der}_{R_o}(R,M)$, as a functor of M, is left exact: If $0 \to M \overset{\alpha}{\to} N \overset{\beta}{\to} P$ is an exact sequence of R-modules, then

$$0 \to \mathrm{Der}_{R_o}(R,M) \xrightarrow{\mathrm{Der}_{R_o}(R,\alpha)} \mathrm{Der}_{R_o}(R,N) \xrightarrow{\mathrm{Der}_{R_o}(R,\beta)} \mathrm{Der}_{R_o}(R,P)$$

is also exact.

This statement can easily be checked directly, but it also follows from a later proposition (3.22).

Suppose now that S/R_o is another algebra, and $\varphi : S \to R$ is an R_o-algebra homomorphism. Then for each R_o-derivation $d : R \to M$ the composition $d \circ \varphi$ is an R_o-derivation from S to M, if M is considered as an S-module via φ. In fact, it suffices to show the product formula: For $s,s' \in S$ we have
$$d(\varphi(ss')) = d(\varphi(s) \cdot \varphi(s')) = \varphi(s) \cdot d\varphi(s') + \varphi(s') \cdot d\varphi(s) =$$
$$= s \cdot d\varphi(s') + s' \cdot d\varphi(s).$$
Therefore we obtain an S-linear map

$$\mathrm{Der}_{R_o}(\varphi,M) : \mathrm{Der}_{R_o}(R,M) \to \mathrm{Der}_{R_o}(S,M) \qquad (d \mapsto d \circ \varphi).$$

1.15. Proposition. The sequence
$$0 \to \mathrm{Der}_S(R,M) \overset{j}{\to} \mathrm{Der}_{R_o}(R,M) \xrightarrow{\mathrm{Der}_{R_o}(\varphi,M)} \mathrm{Der}_{R_o}(S,M)$$
is exact. (Any $d \in \mathrm{Der}_S(R,M)$ is also an R_o-derivation from R to M, and j is defined to be the corresponding inclusion mapping).

This can easily be checked directly, but the statement also follows from a later proposition (3.25).

<u>1.16. Example.</u>   Let X and Y be C^∞-manifolds and $F : X \to Y$ a C^∞-mapping. F induces for any $x \in X$ an $\mathbb{R}$-algebra homomorphism

$$\varphi_x : \mathscr{E}_{F(x)} \to \mathscr{E}_x \qquad ([g] \mapsto [g \circ F]).$$

The induced mapping

$$\mathrm{Der}_{\mathbb{R}}(\varphi_x,\mathbb{R}) : \mathrm{Der}_{\mathbb{R}}(\mathscr{E}_x,\mathbb{R}) \to \mathrm{Der}_{\mathbb{R}}(\mathscr{E}_{F(x)},\mathbb{R})$$

maps the tangent space $T_x(X) = \mathrm{Der}_{\mathbb{R}}(\mathscr{E}_x,\mathbb{R})$ into the tangent space $T_{F(x)}(Y) = \mathrm{Der}_{\mathbb{R}}(\mathscr{E}_{F(x)},\mathbb{R})$. In analysis this mapping is denoted $T_x(F)$ or dF_x, and is called the differential of F in x (or the linear approximation of F in x).

Next we relate R_o-derivations $d : R \to M$ to R_o-algebra homomorphisms of R into the algebra $R \ltimes M$ of "dual numbers". Recall that this algebra is $R \oplus M$ with the multiplication given by

$$(r_1,m_1) \cdot (r_2,m_2) := (r_1 r_2, r_1 m_2 + r_2 m_1) \quad (r_i \in R, m_i \in M).$$

$i : R \to R \ltimes M \ (r \mapsto (r,0))$ is an injective and $p : R \ltimes M \to R$ $((r,m) \to r)$ a surjective R_o-homomorphism with $p \circ i = id_R$.

<u>1.17. Proposition.</u>   a) For any R_o-derivation $d : R \to M$

$$\tilde{d} : R \to R \ltimes M \qquad (r \mapsto (r,dr))$$

is an R_o-algebra-homomorphism with $p \circ \tilde{d} = id_R$.
b) For any R_o-algebra-homomorphism $h : R \to R \ltimes M$ with $p \circ h = id_R$ there is a unique R_o-derivation $d : R \to M$ with $h = \tilde{d}$.

Proof.   a) Obiously $\tilde{d}$ is R_o-linear and $p \circ \tilde{d} = id_R$. Moreover, for $a,b \in R$ we have
$$\tilde{d}(ab) = (ab,d(ab)) = (ab,adb+bda) = (a,da) \cdot (b,db) = \tilde{d}a \cdot \tilde{d}b$$

thus $\tilde{d}$ is a R_o-algebra-homomorphism.

b) Given h, put h(a) = (a,a') with a' $\in$ M for any a $\in$ R.
Let d : R $\to$ M be the mapping a $\mapsto$ a'. Clearly, d is R_o-linear. From h(ab) = h(a)$\cdot$h(b) = (a,a')(b,b') =
(ab,ab'+ba') we see that d is a derivation and h = $\tilde{d}$.

1.18. Definition.   An R_o-derivation d : R $\to$ M is called
<u>universal</u>, if the following condition is satisfied: For any
R_o-derivation δ : R $\to$ N of R into an R-module N there is
one and only one linear mapping ℓ : M $\to$ N with δ = $\ell \circ$ d.

In other words: If one knows a universal derivation
d : R $\to$ M of R/R_o, then any other derivation δ of R/R_o can
be written as a "specialization" $\ell \circ$ d of d with a unique
linear mapping ℓ.

As usual, if d_1 : R $\to$ M_1 and d_2 : R $\to$ M_2 are universal
derivations of R/R_o, then there is a unique R-linear map
ℓ : M_1 $\to$ M_2 such that d_2 = $\ell \circ d_1$, and ℓ is an isomorphism.
So we can say that there is exactly one universal deriva-
tion of R/R_o up to canonical isomorphism.

1.19. Theorem.   For any algebra R/R_o there exists a universal derivation. In fact, the derivation described in
example 1.7 is universal.

Proof.   We shall use the notations of 1.7. Let δ : R $\to$ N be
an arbitrary derivation of R/R_o and $\tilde{\delta}$: R $\to$ R $\ltimes$ N the R_o-homomorphism associated with it according to 1.17: We have
$\tilde{\delta}$(a) = (a,δa) for each a $\in$ R. According to the universal
property of R $\underset{R_o}{\otimes}$ R the injections i : R $\to$ R $\ltimes$ N (a $\mapsto$ (a,0))

and $\tilde{\delta}$ induce an R_o-homomorphism

$h : R \underset{R_o}{\otimes} R \to R \ltimes N$ with

$h(a \otimes b) = \tilde{\delta}(a) \cdot i(b) = (a,\delta a) \cdot (b,0) = (ab,b\delta a)$ for all

$a,b \in R$. For $x = r \otimes 1 - 1 \otimes r \in I$ $(r \in R)$ we obtain

$h(x) = (r,\delta r) - (r,0) = (0,\delta r)$, and hence $h(I^2) = 0$. There-

fore h induces an R-linear mapping $\ell : I/I^2 \to N$ with

$\ell(r \otimes 1 - 1 \otimes r + I^2) = \ell(dr) = \delta r$ for all $r \in R$.

Since the R-module I/I^2 is generated by $\{dr\}_{r \in R}$, we obtain

$\delta = \ell \circ d$, and there can be only one such mapping ℓ. Hence d

is universal.

<u>1.20. Definition.</u>   If $d : R \to M$ is a universal derivation

of R/R_o, then the R-module M (which is unique up to canonical

isomorphism) is denoted by Ω^1_{R/R_o} and is called <u>the module of</u>

<u>(Kähler) differentials</u> of R/R_o. The universal derivation

of R/R_o is sometimes denoted by d_{R/R_o}.

The question, which relations exist between the proper-

ties of an algebra R/R_o and its module of differentials

Ω^1_{R/R_o}, is a major theme of these lectures.

<u>1.21. Remarks.</u>   a) As was shown in the proof of 1.19 we have

$$\Omega^1_{R/R_o} \cong I/I^2$$

where I is the kernel of $R \underset{R_o}{\otimes} R \to R$ $(a \otimes b \mapsto ab)$.

b) For a derivation $\delta : R \to N$ let $R\delta R$ denote the submodule

of N generated by $\{\delta r\}_{r \in R}$. Thus $R\delta R$ is the set of all sums

$\sum_{i=1}^{n} r_i \delta r_i'$ $(r_i, r_i' \in R, n \in \mathbb{N})$. If $d : R \to \Omega^1_{R/R_o}$ is the univer-

sal derivation of R/R_o, we have $\Omega^1_{R/R_o} = RdR$, because I is

generated by $\{r \otimes 1 - 1 \otimes r\}_{r \in R}$.

We can "compute" Ω^1_{R/R_O} already in some special cases.

__1.22. Examples.__ a) Let K/K_O be a separably algebraic field extension. Then $\Omega^1_{K/K_O} = O$, because any derivation of K/K_O is trivial, as was shown in 1.10.

b) Similarly, for a perfect field K of characteristic $>O$ we have $\Omega^1_{K/\mathbf{Z}} = O$ by 1.11.

c) Let R_O be a ring and $R := R_O[\{X_\lambda\}_{\lambda \in \Lambda}]$ the polynomial algebra over R_O in a family of indeterminates X_λ ($\lambda \in \Lambda$). Form the free R-module on the family of indeterminates $\{dX_\lambda\}_{\lambda \in \Lambda}$

$$M := \bigoplus_{\lambda \in \Lambda} R dX_\lambda$$

and define for each $F \in R$ the "formal differential" as in 1.4 by

$$dF = \sum_{\lambda \in \Lambda} \frac{\partial F}{\partial X_\lambda} dX_\lambda.$$

It is easy to see that d is a derivation of R/R_O, we show that it is also universal.

In fact, if $\delta : R \to N$ is an arbitrary derivation of R/R_O, then $\delta F = \sum_{\lambda \in \Lambda} \frac{\partial F}{\partial X_\lambda} \delta X_\lambda$ by 1.9g). The R-linear mapping $\ell : M \to N$ with $\ell(dX_\lambda) = \delta X_\lambda$ satisfies the equation $\ell \circ d = \delta$, and there can be only one such mapping ℓ, since $\{dX_\lambda\}_{\lambda \in \Lambda}$ is a basis of M.

d) For $R = R_O[X_1,\ldots,X_n]$ consider the derivation
$$d : R \to \bigoplus_{i=1}^{n} R dX_i \; , \quad F \mapsto \sum_{i=1}^{n} \frac{\partial F}{\partial X_i} dX_i$$
of example 1.4. Although this example is very similar to that of the polynomial ring, d is not always universal, as will be seen later (5.5a).

<u>1.23. Proposition.</u> For an R-module M the canonical R-linear mapping

$$\mathrm{Hom}_R(\Omega^1_{R/R_o}, M) \to \mathrm{Der}_{R_o}(R,M) \qquad (\ell \mapsto \ell \circ d_{R/R_o})$$

is an isomorphism. In particular, there is a canonical isomorphism

$$\mathrm{Der}_{R_o}(R) \cong \mathrm{Hom}_R(\Omega^1_{R/R_o}, R)$$

which identifies the derivation module of R/R_o canonically with the dual of the differential module of R/R_o.

The first statement of the proposition is only a reformulation of the universal property of d_{R/R_o}. (In a different language: Ω^1_{R/R_o} "represents" the functor $\mathrm{Der}_{R_o}(R,-)$).

Given an R_o-derivation $d : R \to R$ and a ring extension S/R, one may ask whether d can be extended to a derivation $D : S \to S$ such that $D|_R = d$. This is only rarely possible. However, extensions in the sense of the following definition always exist.

Let $d: R \to M$ be a derivation of R/R_o with $M = RdR$, and let S/R be an algebra. We regard any S-module as an R-module via $R \to S$.

<u>1.24. Definition.</u> A derivation $D : S \to N$ of S/R_o into an S-module N is called an <u>extension</u> of d, if there exists an R-linear mapping $\ell : M \to N$ such that

$$
\begin{array}{ccc}
R & \xrightarrow{\ \rho\ } & S \\
\downarrow{\scriptstyle d} & & \downarrow{\scriptstyle D} \\
M & \xrightarrow[\ \ell\ ]{} & N
\end{array}
$$

is commutative. An extension D of d is called <u>universal</u>, if any extension $\Delta : S \to N'$ of d can be uniquely written as a specialization of D, that is: There exists a unique S-linear map $\alpha : N \to N'$ such that $\Delta = \alpha \circ D$.

We shall see later (3.20) that, in the situation of 1.24, a universal extension of d always exists. This will also give a new proof for the existence of a universal derivation of R/R_O: By definition such derivation is a universal extension of the trivial derivation of R_O.

Exercises

1) Let R be a ring, $d : R \to M$ a derivation into an R-module M, $N \subset R$ a multiplicatively closed subset, $I \subset R$ an ideal.

   a) There is a derivation $d_N : R_N \to M_N$ given by the formula

   $$d_N\left(\frac{r}{n}\right) = \frac{ndr - rdn}{n^2} \qquad \left(\frac{r}{n} \in R_N\right)$$

   such that the diagram

   $$\begin{array}{ccc} R & \longrightarrow & R_N \\ \downarrow{\scriptstyle d} & & \downarrow{\scriptstyle d_N} \\ M & \longrightarrow & M_N \end{array}$$

   is commutative.

   b) There is a derivation $\bar{d} : R/I \to M/RdI + IM$ given by the formula

   $$\bar{d}(r+I) = dr + RdI + IM \quad (r \in R)$$

   such that the diagram

   $$\begin{array}{ccc} R & \longrightarrow & R/I \\ \downarrow{\scriptstyle d} & & \downarrow{\scriptstyle \bar{d}} \\ M & \longrightarrow & M/RdI + IM \end{array}$$

   is commutative (RdI is the submodule of M generated by all dx, $x \in I$).

17

2) Let R_o be a ring, $(\alpha_1,\ldots,\alpha_n) \in \mathbf{Z}^n$, $d \in \mathbf{Z}$. A polynomial
$F = \Sigma a_{\nu_1\ldots\nu_n} X_1^{\nu_1}\ldots X_n^{\nu_n} \in R_o[X_1,\ldots,X_n]$ is called $\underline{\text{quasi-}}$
$\underline{\text{homogeneous}}$ of weight $(\alpha_1,\ldots,\alpha_n)$ and degree d, if
$a_{\nu_1\ldots\nu_n} = 0$ whenever $\sum_{i=1}^{n} \nu_i\alpha_i \neq d$.

a) For such polynomial the $\underline{\text{Euler relation}}$

$$\sum_{i=1}^{n} \alpha_i X_i \frac{\partial F}{\partial X_i} = d \cdot F$$

is satisfied.

b) Conversely, if $\mathbb{Q} \subset R_o$ and F satisfies Euler's relation, then F is quasihomogeneous of weight $(\alpha_1,\ldots,\alpha_n)$
and degree d.

3) Let R be a ring, $F \in R$, and $\wp \in \mathrm{Spec}(R)$. We say that F
has $\underline{\text{order}}$ m at $\wp$, if $F \in \wp^m R_\wp$, $F \notin \wp^{m+1} R_\wp$. Suppose R_o is
a ring with $\mathbb{Q} \subset R_o$ and $R := R_o[X_1,\ldots,X_n]$. Assume
$\wp \in \mathrm{Spec}(R)$ contains a polynomial F of degree m whose
order at $\wp$ is also m. Then $\wp$ contains a polynomial of
degree 1. (Hint: Consider the partial derivatives of F).

4) Let R be a ring with $\mathbb{Q} \subset R$. For each $\wp \in \mathrm{Min}(R)$ and
each derivation $d : R \to R$ we have $d(\wp) \subset \wp$
(Hint: Localize at $\wp$ and make use of the fact that
$\wp R_\wp$ consists of nilpotent elements only).

5) Let R be a ring whose characteristic is a prime number
p. If $d : R \to R$ is a derivation, then so is
$d^p := d \circ\ldots\circ d$ (p factors).

6) Let R be a ring and $\delta : R \to R$ a derivation. Let
$N \subset R[X]$ be multiplicatively closed and $\alpha \in R[X]_N$. There
is precisely one derivation $\delta_\alpha : R[X]_N \to R[X]_N$ with
$\delta_\alpha(X) = \alpha$ such that the diagram

$$
\begin{array}{ccc}
R & \longrightarrow & R[X]_N \\
\delta \downarrow & & \downarrow \delta_\alpha \\
R & \longrightarrow & R[X]_N
\end{array}
$$

is commutative.

7) Let L/K be a separably algebraic field extension and
$\delta : K \to K$ a derivation. Then there is exactly one deri-
vation $d : L \to L$ with $d|_K = \delta$.
Hint: Using Zorn's Lemma one can reduce to the case of
a simple extension $L = K[x]$. Let $f \in K[X]$ be the minimal
polynomial of x over K. There is a K-epimorphism
$K[X]_{(f)} \to L$ with kernel $fK[X]_{(f)}$. Choose $\alpha \in K[X]_{(f)}$
such that for the derivation $\delta_\alpha : K[X]_{(f)} \to K[X]_{(f)}$ of
exercise 6) the condition $\delta_\alpha(fK[X]_{(f)}) \subset fK[X]_{(f)}$ is
satisfied. Now apply exercise 1b).

8) Let R_0 be a ring, $(G,+)$ an abelian group, and
$R := R_0[G] = \bigoplus_{g \in G} R_0 x_g$ the group algebra of G over R_0.
Put $M := R \underset{\mathbb{Z}}{\otimes} G$.

a) The R_0-linear map $d : R \to M$ with $dx_g = x_g \otimes g$ for all
$g \in G$ is a derivation.

b) d is universal.

c) For any R-module N there is an isomorphism

$$
\mathrm{Der}_{R_0}(R_0[G],N) \cong \mathrm{Hom}_{\mathbb{Z}}(G,N)
$$

which is functorial in N.

9) (Abelian groups with isomorphic group rings are isomorphic). Under the assumptions of exercise 8) consider the
ring-epimorphism

$$\varepsilon : \mathbf{Z}[G] \to \mathbf{Z} \qquad (\Sigma n_g x_g \mapsto \Sigma n_g).$$

Any abelian group G' can be made a $\mathbf{Z}[G]$-module via ε.
Let G and H be abelian groups such that there is an isomorphism of $\mathbf{Z}$-algebras $\mathbf{Z}[G] \cong \mathbf{Z}[H]$.

a) For any abelian group G' there is an isomorphism

$$\mathrm{Hom}_{\mathbf{Z}}(G,G') \cong \mathrm{Hom}_{\mathbf{Z}}(H,G'),$$

which is functorial in G'.

b) Conclude that $G \cong H$.

10) Let R/R_o and S/R be algebras and I an ideal of S with
$I^2 = 0$. Let $\rho : R \to S$ be the structure homomorphism and
$\varepsilon : S \to S/I$ the canonical epimorphism.

a) For each R_o-homomorphism $\alpha : R \to S$ for which

(*)

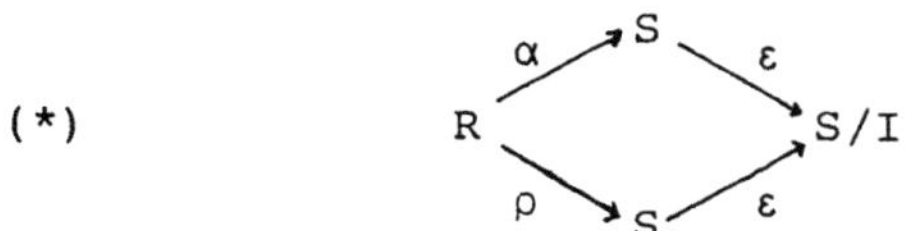

is commutative the map

$$\tilde{\alpha} : R \ltimes I \to S \qquad (r,i) \mapsto \alpha(r)+i$$

is an R_o-homomorphism.

b) The R_o-derivations $\delta : R \to I$ are in natural one-to-
one correspondence with the R_o-homomorphisms $\alpha : R \to S$
that make (*) commutative (the "liftings" of $\varepsilon \circ \rho$).

§ 2. Differential Algebras

A method to study algebras is to study their differential algebras. This notion, which is analogous to the notion of the algebra of differential forms in analysis, will be introduced here and some of its basic properties will be established.

Let R/R_o be an algebra.

2.1. Definition. A <u>differential algebra</u> of R/R_o is an associative (not necessarily commutative) graded R-algebra $\Omega = \bigoplus_{n \in \mathbb{N}} \Omega^n$, on which an R_o-linear map $d : \Omega \to \Omega$ of degree 1 is given (i.e. $d\Omega^n \subset \Omega^{n+1}$) such that the following axioms are satisfied:

a) $\Omega^o = R$ and R is contained in the center of Ω.

b) $\Omega = R[dR]$ (i.e. as an R-algebra Ω is generated by the elements dr ($r \in R$)).

c) For all $r,r' \in R$ we have $d(rr') = rdr' + r'dr$.

d) For all $r,r_1,\ldots,r_m \in R$ we have $d(rdr_1\ldots dr_m) = drdr_1\ldots dr_m$.

e) $drdr = 0$ for each $r \in R$.

The mapping d is called the <u>differentiation</u> of Ω and the elements of Ω^n are called <u>n-forms</u>. For every ring R a differential algebra of $R/\mathbb{Z}$ is called an <u>absolute differential algebra</u>.

Clearly any differential algebra of R/R_o is also one of R/R_o', if R_o' is the image of R_o in R.

From the axioms we can derive the following properties of a differential algebra (Ω,d) of R/R_o.

2.2. Rules.

a) The restriction $d : R \to \Omega^1$ of the differentiation d to elements of degree 0 is a derivation of R/R_0.

b) Each $\omega \in \Omega$ is a finite sum of elements $rdr_1 \ldots dr_m$ $(r,r_1,\ldots,r_m \in R)$. Such elements have degree m.

c) For all $r,r' \in R$ we have $drdr' + dr'dr = 0$.

d) For $\omega = \Sigma rdr_1 \ldots dr_m$ we have $d\omega = \Sigma drdr_1 \ldots dr_m$.

e) Ω is anticommutative, i.e. for $\omega_m \in \Omega^m$, $\omega_n \in \Omega^n$

$$\omega_m \cdot \omega_n = (-1)^{m \cdot n} \omega_n \cdot \omega_m.$$

Each homogeneous left (right) ideal of Ω is a two-sided ideal.

f) d is an antiderivation, i.e. for $\omega_m \in \Omega^m$, $\omega \in \Omega$

$$d(\omega_m \cdot \omega) = d\omega_m \cdot \omega + (-1)^m \omega_m \cdot d\omega.$$

g) $d \circ d = 0$, i.e. (Ω,d) is a complex of R_0-modules

$$\Omega^0 \overset{d}{\to} \Omega^1 \overset{d}{\to} \Omega^2 \to \ldots.$$

h) $Z(\Omega) := \ker d$ is a graded subring of Ω.

j) $B(\Omega) := \operatorname{im} d$ is a two-sided homogeneous ideal of $Z(\Omega)$.

Proofs. a) follows from 2.1c) and the fact that d is R_0-linear and of degree 1. b) is a consequence of 2.1a) and 2.1b).

c) From 2.1e) we obtain $0 = d(r+r') \cdot d(r+r') = (dr+dr') \cdot (dr+dr') = drdr' + dr'dr$.

d) follows from 2.1d), because d is linear.

e) Since the multiplication on Ω is bilinear, we may assume $\omega_m = rdr_1 \ldots dr_m$, $\omega_n = sds_1 \ldots ds_n$ $(r,r_1,\ldots,r_m,s,s_1,\ldots,s_n \in R)$. From 2.2c) we conclude $\omega_m \cdot \omega_n = rsdr_1 \ldots dr_m ds_1 \ldots ds_n =$

$rs \cdot (-1)^{m \cdot n} ds_1 \ldots ds_n dr_1 \ldots dr_m = (-1)^{m \cdot n} \omega_n \cdot \omega_m$. The second statement in e) follows from

2.3. Lemma. Let I be a homogeneous left (right) ideal in a graded anticommutative algebra A. Then I is a two-sided ideal of A.

If $A \cdot I \subset I$, then also $I \cdot A \subset I$. To show this, let $\Sigma x_m \in I$, $\Sigma r_n \in A$ with $x_i, r_i \in A^i$ be given. Since I is homogeneous, we have $x_i \in I$ for all i. Then $(\Sigma x_m)(\Sigma r_n) = \Sigma\limits_{\rho}(\sum\limits_{m+n=\rho} x_m r_n) = \Sigma\limits_{\rho}(\sum\limits_{m+n=\rho} (-1)^{m \cdot n} r_n x_m)$ and this element belongs to I, since $r_n x_m \in I$ for all n,m by assumption.

f) Since d is R_o-linear and the multiplication on Ω bilinear, we may assume that $\omega_m = rdr_1 \ldots dr_m$, $\omega = sds_1 \ldots ds_n$ $(r, r_1, \ldots, r_m, s, s_1, \ldots, s_n \in R)$. By 2.1c) and d) we have
$d(\omega_m \omega) = d(rsdr_1 \ldots dr_m ds_1 \ldots ds_n) = (rds + sdr)dr_1 \ldots dr_m ds_1 \ldots ds_n = (drdr_1 \ldots dr_m)(sds_1 \ldots ds_n) + (-1)^m (rdr_1 \ldots dr_m)(dsds_1 \ldots ds_n) = d\omega_m \cdot \omega + (-1)^m \omega_m d\omega$.

g) For $\omega = rdr_1 \ldots dr_m$ we get $(d \circ d)(\omega) = d(1 \cdot drdr_1 \ldots dr_m) = d(1) \cdot drdr_1 \ldots dr_m = O$, since $d(1) = O$ by 1.9a).

h) For $\omega = \Sigma \omega_n$ ($\omega_n \in \Omega^n$) we have $d\omega = \Sigma d\omega_n = O$ if and only if $d\omega_n = O$ for all $n \in \mathbb{N}$, since d is of degree 1. Hence $Z(\Omega)$ is a graded subgroup of $(\Omega,+)$. Moreover, for $\omega, \omega' \in Z(\Omega)$ one sees from 2.2f) that also $\omega \cdot \omega' \in Z(\Omega)$, hence $Z(\Omega)$ is a graded subring of Ω.

j) From $d \circ d = O$ we conclude that $B(\Omega) \subset Z(\Omega)$. Obviously $B(\Omega)$ is a graded subgroup of $(Z(\Omega),+)$. By lemma 2.3 it is enough to show $Z(\Omega) \cdot B(\Omega) \subset B(\Omega)$. But for $\omega_m \in Z(\Omega)$, $d\omega_n \in B(\Omega)$ ($\omega_m \in \Omega^m, \omega_n \in \Omega^n$) we have by 2.2f)
$\omega_m \cdot d\omega_n = (-1)^m d(\omega_m \cdot \omega_n) + (-1)^{m+1} d\omega_m \cdot \omega_n = d((-1)^m \omega_m \cdot \omega_n) \in B(\Omega)$.

2.4. Definition. The elements of $Z(\Omega)$ are called <u>closed</u> elements of Ω, those of $B(\Omega)$ <u>exact</u>. The graded algebra

$$H_{DR}(\Omega) := Z(\Omega)/B(\Omega)$$

is called the <u>cohomology</u> of Ω (<u>deRham-cohomology</u>).

We have $H_{DR}(\Omega) = \underset{n\in\mathbf{Z}}{\oplus} H_{DR}^n(\Omega)$, where $H_{DR}^n(\Omega) = Z^n(\Omega)/B^n(\Omega)$ with $Z^n(\Omega) := Z(\Omega) \cap \Omega^n$, $B^n(\Omega) := B(\Omega) \cap \Omega^n$. $H_{DR}(\Omega)$ is an important invariant of the differential algebra Ω (and of the algebra R/R_o). In this text only very little will be said about $H_{DR}(\Omega)$.

2.5. Examples of differential algebras.

a) <u>The trivial differential algebra of a ring R.</u>
It is given by $\Omega = \Omega^o = R$, $\Omega^n = 0$ for $n > 0$. Obviously d has to be the zero map.

b) <u>The algebra of C^∞-differential forms on a C^∞-manifold.</u>
Let R be the ring of C^∞-functions $f : X \to \mathbb{R}$ on a C^∞-manifold X and Ω the algebra of C^∞-differential forms on X. Let $d : \Omega \to \Omega$ be the Cartan differentiation of differential forms. Then with $R_o := \mathbb{R}$ the pair (Ω,d) is a differential algebra of R/R_o in the sense of definition 2.1, as is shown in analysis. Poincaré's lemma states that for a contractible manifold $H_{DR}^n(\Omega) = 0$ for $n \geq 1$.

c) Let R/R_o be an algebra and $d : R \to M$ an R_o-derivation of R into an R-module M with $M = RdR$. Let $\Omega := R \ltimes M$ be the ring of dual numbers and put $\Omega^o := R$, $\Omega^1 := M$, $\Omega^i := 0$ for $i > 1$. Extend the given derivation $d : \Omega^o \to \Omega^1$ by zero to Ω. Then (Ω,d) is a differential algebra of R/R_o. We call (Ω,d) <u>the differential algebra associated with the derivation d.</u>

This example, together with 2.2a), shows the close connection between differential algebras and derivations of R/R_o.

d) Let R_o be a ring, $R := R_o[X_1,\ldots,X_n]$ a polynomial algebra over R_o and $\Omega^1 := RdX_1 \oplus \ldots \oplus RdX_n$ the module of formal differentials of R/R_o (1.4). Let $\Omega := \Lambda\Omega^1$ be the exterior algebra of the R-module Ω^1 and put $\Omega^p := \Lambda^p\Omega^1$.

Then any $\omega \in \Omega$ may be written uniquely as $\omega = \sum_{p\in\mathbb{N}} \omega_p$ with

$$\omega_p = \sum_{1\le\nu_1<\ldots<\nu_p\le n} r_{\nu_1\ldots\nu_p} dX_{\nu_1} \wedge\ldots\wedge dX_{\nu_p} \qquad (r_{\nu_1\ldots\nu_p} \in R).$$

Define $d : \Omega \to \Omega$ by $d\omega = \sum_{p\in\mathbb{N}} d\omega_p$ with

$$d\omega_p = \sum_{1\le\nu_1<\ldots<\nu_p\le n} dr_{\nu_1\ldots\nu_p} \wedge dX_{\nu_1} \wedge\ldots\wedge dX_{\nu_p}$$

where $dr_{\nu_1\ldots\nu_p} = \sum_{i=1}^{n} \dfrac{\partial r_{\nu_1\ldots\nu_p}}{\partial X_i} dX_i$ is the formal differential of $r_{\nu_1\ldots\nu_p}$ (1.4). Then d is an R_o-linear map of degree 1 and it is clear that the axioms of 2.1 are satisfied with the possible exception of 2.1d).

In order to show 2.1d), one may prove first that $d \circ d = 0$ and that the formula

$$(1) \qquad d(\omega_k \wedge \omega_\ell) = d\omega_k \wedge \omega_\ell + (-1)^k \omega_k \wedge d\omega_\ell$$

$(\omega_k \in \Omega^k, \omega_\ell \in \Omega^\ell)$ is valid, then 2.1d) follows easily.

For $r \in R$ we have $d(dr) = d\left(\sum_{i=1}^{n} \dfrac{\partial r}{\partial X_i} dX_i \right) =$

$$\sum_{i=1}^{n} \left(\sum_{j=1}^{n} \dfrac{\partial^2 r}{\partial X_j \partial X_i} dX_j \right) \wedge dX_i = \sum_{i<j} \left(\dfrac{\partial^2 r}{\partial X_i \partial X_j} - \dfrac{\partial^2 r}{\partial X_j \partial X_i} \right) dX_i \wedge dX_j = 0,$$

since $\dfrac{\partial^2 r}{\partial X_i \partial X_j} = \dfrac{\partial^2 r}{\partial X_j \partial X_i}$. From this we get $d(d\omega) = 0$ for all $\omega \in \Omega$.

Moreover, for $\omega_k = rdX_{\nu_1} \wedge\ldots\wedge dX_{\nu_k}$, $\omega_\ell = sdX_{\mu_1} \wedge\ldots\wedge dX_{\mu_\ell}$

25

$(r, s \in R, 1 \leq \nu_1 < \ldots < \nu_k \leq n, 1 \leq \mu_1 < \ldots < \mu_\ell \leq n)$ we obtain $d(\omega_k \wedge \omega_\ell) = d(rsdX_{\nu_1} \wedge \ldots \wedge dX_{\nu_k} \wedge dX_{\mu_1} \wedge \ldots \wedge dX_{\mu_\ell}) =$ $(sdr + rds) \wedge dX_{\nu_1} \wedge \ldots \wedge dX_{\nu_k} \wedge dX_{\mu_1} \wedge \ldots \wedge dX_{\mu_\ell} =$ $d\omega_k \wedge \omega_\ell + (-1)^k \omega_k \wedge d\omega_\ell$ and, since d is linear, formula (1) follows in the general case.

We have therefore shown that (Ω, d) is a differential algebra of $R_o[X_1, \ldots, X_n]/R_o$. In a completely analogous way one can also construct a differential algebra of $R_o[\![X_1, \ldots, X_n]\!]/R_o$.

e) Let R_o be a field and R a separably algebraic extension field of R_o. Then any differential algebra Ω of R/R_o is trivial. This is a consequence of the fact that the derivation $d : R \to \Omega^1$ is trivial (by 1.10) and that $\Omega = R[dR]$. Similarly any differential algebra of a perfect field R of characteristic $p > 0$ over any subring $R_o \subset R$ is trivial by 1.11.

We shall now discuss homomorphisms of differential algebras. Let (Ω, d) be a differential algebra of an algebra R/R_o and (Ω', d') a differential algebra of an algebra S/R_o.

<u>2.6. Definition.</u>   A mapping $\varphi : \Omega \to \Omega'$ is called <u>a homomorphism of differential algebras</u>, if the following are true:
a) φ is a homomorphism of R_o-algebras
(i.e. φ is R_o-linear and a ring homomorphism),
b) φ is compatible with the grading (i.e. $\varphi(\Omega^n) \subset \Omega'^n$ for all $n \in \mathbb{N}$).
c) φ is compatible with differentiation (i.e. $\varphi \circ d = d' \circ \varphi$).

We write φ^n for $\varphi|_{\Omega^n} : \Omega^n \to \Omega'^n$. It is clear that $\varphi^0 : R \to S$ is a homomorphism of R_o-algebras.

2.7. <u>Example.</u>   Let X and Y be two C^∞-manifolds and Ω resp.

Ω' their algebras of C^∞-differential forms. Let $f : X \to Y$

be a C^∞-diffeomorphism. In analysis the "pullback"

$f_* : \Omega' \to \Omega$ of differential forms is defined. For a C^∞-

function $\varphi : Y \to \mathbb{R}$ we have $f_*(\varphi) = \varphi \circ f$ and

$f_*(d'\varphi) = d(f_*\varphi) = d(\varphi \circ f)$. Moreover, f_* is a homomorphism

of graded $\mathbb{R}$-algebras. (These conditions define f_* uniquely).

Hence f_* is an example of a homomorphism of differential

algebras (here even an isomorphism). Another example is

given by the "restriction of differential forms" to an open

submanifold of a manifold.

An R_o-differential algebra is a differential algebra of

some algebra R/R_o. Since the composition of two homomorphisms

of R_o-differential algebras is also one, the R_o-differential

algebras form, together with the homomorphisms of differen-

tial algebras, a category $\mathcal{D}_{R_o}$. Let $\mathcal{A}_{R_o}$ denote the category

of R_o-algebras.

2.8. Rules.

a) Any homomorphism $\varphi : \Omega \to \Omega'$ of R_o-differential algebras

is uniquely determined by its restriction $\varphi^o : \Omega^o \to \Omega'^o$ to

its elements of degree O. Namely, we have the formula

$$\varphi(\Sigma r dr_1 \ldots dr_n) = \Sigma \varphi^o(r) \cdot d'\varphi^o(r_1) \ldots d'\varphi^o(r_n).$$

In particular, if Ω is a differential algebra of R/R_o,

Ω' one of S/R_o and $\rho : R \to S$ a homomorphism of R_o-algebras,

there is <u>at most</u> one homomorphism $\varphi : \Omega \to \Omega'$ of differential

algebras with $\varphi^o = \rho$.

b) A diagram in $\mathcal{D}_{R_o}$ is commutative if and only if the

corresponding diagram in $\mathcal{A}_{R_o}$, given by restricting all

algebras and homomorphisms to the elements of degree 0, is
commutative.

c) If R'/R_0 is a subalgebra of an algebra R/R_0 and (Ω, d) a
differential algebra of R/R_0, then $\Omega' := R'[dR']$, the subalgebra of Ω generated over R' by the elements dr' ($r' \in R'$)
forms together with $d' := d|_{\Omega'}$ a differential algebra of
R'/R_0, and the canonical injection $\Omega' \to \Omega$ is a homomorphism
of differential algebras.

d) If $\varphi : \Omega \to \Omega'$ is a homomorphism in $\mathcal{D}_{R_0}$, then im φ is a
differential algebra of $\varphi^0(R)/R_0$. ker $\varphi =: I$ is a homoge-
neous ideal of Ω, which is, moreover, <u>differentially closed</u>,
i.e. $dI \subset I$.

e) For φ as in d) we have $\varphi(Z(\Omega)) \subset Z(\Omega')$ and
$\varphi(B(\Omega)) \subset B(\Omega')$. Hence φ induces a homomorphism of graded
R_0-algebras

$$H_{DR}(\varphi) : H_{DR}(\Omega) \to H_{DR}(\Omega') \quad (\omega + B(\Omega) \mapsto \varphi(\omega) + B(\Omega')).$$

Proofs. The formula in a) follows from the fact that φ is
a ring homomorphism and compatible with differentiation.
The other statements in a) are then clear and so is b).
c) The elements of Ω' are of the form $\Sigma r' dr'_1 \ldots dr'_n$ with
$r', r'_1, \ldots, r'_n \in R'$. Therefore $d(\Omega') \subset \Omega'$ and d' is defined.
The axioms of a differential algebra are satisfied in Ω',
since they hold in Ω.
d) The statement about im φ follows from c). Since φ is
compatible with the grading, $I = $ ker φ is a homogeneous
two-sided ideal of Ω. Moreover, for $\omega \in I$ we have
$\varphi(d\omega) = d'(\varphi(\omega)) = d'(0) = 0$, hence $d\omega \in I$ and $dI \subset I$.
e) follows similarly from the fact that φ is compatible

with differentiation.

Now let R/R_o be an algebra and (Ω, d) a differential algebra of R/R_o.

2.9. **Proposition**   (Existence of residue class differential
algebras). If I is a homogeneous differentially closed ideal
of Ω, put $\Omega' := \Omega/I$ and define $d' : \Omega' \to \Omega'$ by
$d'(\omega+I) := d\omega + I$. Then:
a) (Ω',d') is a differential algebra of $R/I \cap R$ over R_o and
$\varepsilon : \Omega \to \Omega'$ $(\omega \mapsto \omega+I)$ a homomorphism of differential algebras.
b) Let (Ω'',d'') an arbitrary object in $\mathcal{D}_{R_o}$ and $\varphi : \Omega \to \Omega''$ a
morphism of $\mathcal{D}_{R_o}$ with $I \subset \ker \varphi$, then there is (exactly one)
morphism $h : \Omega' \to \Omega''$ with $\varphi = h \circ \varepsilon$.

Proof. a) Since I is homogeneous, Ω' is graded with
$\Omega'^n = \Omega^n/I \cap \Omega^n$ for all $n \in \mathbb{N}$. In particular, $\Omega'^o = R/I \cap R$.
Since $dI \subset I$, the mapping d' is well-defined. Now one checks
easily that the axioms for a differential algebra are
satisfied in Ω' and that ε is a morphism of $\mathcal{D}_{R_o}$.
b) By the homomorphism theorem for algebras there is
precisely one R_o-algebra homomorphism $h : \Omega' \to \Omega''$ with
$\varphi = h \circ \varepsilon$. Since φ and ε are compatible with the gradings,
so is h. For $\omega \in \Omega$ we have $h(d'(\omega+I)) = h(d\omega+I) = h(\varepsilon(d\omega))$
$= \varphi(d\omega) = d''\varphi(\omega) = (d'' \circ h)(\omega+I)$, hence $h \circ d' = d'' \circ h$ and h
is a homomorphism of differential algebras.

2.10. **Example.**   If $I := \bigoplus_{\nu \geq m+1} \Omega^\nu$ for some $m \in \mathbb{N}$, then
$\Omega/I = \Omega^o \oplus \ldots \oplus \Omega^m$ is a differential algebra of R/R_o.

2.11. Theorem. The category $\mathcal{D}_{R_o}$ is complete and cocomplete.

Remember that a category $\mathcal{C}$ is called complete (cocomplete), if any diagram in $\mathcal{C}$ has a limit (colimit) in $\mathcal{C}$.

We first prove the <u>existence of products</u> in $\mathcal{D}_{R_o}$:

Let $\{(\Omega_\lambda, d_\lambda)\}_{\lambda \in \Lambda}$ be a familiy of R_o-differential algebras and let $\tilde{\Omega} := \prod_{\lambda \in \Lambda} \Omega_\lambda$ be their direct product. For $(\omega_\lambda)_{\lambda \in \Lambda} \in \tilde{\Omega}$ define $\tilde{d}(\omega_\lambda)_{\lambda \in \Lambda}$ to be $(d_\lambda \omega_\lambda)_{\lambda \in \Lambda}$. $\tilde{\Omega}$ contains $R := \prod_{\lambda \in \Lambda} \Omega_\lambda^o$ as a subring of its center. Let $\Omega := R[\tilde{d}R]$ be the subring of $\tilde{\Omega}$ generated over R by all elements $\tilde{d}r$ ($r \in R$). Any element of Ω is a finite sum of elements $r\tilde{d}r_1 \ldots \cdot \tilde{d}r_m$ ($r, r_1, \ldots, r_m \in R$) and $\tilde{d}(r\tilde{d}r_1 \ldots \tilde{d}r_m) = \tilde{d}r\tilde{d}r_1 \cdot \ldots \cdot \tilde{d}r_m$. Hence $\tilde{d}(\Omega) \subset \Omega$. Let d be the restriction of $\tilde{d}$ to Ω. We want to show that (Ω, d) is a differential algebra of R/R_o, which has the universal property of the product of the differential algebras $(\Omega_\lambda, d_\lambda)$.

We call an element $(\omega_\lambda)_{\lambda \in \Lambda} \in \Omega$ ($\omega_\lambda \in \Omega_\lambda$ for all λ) homogeneous of degree p, if $\omega_\lambda \in \Omega_\lambda^p$ for all λ. Let Ω^p be the set of all such elements of Ω. Then $\Omega^o = R$, and from $\Omega = R[dR]$ it is easily seen that $\Omega = \bigoplus_{p \in \mathbb{N}} \Omega^p$, $\Omega^p \cdot \Omega^q \subset \Omega^{p+q}$. In other words, Ω is a graded R-algebra and $d : \Omega \to \Omega$ is of degree one. The axioms 2.1a),c)-e) of a differential algebra are also satisfied in Ω, they can be checked componentwise.

Let $p_\lambda : \Omega \to \Omega_\lambda$ be the composition of the inclusion $\Omega \hookrightarrow \tilde{\Omega}$ with the projection $\tilde{\Omega} \to \Omega_\lambda$. Clearly, this is a homomorphism of differential algebras.

Now let Ω^* be an arbitrary R_o-differential algebra and $\varphi_\lambda : \Omega^* \to \Omega_\lambda$ homomorphisms of differential algebras ($\lambda \in \Lambda$). Define $\varphi : \Omega^* \to \tilde{\Omega}$ by $\varphi(\omega) = (\varphi_\lambda(\omega))_{\lambda \in \Lambda}$ for all $\omega \in \Omega^*$. Then

$\varphi(\Omega^*) \subset \Omega$ and $\varphi_\lambda = p_\lambda \circ \varphi$ for all λ. Moreover, the restriction $\varphi : \Omega^* \to \Omega$ is a homomorphism of differential algebras. By 2.8a) there can be only one such homomorphism φ with $\varphi_\lambda = p_\lambda \circ \varphi$ for all λ, since φ° has to coincide with the ring homomorphism $\Omega^{*\circ} \to \prod_{\lambda \in \Lambda} \Omega_\lambda^\circ$ $(\omega^* \mapsto (\varphi_\lambda(\omega^*)))$. We write $\prod_{\lambda \in \Lambda} \Omega_\lambda$ for the product of differential algebras $(\Omega_\lambda, d_\lambda)$.

Next we show the <u>existence of equalizers</u> in $\mathcal{D}_{R_\circ}$. Let (Ω_1, d_1) and (Ω_2, d_2) be two objects of $\mathcal{D}_{R_\circ}$ and $\alpha, \beta : \Omega_1 \to \Omega_2$ two morphisms. Then $R := \{r \in \Omega_1^\circ \mid \alpha(r) = \beta(r)\}$ is an $R_\circ$-subalgebra of Ω_1°. By 2.8c), $\Omega := R[d_1 R]$ is a differential algebra of $R/R_\circ$ and the inclusion $i : \Omega \hookrightarrow \Omega_1$ a homomorphism of differential algebras with $\alpha \circ i = \beta \circ i$. If Ω^* is an arbitrary object of $\mathcal{D}_{R_\circ}$ and $j : \Omega^* \to \Omega_1$ a morphism with $\alpha \circ j = \beta \circ j$, then clearly $j(\Omega^{*\circ}) \subset \Omega^\circ$ and hence $j(\Omega^*) \subset \Omega$. Therefore Ω is an equalizer of α and β.

As is well known, the existence of products and equalizers in a category ensure the existence of arbitrary limits in that category.

In order to prove the existence of colimits, we first show the <u>existence of the coproduct of two differential algebras</u>, which is also called the <u>tensor product</u> of differential algebras.

For objects (Ω_1, d_1) and (Ω_2, d_2) of $\mathcal{D}_{R_\circ}$ let $\Omega := \Omega_1 \otimes_{R_\circ} \Omega_2$ be the tensor product of the $R_\circ$-modules Ω_1 and Ω_2. Then $\Omega = \bigoplus_{p \in \mathbb{N}} \Omega^p$ with $\Omega^p := \bigoplus_{r+s=p} \Omega_1^r \otimes_{R_\circ} \Omega_2^s$. There is a unique multiplication on Ω such that for $a_m \in \Omega_1^m$, $\bar{a}_r \in \Omega_1^r$, $b_n \in \Omega_2^n$, $\bar{b}_s \in \Omega_2^s$

$$(2) \qquad (a_m \otimes b_n)(\bar{a}_r \otimes \bar{b}_s) = (-1)^{n \cdot r} a_m \bar{a}_r \otimes b_n \bar{b}_s$$

With this multiplication, Ω becomes a graded, associative and anticommutative R_o-algebra. $R := \Omega^o = \Omega_1^o \otimes_{R_o} \Omega_2^o$ becomes

a commutative R_o-algebra, the tensor product of Ω_1^o and Ω_2^o as

R_o-algebras, and Ω^o is contained in the center of Ω.

Similarly, there is a unique R_o-linear map $d : \Omega \to \Omega$

such that for $a_m \in \Omega_1^m$, $b_n \in \Omega_2^n$

$$(3) \qquad d(a_m \otimes b_n) = d_1 a_m \otimes b_n + (-1)^m a_m \otimes d_2 b_n$$

and d is of degree 1. By (3) we have $d(a_m \otimes 1) = d_1 a_m \otimes 1$,

$d(1 \otimes b_n) = 1 \otimes d_2 b_n$, and since $\Omega_1 = R_1[d_1 R_1]$, $\Omega_2 = R_2[d_2 R_2]$

with $R_i := \Omega_i^o$ $(i=1,2)$, we obtain $\Omega = R[dR]$.

Using (2) and (3) an easy calculation shows that

$$d((a_m \otimes b_n)(\bar{a}_r \otimes \bar{b}_s)) = d(a_m \otimes b_n) \cdot (\bar{a}_r \otimes \bar{b}_s) + (-1)^{m+n}(a_m \otimes b_n) \cdot d(\bar{a}_r \otimes \bar{b}_s).$$

In particular, the axiom 2.1c) is satisfied for (Ω, d), and

also the formula 2.2f) holds, which says that d is an anti-

derivation on Ω.

From (3) we can conclude that $d \circ d = 0$. This fact and

the formula 2.2f) imply that 2.1d) is satisfied too.

Finally for $r_i^j \in R_i$ $(i=1,2)$ we have $d(\Sigma r_1^j \otimes r_2^j) \cdot d(\Sigma r_1^j \otimes r_2^j) =$

$\sum_{j,k} (d_1 r_1^j \otimes r_2^j + r_1^j \otimes d_2 r_2^j)(d_1 r_1^k \otimes r_2^k + r_1^k \otimes d_2 r_2^k) = 0$ by strait-

forward computation, hence 2.1e).

We have shown that (Ω, d) is a differential algebra of

$R_1 \otimes_{R_o} R_2 / R_o$. (2) and (3) moreover imply that the canonical

mappings

$$\alpha : \Omega_1 \to \Omega \qquad (a \mapsto a \otimes 1)$$

$$\beta : \Omega_2 \to \Omega \qquad (b \mapsto 1 \otimes b)$$

are morphisms of $\mathcal{D}_{R_o}$.

Now let (Ω^*, d^*) be an arbitrary R_o-differential algebra

and $\varphi : \Omega_1 \to \Omega^*$, $\psi : \Omega_2 \to \Omega^*$ homomorphisms of differential

algebras. There is a unique R_o-linear map $h : \Omega_1 \otimes_{R_o} \Omega_2 \to \Omega^*$

with $h(a \otimes b) = \varphi(a) \cdot \psi(b)$ for $a \in \Omega_1$, $b \in \Omega_2$. h is compatible with the gradings and (2) yields the formula

$$h((a_m \otimes b_n)(\overline{a}_r \otimes \overline{b}_s)) = h(a_m \otimes b_n) \cdot h(\overline{a}_r \otimes \overline{b}_s).$$

Similarly, using (3), an easy computation shows that

$$h(d(a_m \otimes b_n)) = d*(h(a_m \otimes b_n)).$$ Hence h is a homomorphism of differential algebras.

From the definition of h it is clear that $\varphi = h \circ \alpha$ and $\psi = h \circ \beta$. That only one such h can exist is already a consequence of the universal property of the tensorproduct of R_o-modules. We write $\Omega_1 \otimes_{R_o} \Omega_2$ for the coproduct of two differential algebras (Ω_1, d_1), (Ω_2, d_2).

Since the coproduct of two objects of $\mathcal{D}_{R_o}$ exists, finite coproducts exist too. Next we show the <u>existence of direct limits</u> in $\mathcal{D}_{R_o}$.

Let $\{\Omega_\lambda, \varphi_{\lambda'}^\lambda\}$ be a direct system in $\mathcal{D}_{R_o}$ over a directed set Λ. This means that Λ is partially ordered and that for $\lambda, \lambda' \in \Lambda$ there is $\lambda'' \in \Lambda$ with $\lambda \le \lambda''$, $\lambda' \le \lambda''$. Moreover, for each $(\lambda, \lambda') \in \Lambda \times \Lambda$ with $\lambda \le \lambda'$ there is a morphism $\varphi_{\lambda'}^\lambda : \Omega_\lambda \to \Omega_{\lambda'}$ such that $\varphi_\lambda^\lambda = \mathrm{id}_{\Omega_\lambda}$ for all $\lambda \in \Lambda$ and $\varphi_{\lambda''}^{\lambda'} \circ \varphi_{\lambda'}^\lambda = \varphi_{\lambda''}^\lambda$ for $\lambda \le \lambda' \le \lambda''$.

We first construct the direct limit of $\{\Omega_\lambda, \varphi_{\lambda'}^\lambda\}$ in the category of R_o-modules: In $\underset{\lambda \in \Lambda}{\oplus} \Omega_\lambda$ consider the submodule U generated by all elements

$$\omega_\lambda - \varphi_{\lambda'}^\lambda(\omega_\lambda) \quad \text{with } \omega_\lambda \in \Omega_\lambda \text{ and } \lambda \le \lambda'.$$

Put $\Omega := \underset{\lambda \in \Lambda}{\oplus} \Omega_\lambda / U$, and let $i^\lambda : \Omega_\lambda \to \Omega$ be the composition of the canonical injection $\Omega_\lambda \to \underset{\lambda \in \Lambda}{\oplus} \Omega_\lambda$ and the canonical epimorphism $\underset{\lambda \in \Lambda}{\oplus} \Omega_\lambda \to \Omega$. By construction $i^\lambda = i^{\lambda'} \circ \varphi_{\lambda'}^\lambda$ for all $\lambda \le \lambda'$. Since Λ is directed, we have $\Omega = \underset{\lambda \in \Lambda}{\cup} i^\lambda(\Omega_\lambda)$. Moreover, $\Omega = \underset{n \in \mathbb{N}}{\oplus} \Omega^n$ with $\Omega^n = \underrightarrow{\lim}\, \Omega_\lambda^n$, which is also the image of $\bigoplus_{\lambda \in \Lambda} \Omega_\lambda^n$ in Ω.

For given $\omega, \omega' \in \Omega$ we can choose a $\lambda \in \Lambda$ and elements $\omega_\lambda, \omega_\lambda' \in \Omega_\lambda$ with $\omega = i^\lambda(\omega_\lambda)$, $\omega' = i^\lambda(\omega_\lambda')$. If we define $\omega \cdot \omega' := i^\lambda(\omega_\lambda \cdot \omega_\lambda')$, then this product is independent of the choice of λ and $\omega_\lambda, \omega_\lambda'$. Similarly, we define $d\omega := i^\lambda(d_\lambda \omega_\lambda)$, if d_λ is the differentiation of Ω_λ. It is now easy to check that Ω, with the above grading, multiplication, and differentiation is a differential algebra, and that $i^\lambda : \Omega_\lambda \to \Omega$ is a homomorphism of differential algebras for all $\lambda \in \Lambda$.

Suppose Ω^* is an arbitrary differential algebra and $j^\lambda : \Omega_\lambda \to \Omega^*$ are morphisms in $\mathcal{D}_{R_0}$ such that $j^\lambda = j^{\lambda'} \circ \varphi_\lambda^{\lambda'}$, for $\lambda \leq \lambda'$. Then there is precisely one homomorphism of R_0-modules $h : \Omega \to \Omega^*$ with $j^\lambda = h \circ i^\lambda$ for all $\lambda \in \Lambda$ (induced by the canonical mapping $\bigoplus_{\lambda \in \Lambda} \Omega_\lambda \to \Omega^*$). It is easily seen that h is already a homomorphism of differential algebras. We denote the direct limit of $\{\Omega_\lambda, \varphi_\lambda^{\lambda'}\}$ in $\mathcal{D}_{R_0}$ by $\varinjlim \Omega_\lambda$.

Since arbitrary coproducts are direct limits of finite coproducts, we have also shown the <u>existence of coproducts</u> in $\mathcal{D}_{R_0}$. In order to prove the existence of arbitrary colimits, and therefore finish the proof of theorem 2.11, it will be enough to show the <u>existence of coequalizers</u> in $\mathcal{D}_{R_0}$.

As in the existence proof for equalizers let $\alpha, \beta : \Omega_1 \to \Omega_2$ be two morphisms in $\mathcal{D}_{R_0}$. Consider the (two-sided ideal) I of Ω_2 generated by all (homogeneous) elements $\alpha(\omega_n) - \beta(\omega_n)$ with $\omega_n \in \Omega_1^n$ ($n \in \mathbb{N}$). For $\omega_m' \in \Omega_2^m$ we have

$$d_2(\omega_m' \cdot (\alpha(\omega_n) - \beta(\omega_n))) =$$
$$d_2\omega_m'(\alpha(\omega_n) - \beta(\omega_n)) + (-1)^m \omega_m'(\alpha(d_1\omega_n) - \beta(d_1\omega_n))$$

hence I is homogeneous and differentially closed. Therefore the residue class differential algebra $\Omega := \Omega_2/I$ exists (2.9). For the canonical epimorphism $\varepsilon : \Omega_2 \to \Omega_2/I$ one obviously has $\varepsilon \circ \alpha = \varepsilon \circ \beta$.

If $\varphi : \Omega_2 \to \Omega^*$ is an arbitrary morphism in $\mathcal{D}_{R_O}$ with $\varphi \circ \alpha = \varphi \circ \beta$, then $I \subset \ker \varphi$ and hence there is exactly one morphism $h : \Omega \to \Omega^*$ in $\mathcal{D}_{R_O}$ with $\varphi = h \circ \varepsilon$ by the universal property of Ω_2/I. Therefore Ω is the coequalizer of α and β.

Observe that, if $\Omega \in \mathcal{D}_{R_O}$ is a limit (colimit) of differential algebras Ω_λ, then Ω^O is the corresponding limit (colimit) of the R_O-algebras Ω_λ^O in $\mathcal{A}_{R_O}$.

Exercises

1) Let (Ω, d) be a differential algebra of an algebra R/R_O.

   a) For each family $\{r_\lambda\}_{\lambda \in \Lambda}$ of elements of R the ideal generated by $\{dr_\lambda\}_{\lambda \in \Lambda}$ in Ω is homogeneous and differentially closed.

   b) If the R-module Ω^1 can be generated by n elements, then $\Omega^m = O$ for $m > n$.

2) Let R_O be a ring with $\mathbb{Q} \subset R$, and let $R := R_O[X_1, \ldots, X_n]$. Let Ω be the differential algebra of R/R_O described in example 2.5d) with $\Omega^1 = RdX_1 \oplus \ldots \oplus RdX_n$. For each $\omega = \sum_{i=1}^n f_i dX_i \in \Omega^1$ $(f_i \in R)$ with $d\omega = O$ there is $g \in R$ with $dg = \omega$.

3) Let R_O be a ring, $R := R_O[X_1, X_2]$, $M := RdX_1 \oplus RdX_2$, and $d : R \to M$ the derivation with $df = \frac{\partial f}{\partial X_1}dX_1 + \frac{\partial f}{\partial X_2}dX_2$ for all $f \in R$. Let U be the R-submodule of M generated by

$X_1 dX_2$, $\Omega^1 := M/U$, and let $\delta : R \to \Omega^1$ be the composition of d with the canonical epimorphism $M \to M/U$. Put $\Omega := \wedge\Omega^1$ (the exterior algebra of the R-module Ω^1). Show that it is <u>not</u> possible to extend $\delta : R \to \Omega^1$ to a mapping $\delta : \Omega \to \Omega$ in such a way that (Ω,δ) becomes a differential algebra of R/R_o.

4) A limit of a diagram

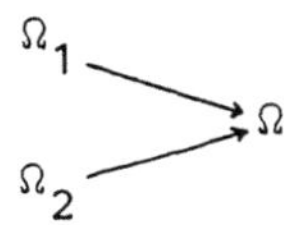

is written $\Omega_1 \underset{\Omega}{\prod} \Omega_2$ and called <u>fiber-product</u> (pull-back).
A colimit of a diagram

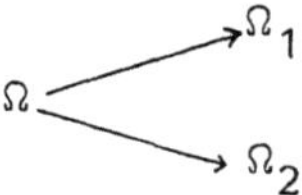

is written $\Omega_1 \underset{\Omega}{\amalg} \Omega_2$ and called <u>fiber-sum</u> (push-out). Give a construction of $\Omega_1 \underset{\Omega}{\prod} \Omega_2$ and $\Omega_1 \underset{\Omega}{\amalg} \Omega_2$ in the category of R_o-differential algebras.

§ 3. Universal Extension of a Differential Algebra

We first prove the existence of a "universal" differential algebra of R/R_o , of which any other differential algebra of R/R_o is a homomorphic image. Later the notion of universal differential algebra will be generalized to that of a "universal extension" of a differential algebra.

Let R/R_o and S/R_o be two algebras and $\rho : R \to S$ an R_o-algebra-homomorphism. If (Ω,d) is a differential algebra of R/R_o and (Ω',d') a differential algebra of S/R_o, there is at most one homomorphism of differential algebras $\varphi : \Omega \to \Omega'$ with $\varphi^o = \rho$ (2.8a)). We call such φ a $\underline{\rho\text{-homomorphism}}$. It satisfies the formula

$$(1) \qquad \varphi(\Sigma r dr_1 \ldots dr_n) = \Sigma \rho(r) d' \rho(r_1) \ldots d' \rho(r_n).$$

In the special case $S = R$, $\rho = id_R$, a ρ-homomorphism is also called $\underline{R\text{-homomorphism}}$.

$\underline{3.1.\ Lemma.}$ a) A ρ-homomorphism $\varphi : \Omega \to \Omega'$ is surjective if and only if ρ is. In particular, any R-homomorphism is surjective.

b) If Ω and Ω' are differential algebras of R/R_o and $\varphi : \Omega \to \Omega'$, $\psi : \Omega' \to \Omega$ R-homomorphisms, then φ and ψ are isomorphisms, inverse to each other.

a) is an immediate consequence of (1). In order to show b) one has simply to observe that an R-homomorphism $\Omega \to \Omega$ is necessarily the identity map.

<u>3.2. Theorem</u> (Existence of the universal differential algebra). For each algebra R/R_o there is a differential algebra $(\Omega_{R/R_o}, d)$ of R/R_o having the following property: For any homomorphism $\rho : R \to S$ of R_o-algebras and any differential algebra (Ω, δ) of S/R_o there is a (unique) ρ-homomorphism $h : \Omega_{R/R_o} \to \Omega$.

Proof. For each $r \in R$ we choose an indeterminate Δr and form the free R-algebra $R[\Delta R]$ on $\Delta R := \{\Delta r \mid r \in R\}$. This is the <u>non-commutative</u> polynomial algebra in the indeterminates Δr, where the elements of R, however, commute with the Δr. Any element of $R[\Delta R]$ is a finite sum of "monomials" $r_o \Delta r_1 \cdots \Delta r_n$ $(r_o, r_1, \ldots, r_n \in R)$. If we give the elements of R degree 0 and Δr $(r \in R)$ degree 1, then $R[\Delta R]$ becomes a graded R-algebra: $R[\Delta R] = R \oplus R\Delta R \oplus R\Delta R\Delta R \oplus \ldots$.

Let I be the two-sided ideal of $R[\Delta R]$ generated by all elements

$$
\begin{aligned}
&\Delta(r+r') - \Delta r - \Delta r' \\
&\Delta(r \cdot r') - r\Delta r' - r'\Delta r \\
&\Delta(r_o \cdot 1) \\
&\Delta r \cdot \Delta r
\end{aligned}
$$

(2)

with $r, r' \in R$, $r_o \in R_o$. Since I is generated by homogeneous elements of $R[\Delta R]$, it is a homogeneous ideal. Put

$$\Omega_{R/R_o} := R[\Delta R]/I.$$

This is a graded R-algebra with $\Omega^o_{R/R_o} = R$, since $I \cap R = (0)$. As an R-algebra, Ω_{R/R_o} is generated by the elements

$$dr := \Delta r + I \quad (r \in R)$$

because the Δr generate $R[\Delta R]$. Since the elements (2) are in I, we have in Ω_{R/R_o} relations

$$d(r+r') = dr + dr'$$
$$d(r\cdot r') = rdr' + r'dr$$
$$(2')\qquad d(r_o\cdot 1) = 0$$
$$dr\cdot dr = 0$$

for all $r, r' \in R$, $r_o \in R_o$. From the last relation we can easily conclude that Ω_{R/R_o} is anticommutative, which is equivalent with $\Delta r \Delta r' + \Delta r' \Delta r$ being an element of I for all $r, r' \in R$.

Let $\Delta : R[\Delta R] \to R[\Delta R]$ be the $\mathbb{Z}$-linear map with $\Delta(r\Delta r_1 \ldots \Delta r_n) = \Delta r \Delta r_1 \ldots \Delta r_n$ for all $r, r_1, \ldots, r_n \in R$. We wish to show $\Delta(I) \subset I$.

If α is one of the elements (2) and if $\beta, \beta' \in R[\Delta R]$, we have to show that $\Delta(\beta\cdot\alpha\cdot\beta') \in I$. It is enough to show this, if β and β' are of the form $r\Delta r_1 \ldots \Delta r_n$. Using that $\Delta r \Delta r' \equiv -\Delta r' \Delta r$ mod I for $r, r' \in R$, it suffices to show $\Delta(\tilde{r}\cdot\alpha) \in I$ for all $r \in R$. This is only in question, if $\alpha = \Delta(r\cdot r') - r\Delta r' - r'\Delta r$. But in this case

$$\Delta(\tilde{r}\alpha) = \Delta\tilde{r}\cdot\Delta(r\cdot r') - \Delta(\tilde{r}\cdot r)\Delta r' - \Delta(\tilde{r}r')\Delta r \equiv$$
$$r\Delta\tilde{r}\Delta r' + r'\Delta\tilde{r}\Delta r - \tilde{r}\Delta r\Delta r' - r\Delta\tilde{r}\Delta r' - \tilde{r}\Delta r'\Delta r - r'\Delta\tilde{r}\Delta r \equiv$$
$$-\tilde{r}(\Delta r\Delta r' + \Delta r'\Delta r) \equiv 0 \text{ mod } I. \text{ Hence we have in fact } \Delta(I) \subset I.$$

Δ induces a $\mathbb{Z}$-linear mapping

$$d : \Omega_{R/R_o} \to \Omega_{R/R_o} \qquad (d(\beta+I) = \Delta\beta+I \text{ for } \beta \in R[\Delta R]).$$

In particular, $r \in R$ is mapped onto the element, which was denoted dr already at the beginning of the proof. The formulas (2') show that d is R_o-linear, and from the definition of Δ we obtain $d(rdr_1 \ldots dr_n) = drdr_1 \ldots dr_n$ for $r, r_1, \ldots, r_n \in R$.

Putting everything together, we have shown that $(\Omega_{R/R_O}, d)$ is a differential algebra of R/R_O.

If $\rho : R \to S$ and Ω are given as in the theorem, there is a ring homomorphism $\chi : R[\Delta R] \to \Omega$ with $\chi(r) = \rho(r)$ and $\chi(\Delta r) = \delta\rho(r)$ for all $r \in R$. χ is compatible with the gradings and maps the elements (2) onto O. Hence $\chi(I) = O$, and we have an induced homomorphism of graded algebras $h : \Omega_{R/R_O} \to \Omega$ with $h(r) = \rho(r)$ and $h(dr) = \delta\rho(r) = \delta h(r)$ for all $r \in R$. The last formula implies that h commutes with differentiation, and hence is a ρ-homomorphism of differential algebras.

3.3. Definition. $(\Omega_{R/R_O}, d)$ is called the <u>universal differential algebra</u> (deRham algebra) of the algebra R/R_O.

This concept is due to Kähler [Kä]. Ω_{R/R_O} is uniquely determined (up to canonical isomorphism) by the following universal property:

3.4. Corollary. For any differential algebra Ω of R/R_O there is an R-homomorphism $h : \Omega_{R/R_O} \to \Omega$. In particular, $\Omega = \Omega_{R/R_O}/I$ with a homogeneous differentially closed ideal I of Ω_{R/R_O}.

It is clear that $\Omega_{R/R_O} = \Omega_{R/R_O'}$, where R_O' is the image of R_O in R. Moreover, the universal differential algebra defines a covariant functor $\mathcal{A}_{R_O} \to \mathcal{D}_{R_O}$:

3.5. Corollary. If $\rho : R \to S$ is a homomorphism of R_O-algebras, then there is a ρ-homomorphism $\rho^* : \Omega_{R/R_O} \to \Omega_{S/R_O}$

$(\rho*(\Sigma r_0 dr_1 \ldots dr_n) = \Sigma \rho(r_0) d\rho(r_1) \ldots d\rho(r_n))$.

A relation between differential algebras and exterior algebras is given in the next proposition.

3.6. Proposition.   Let (Ω,d) be a differential algebra.

a) If $\omega \in \Omega$ is homogeneous of odd degree, then $\omega^2 = 0$.

b) There is a canonical epimorphism of graded R-algebras
$\varepsilon : \Lambda\Omega^1 \to \Omega$ $(\varepsilon(a_1 \wedge \ldots \wedge a_n) = a_1 \cdot \ldots \cdot a_n$ for $a_1, \ldots, a_n \in \Omega^1)$.

Proof.   a) For $\omega = rdr_1 \ldots dr_m$ $(m > 0)$ we have $\omega^2 = 0$, since $dr_1 dr_1 = 0$ and Ω is anticommutative. If m is odd, and $\omega, \omega' \in \Omega$ are homogeneous of degree m, then $\omega\omega' + \omega'\omega = 0$ by 2.2e). Now let $\omega = \sum_{i=1}^{t} \omega_i$, where the ω_i are of the form $rdr_1 \ldots dr_m$ with a fixed odd m. Then $\omega^2 = \sum_{i,j=1}^{t} \omega_i \omega_j = \sum_{i<j} (\omega_i \omega_j + \omega_j \omega_i) = 0$.

b) Since $\omega^2 = 0$ for each $\omega \in \Omega^1$ by a), the universal property of the exterior algebra $\Lambda\Omega^1$ allows us to extend the canonical injection $\Omega^1 \to \Omega$ to a homomorphism of R-algebras $\varepsilon : \Lambda\Omega^1 \to \Omega$ with $\varepsilon(a_1 \wedge \ldots \wedge a_n) = a_1 \cdot \ldots \cdot a_n$ for $a_1, \ldots, a_n \in \Omega^1$. Since Ω is generated as an R-algebra by Ω^1, this homomorphism is surjective.

3.7. Definition.   Under the assumptions of 3.6, Ω is called an **exterior differential algebra**, if the canonical epimorphism ε is an isomorphism.

3.8. Proposition.   a) Ω_{R/R_0} is an exterior differential algebra.

b) Let $\Omega = \Omega_{R/R_0}/I$ be an arbitrary differential algebra of

R/R_o. Ω is an exterior differential algebra if and only if I is generated by its elements of degree 1.

Proof. a) We construct an inverse to $\varepsilon : \Lambda\Omega^1_{R/R_o} \to \Omega_{R/R_o}$. Write $\Omega_{R/R_o} = R[\Delta R]/I$ as in the proof of 3.2. Let $h : R[\Delta R] \to \Lambda\Omega^1_{R/R_o}$ be the ring homomorphism with $h(r) = r$, $h(\Delta r) = dr$ for all $r \in R$. Then $h(r\Delta r_1 \ldots \Delta r_n) = rdr_1 \wedge \ldots \wedge dr_n$ for $r,r_1,\ldots,r_n \in R$. In $\Lambda\Omega^1_{R/R_o}$ the formulas $d(r+r') = dr+dr'$, $d(r \cdot r') = rdr'+r'dr$, $d(r_o \cdot 1) = 0$ and $dr \wedge dr = 0$ are true for $r,r' \in R$, $r_o \in R_o$. Therefore $h(I) = 0$, and h induces a ring homomorphism

$$\eta : \Omega_{R/R_o} \to \Lambda\Omega^1_{R/R_o} \qquad (rdr_1 \ldots dr_m \mapsto rdr_1 \wedge \ldots \wedge dr_m)$$

which is obviously an inverse to ε.

b) Consider the commutative diagram of canonical homomorphisms

$$
\begin{array}{ccc}
\Lambda\Omega^1_{R/R_o} & \xrightarrow{\ \alpha\ } & \Lambda\Omega^1 \\
{\scriptstyle\varepsilon}\downarrow & & \downarrow{\scriptstyle\varepsilon'} \\
\Omega_{R/R_o} & \xrightarrow[\ \beta\ ]{} & \Omega
\end{array}
$$

Since $\Omega^1 = \Omega^1_{R/R_o}/I^1$, we have $\Lambda\Omega^1 = \Lambda\Omega^1_{R/R_o}/(I^1)$, and $\ker(\alpha)$ is generated by I^1. Therefore ε' is an isomorphism if and only if $I = \ker(\beta)$ is also generated by I^1.

Next we shall exhibit the relation between Ω_{R/R_o} and the module of Kähler differentials, introduced in 1.20.

3.9. Proposition. Let $(\Omega_{R/R_o},d)$ be the universal differential algebra of R/R_o. The R_o-derivation $d : R \to \Omega^1_{R/R_o}$ is universal, hence Ω^1_{R/R_o} can be canonically identified with the module of Kähler differentials of R/R_o (which was in § 1 already denoted Ω^1_{R/R_o}).

Proof. Let $\delta : R \to M$ be a derivation of R/R_o into an R-module M. Let $R\delta R$ denote the submodule of M generated by $\{\delta r \mid r \in R\}$. Make $R \ltimes R\delta R$ into a differential algebra as in 2.5c). By 3.4 there is an R-homomorphism $\Omega_{R/R_o} \to R \ltimes R\delta R$, in particular an R-linear map $\ell : \Omega^1_{R/R_o} \to R\delta R$ with $\ell \circ d = \delta$. Since Ω^1_{R/R_o} is generated as an R-module by $\{dr \mid r \in R\}$, there can be only one such ℓ. Hence $d : R \to \Omega^1_{R/R_o}$ is universal.

We see that Ω_{R/R_o} is the exterior algebra of the module of Kähler differentials, with the universal derivation $d : R \to \Omega^1_{R/R_o}$ extended to all of Ω_{R/R_o}.

3.10. Examples.

a) The differential algebra described in 2.5d) is the universal differential algebra of the polynomial algebra $R_o[X_1, \ldots, X_n]/R_o$:

$$\Omega_{R_o[X_1, \ldots, X_n]/R_o} = \Lambda(RdX_1 \oplus \ldots \oplus RdX_n).$$

b) If L/K is a separable algebraic field extension, then $\Omega_{L/K}$ is the trivial differential algebra of L.

c) If K is a perfect field of characteristic $p > 0$, then $\Omega_{K/\mathbb{Z}}$ is trivial.

For many purposes it is useful to generalize the notion of universal differential algebra. We fix a pair $(R/R_o, \Omega)$, where R/R_o is an algebra and Ω a differential algebra of R/R_o. Let $\rho : R \to S$ be a homomorphism of R_o-algebras.

3.11. __Definition.__   A __universal ρ-extension of Ω__ is a pair $(\Omega_\rho, \varphi_\rho)$, where Ω_ρ is a differential algebra of S/R_o and $\varphi_\rho : \Omega \to \Omega_\rho$ a ρ-homomorphism, such that the following universal property is satisfied: Let $\sigma : S \to T$ be any morphism in $\mathcal{A}_{R_o}$, Ω^* a differential algebra of T/R_o and $\psi : \Omega \to \Omega^*$ a $\sigma \circ \rho$-homomorphism. Then there is a σ-homomorphism $h : \Omega_\rho \to \Omega^*$.

If the universal property is satisfied, then $\psi = h \cdot \varphi_\rho$ by 2.8b). Moreover, $(\Omega_\rho, \varphi_\rho)$ is unique up to canonical isomorphism. To obtain uniqueness it is already enough to require the universal property in case $T = S$, $\sigma = \mathrm{id}_S$.

When it is clear, which ρ we mean, we usually write Ω_S for Ω_ρ and call Ω_S the __universal S-extension of Ω.__ If $\Omega = R$ is the trivial differential algebra, then the mapping

$$\varphi_\rho : R \to \Omega_{S/R} \qquad (\varphi_\rho(r) = \rho(r) \text{ for all } r \in R)$$

has the universal property of 3.11 (by 3.2), hence the universal differential algebra $\Omega_{S/R}$ is the universal S-extension of the trivial differential algebra of R.

3.12. __Theorem.__   In the situation of 3.11 the universal ρ-extension $(\Omega_\rho, \varphi_\rho)$ exists.

Proof.   Write $\Omega = \Omega_{R/R_o}/I$ with a homogeneous differentially closed ideal I of Ω_{R/R_o} (3.4). Let $I \cdot \Omega_{S/R_o}$ denote the extension ideal of I under the functorial homomorphism $\Omega_{R/R_o} \to \Omega_{S/R_o}$. It is generated by homogeneous elements. By the following lemma it is also differentially closed.

3.13. __Lemma.__   If an ideal I of a differential algebra Ω is generated by homogeneous elements ω_λ and their derivatives $d\omega_\lambda$

then I is differentially closed.

For each homogeneous $\omega \in \Omega$ we have
$d(\omega\omega_\lambda) = d\omega \cdot \omega_\lambda \pm \omega d\omega_\lambda \in I$ and $d(\omega d\omega_\lambda) = d\omega d\omega_\lambda \in I$, from which the statement of the lemma follows.

Define $\Omega_\rho := \Omega_{S/R_O}/I\Omega_{S/R_O}$. Since $I\Omega_{S/R_O}$ does not contain elements of degree O, Ω_ρ is a differential algebra of S/R_O The mapping $\varphi_\rho : \Omega \to \Omega_\rho$ induced by $\Omega_{R/R_O} \to \Omega_{S/R_O}$ is a ρ-homomorphism.

For ψ as in 3.11 let $\Omega_{S/R_O} \to \Omega*$ be a σ-homomorphism, which exists according to 3.2, and consider the following diagram in $\mathcal{D}_{R_O}$:

$$
\begin{array}{ccc}
\Omega_{R/R_O} & \longrightarrow & \Omega_{S/R_O} \\
\downarrow & & \downarrow \\
\Omega & \underset{\psi}{\longrightarrow} & \Omega*
\end{array}
$$

It is commutative, as it commutes on elements of degree O. Since I has image O in $\Omega*$, there is an induced σ-homomorphism of $\Omega_\rho = \Omega_{S/R_O}/I\Omega_{S/R_O}$ into $\Omega*$, which is what had to be shown.

We note the following information that was obtained in the proof.

__3.14. Remark.__   Let $\Omega = \Omega_{R/R_O}/I$. Then

$$\Omega_\rho = \Omega_{S/R_O}/I\Omega_{S/R_O}.$$

__3.15. Corollary__ (Functorial property of the universal extension). Given a commutative diagram

$$
\begin{array}{ccc}
R & \overset{\rho}{\longrightarrow} & S \\
\varphi\downarrow & & \downarrow\psi \\
R' & \underset{\rho'}{\longrightarrow} & S'
\end{array}
$$

in $\mathscr{A}_{R_O}$, differential algebras Ω of R/R_O, Ω' of R'/R_O, and a
φ-homomorphism $h : \Omega \to \Omega'$, there is a (unique) ψ-homomorphism
$h_\psi : \Omega_\rho \to \Omega'_{\rho'}$, and the diagram

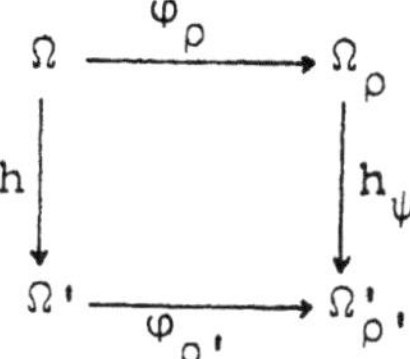

of $\mathscr{D}_{R_O}$ is commutative.

The existence of h_ψ is a consequence of the universal
property of Ω_ρ. The diagram is commutative, since in degree O
it is the original diagram in $\mathscr{A}_{R_O}$.

3.16. Corollary. If Ω is an exterior differential algebra,
so is Ω_ρ.

By 3.8b) I is generated by its elements of degree 1 and so
is $I\Omega_{S/R_O}$ in Ω_{S/R_O}. The claim follows from the formula of
3.14, on account of 3.8b).

We now want to show the <u>transitive law for universal ex-
tensions</u>.

3.17. Proposition. Given a sequence $R \overset{\rho}{\to} S \overset{\sigma}{\to} T$ in $\mathscr{A}_{R_O}$ and
a differential algebra Ω of R/R_O, there is a canonical T-iso-
morphism

$$\Omega_{\sigma \circ \rho} \cong (\Omega_\rho)_\sigma.$$

Proof. The composition of canonical homomorphisms
$\Omega \to \Omega_\rho \to (\Omega_\rho)_\sigma$ is a $\sigma \circ \rho$-homomorphism, hence there is a T-
homomorphism $\alpha : \Omega_{\sigma \circ \rho} \to (\Omega_\rho)_\sigma$ by the universal property of
$\Omega_{\sigma \circ \rho}$. By the functorial property of universal extensions

there is a σ-homomorphism $\Omega_\rho \to \Omega_{\sigma \circ \rho}$ and hence a T-homomorphism $\beta : (\Omega_\rho)_\sigma \to \Omega_{\sigma \circ \rho}$ by the universal property of $(\Omega_\rho)_\sigma$.
By 3.1b), α and β are inverse isomorphisms.

3.18. Corollary. $\quad \Omega_{S/R_o} = (\Omega_{R/R_o})_\rho.$

This corollary is frequently used to compute Ω_{S/R_o}, once
Ω_{R/R_o} is known. A similar formula is given in the next proposition.

3.19. Proposition. Let $\rho : R \to S$ be a homomorphism of R_o-algebras, Ω a differential algebra of R/R_o, and Ω_ρ its uni-
versal ρ-extension. Then there is an S-isomorphism

$$\Omega_{S/R} \cong \Omega_\rho / (\{d\rho(r)\}_{r \in R}).$$

In particular

$$\Omega_{S/R} \cong \Omega_{S/R_o} / (\{d\rho(r)\}_{r \in R}).$$

Proof. The ideal $I := (\{d\rho(r)\}_{r \in R})$ is homogeneous, and by
3.13 it is also differentially closed. Hence Ω_ρ/I is a differential algebra of S/R_o. Since

$$d(r\omega) = d\rho(r) \cdot \omega + rd\omega \qquad (r \in R, \omega \in \Omega_\rho)$$

the differentiation on Ω_ρ/I is R-linear, and so Ω_ρ/I is a
differential algebra of S/R. There is an S-homomorphism
$\alpha : \Omega_{S/R} \to \Omega_\rho/I.$

Conversely, the mapping $\Omega \to \Omega_{S/R}$, which is ρ on $\Omega^o = R$
and O on Ω^n $(n > 0)$, is a ρ-homomorphism. It induces an S-
homomorphism $h : \Omega_\rho \to \Omega_{S/R}$. Since $h(d\rho(r)) = 0$ for all $r \in R$,
h induces an S-homomorphism $\beta : \Omega_\rho/I \to \Omega_{S/R}$, and by 3.1b)
α and β are isomorphisms.

In the rest of this paragraph we shall apply the statements about universal extensions of differential algebras to derivations.

3.20. Proposition. Let R/R_o be an algebra and $\delta : R \to M$ an R_o-derivation into an R-module M with $M = R\delta R$. For each algebra S/R there exists a universal S-extension of δ (see 1.24 for definition).

Proof. Let $\Omega := R \ltimes M$ be the differential algebra associated with δ (see 2.5c)). Let (Ω_S, d) be its universal S-extension and put $N := \Omega_S^1$. From the canonical homomorphism $\Omega \to \Omega_S$ we obtain a commutative diagram

$$
\begin{array}{ccc}
R & \xrightarrow{\ \delta\ } & M = \Omega^1 \\
\downarrow & & \downarrow{\scriptstyle \ell} \\
S & \xrightarrow{\ d\ } & N = \Omega_S^1
\end{array}
$$

where ℓ is R-linear. Therefore d is an extension of δ. To show that it is a universal extension of δ, consider an arbitrary extension $\Delta : S \to N^*$ of δ. Again $\Omega^* := S \ltimes S\Delta S$ is a differential algebra of S/R_o, and since Δ is an extension of δ, there is a ρ-homomorphism $\Omega \to \Omega^*$, where $\rho : R \to S$ is the structure homomorphism of S/R.

By the universal property of Ω_S there is an S-homomorphism $h : \Omega_S \to \Omega^*$. In particular, there is an S-linear map $h^1 : \Omega_S^1 = N \to \Omega^{*1} \subset N^*$ such that $\Delta = h^1 \circ d$. Since Ω_S^1 is generated by $\{ds \mid s \in S\}$ as an S-module, there can be only one such h^1. Hence d is a universal extension of δ.

A universal extension $d : S \to N$ of δ is, of course, unique up to canonical isomorphism. We write $\Omega_{S/\delta}^1$ for N and call

$\Omega^1_{S/\delta}$ the _module of Kähler differentials of S/δ_. If δ is the trivial derivation of R, then $\Omega^1_{S/\delta} = \Omega^1_{S/R}$ is the usual module of Kähler differentials of S/R.

For an S-module N let $\text{Der}_\delta(S,N)$ denote the set of all derivations $\Delta : S \to N$ that extend δ. As is easily seen $\text{Der}_\delta(S,N)$ is a submodule of the S-module $\text{Der}_{R_O}(S,N)$. We sometimes call the elements of $\text{Der}_\delta(S,N)$ _derivations of S/δ_.

<u>3.21. Proposition.</u> a) The mapping

$$u_N : \text{Hom}_S(\Omega^1_{S/\delta},N) \to \text{Der}_\delta(S,N) \qquad (\ell \mapsto \ell \circ d)$$

is an isomorphism of S-modules (d := universal S-extension of δ).
b) For each S-linear map $\alpha : N \to N'$ from N into an S-module N' the diagram

$$
\begin{array}{ccc}
\text{Hom}_S(\Omega^1_{S/\delta},N) & \xrightarrow{\ u_N\ } & \text{Der}_\delta(S,N) \\
\downarrow{\scriptstyle \text{Hom}_S(\Omega^1_{S/\delta},\alpha)} & & \downarrow{\scriptstyle \text{Der}_\delta(S,\alpha)} \\
\text{Hom}_S(\Omega^1_{S/\delta},N') & \xrightarrow{\ u_{N'}\ } & \text{Der}_\delta(S,N')
\end{array}
$$

commutes. Here $\text{Der}_\delta(S,\alpha)(\Delta) = \alpha \circ \Delta$ for all $\Delta \in \text{Der}_\delta(S,N)$.

That u_N is an isomorphism is just another formulation of the universal property of $\Omega^1_{S/\delta}$. The statement in b) can be checked immediately.

<u>3.22. Corollary.</u>   For each exact sequence of S-modules $O \to A \xrightarrow{\alpha} B \xrightarrow{\beta} C$ the following sequence is exact

$$(3) \qquad O \to \text{Der}_\delta(S,A) \xrightarrow{\ \text{Der}_\delta(S,\alpha)\ } \text{Der}_\delta(S,B) \xrightarrow{\ \text{Der}_\delta(S,\beta)\ } \text{Der}_\delta(S,C).$$

This follows from 3.21 and the left exactness of the Hom-functor.

Let Ω be the differential algebra given by δ (2.5c). If we consider $\Omega^1_{S/\delta}$ as the module of elements of degree one of Ω_S, proposition 3.15 immediately yields

3.23. <u>Proposition</u> (Functorial property of $\Omega^1_{S/\delta}$).

Let S/R and T/R be algebras and $d_S : S \to \Omega^1_{S/\delta}$, $d_T : T \to \Omega^1_{T/\delta}$ the universal extensions of δ to S resp. T. Then for any R-algebra homomorphism $\psi : S \to T$ there is exactly one S-linear map ψ^* such that the following diagram commutes:

$$
\begin{array}{ccc}
S & \xrightarrow{\;d_S\;} & \Omega^1_{S/\delta} \\
\psi \downarrow & & \downarrow \psi^* \\
T & \xrightarrow{\;d_T\;} & \Omega^1_{T/\delta}
\end{array}
$$

ψ^* satisfies the formula $\psi^*(d_S s) = d_T(\psi(s))$ for all $s \in S$.

Similarly, proposition 3.19 implies

3.24. <u>Proposition</u>. Under the assumptions of 3.23, there is a T-linear map

$$\beta : \Omega^1_{T/\delta} \to \Omega^1_{T/S} \quad \text{with} \quad \beta(d_T t) = dt \quad \text{for all } t \in T$$

where $d : T \to \Omega^1_{T/S}$ is the universal derivation of T/S and the sequence

$$(4) \qquad T \otimes_S \Omega^1_{S/\delta} \xrightarrow{\alpha} \Omega^1_{T/\delta} \xrightarrow{\beta} \Omega^1_{T/S} \to 0$$

is exact (α is the mapping induced by ψ^*). In particular, there is an exact sequence

$$(4') \qquad T \otimes_S \Omega^1_{S/R_0} \xrightarrow{\alpha} \Omega^1_{T/R_0} \xrightarrow{\beta} \Omega^1_{T/S} \to 0.$$

3.25. <u>Corollary</u>. For each T-module N one has an exact sequence

$$(5) \qquad 0 \to \mathrm{Der}_S(T,N) \xrightarrow{\;\beta^*\;} \mathrm{Der}_\delta(T,N) \xrightarrow{\;\alpha^*\;} \mathrm{Der}_\delta(S,N).$$

Here $\beta*$ is the inclusion mapping, given by the fact that a derivation of T/S is always a T-extension of δ, and $\alpha*$ is the mapping $\Delta \mapsto \Delta \circ \psi$.

Proof. Applying the Hom-functor to (4) we get an exact sequence

$$0 \to \mathrm{Hom}_T(\Omega^1_{T/S}, N) \to \mathrm{Hom}_T(\Omega^1_{T/\delta}, N) \to \mathrm{Hom}_T(T \otimes_S \Omega^1_{S/\delta}, N).$$

By 3.21a) the first module can be identified with $\mathrm{Der}_S(T,N)$, the second with $\mathrm{Der}_\delta(T,N)$ and the third with $\mathrm{Der}_\delta(S,N)$, since $\mathrm{Hom}_T(T \otimes_S \Omega^1_{S/\delta}, N) \cong \mathrm{Hom}_S(\Omega^1_{S/\delta}, N)$.

For applications of the exact sequence (5), see exercise 5).

The question, in which cases the exact sequences (3),(4) and (5) can be extended by 0 on the right (left), is important for many applications. Results of this type will be considered later. More generally it can be shown that for each algebra S/R there are two families $\{T_i(S/R,M)\}_{i \in \mathbb{N}}$ and $\{T^i(S/R,M)\}_{i \in \mathbb{N}}$ of functors of S-modules M having the following properties:

a) The T_i and T^i are S-modules, and $T_0(S/R,M) = \Omega^1_{S/R} \otimes_S M$, $T^0(S/R,M) = \mathrm{Der}_R(S,M)$.

b) If $0 \to M \to N \to P \to 0$ is an exact sequence of S-modules, there are long exact sequences

$$\ldots \to T_i(S/R,M) \to T_i(S/R,N) \to T_i(S/R,P) \to T_{i-1}(S/R,M) \to \ldots$$
$$\ldots \to T_0(S/R,M) \to T_0(S/R,N) \to T_0(S/R,P) \to 0$$

and

$$0 \to T^0(S/R,M) \to T^0(S/R,N) \to T^0(S/R,P) \to T^1(S/R,M) \to \ldots$$

c) If $R \to S \to T$ is a sequence in $\mathcal{A}_{R_0}$, and M is a T-module, there are long exact sequences

$$\cdots \to T_i(S/R,M) \to T_i(T/R,M) \to T_i(T/S,M) \to T_{i-1}(S/R,M) \to \cdots$$

$$\cdots \to T_o(S/R,M) \to T_o(T/R,M) \to T_o(T/S,M) \to 0$$

and

$$0 \to T^o(T/S,M) \to T^o(T/R,M) \to T^o(S/R,M) \to T^1(T/S,M) \to \cdots$$

The T_i and T^i are called <u>cotangent functors</u>. The sequences in b) and c) extend the sequences (3), (4) and (5) in the case that δ is trivial to long exact sequences. There are many "vanishing theorems" for the modules $T_i(S/R,M)$ and $T^i(S/R,M)$, which yield important information about algebras S/R. We are not going to discuss this theory, but we will introduce later in §5 an analogue to the functor T_1 in the very special case of a field extension. For more informations about the T_i and T^i, see the book [And] of M.André (where they are denoted by H_i and H^i).

<u>Exercises</u>

1) (Extensions of gradings to differential algebras)

Let $R = \bigoplus_{q\in\mathbb{N}} R^q$ be a positively graded ring.

a) Show that Ω_{R/R^o} has a natural structure of a bigraded algebra

$$\Omega_{R/R^o} = \bigoplus_{p,q\in\mathbb{N}} \Omega_{R/R^o}^{p,q} \;,\quad \Omega_{R/R^o}^{p,q} \cdot \Omega_{R/R^o}^{p',q'} \subset \Omega_{R/R^o}^{p+p',q+q'},$$

where $\Omega_{R/R^o}^{o,q} = R^q$ and $dr \in \Omega_{R/R^o}^{1,q}$ for each $r \in R^q$.

(Hint: Use the construction of Ω_{R/R^o} given in the proof of 3.2).

b) Prove that there is an induced structure of bigraded algebras on the deRham-cohomology of Ω_{R/R^o}:

$$H_{DR}(\Omega_{R/R^o}) = \bigoplus_{p,q\in\mathbb{N}} H_{DR}^{p,q}(\Omega_{R/R^o}).$$

2) Under the assumptions of 1) Ω^1_{R/R^O} is a graded module over the graded ring R:

$$\Omega^1_{R/R^O} = \bigoplus_{q\in\mathbb{N}} \Omega^{1,q}_{R/R^O} \;,\; R^p \cdot \Omega^{1,q}_{R/R^O} \subset \Omega^{1,p+q}_{R/R^O} \;.$$

Let M be a $\mathbb{Z}$-graded module over R:

$M = \bigoplus_{q\in\mathbb{Z}} M^q$, $R^p \cdot M^q \subset M^{p+q}$. Let $\mathrm{Der}_{R^O}(R,M)^q$ be the set of all R^O-derivations $d: R \to M$ with $d(R^p) \subset M^{p+q}$ for all $p \in \mathbb{N}$. Show that, if the R-module Ω^1_{R/R^O} is generated by finitely many (homogeneous) elements, then

$$\mathrm{Der}_{R^O}(R,M) = \bigoplus_{q\in\mathbb{Z}} \mathrm{Der}_{R^O}(R,M)^q$$

and $\mathrm{Der}_{R^O}(R,M)$ is a graded module over the graded ring R.

3) (Extension of derivations to differential algebras)

Let R/R_O be an algebra.

a) For each R-linear map $\ell : \Omega^1_{R/R_O} \to R$ there is an R-linear map $\delta_\ell : \Omega_{R/R_O} \to \Omega_{R/R_O}$ (of degree -1) satisfying the rule

$$\delta_\ell(dr_1 \cdot \ldots \cdot dr_m) = \sum_{i=1}^{m} (-1)^{i-1} \ell(dr_i) dr_1 \ldots \widehat{dr_i} \ldots dr_m \quad (r_1,..,r_m \in R)$$

Moreover,

$$\delta_\ell(\omega_m \omega) = \delta_\ell \omega_m \cdot \omega + (-1)^m \omega_m \delta_\ell \omega \text{ for } \omega_m \in \Omega^m_{R/R_O},\; \omega \in \Omega_{R/R_O}.$$

b) For $\delta \in \mathrm{Der}_{R_O}(R)$ let $\ell : \Omega^1_{R/R_O} \to R$ be the R-linear map with $\delta = \ell \circ d$. Define

$$\delta(\omega) := (d \circ \delta_\ell + \delta_\ell \circ d)(\omega) \quad \text{for } \omega \in \Omega_{R/R_O}.$$

Show the rules

α) For $\omega = r \in R$ the two meanings of $\delta(r)$ coincide and

$$\delta(dr) = d(\delta r).$$

β) $\delta(\omega_1\omega_2) = \delta\omega_1 \cdot \omega_2 + \omega_1 \cdot \delta\omega_2$ for $\omega_1, \omega_2 \in \Omega_{R/R_o}$.

γ) If R'/R_o is a subalgebra of R/R_o with $\delta(R') \subset R'$, then δ induces a mapping $\delta : \Omega_{R/R'} \to \Omega_{R/R'}$ also satisfying α) and β).

4) (Extension of automorphisms to differential algebras)

Let Ω be a differential algebra of an algebra R/R_o, and let S/R be another algebra. Let G be a group that acts on S as a group of automorphisms of S/R (In other words, there is given a group homomorphism $G \to \text{Aut}(S/R)$). Let $S^G := \{s \in S \mid s^g = s$ for all $g \in G\}$ be the <u>ring of in-</u><u>variants</u> of this action. Show that the action of G can be extended in a natural manner to all of Ω_S and that the image of $\Omega_{S^G} \to \Omega_S$ is contained in $[\Omega_S]^G$, the sub-algebra of "G-invariant forms" of Ω_S. Moreover, there is an induced action of G on $H_{DR}(\Omega_S)$.

5) Given an algebra T/R and a subalgebra S/R of T/R, the following two statements are equivalent:

α) Each derivation of T/S is trivial.

β) Each derivation Δ of T/R is uniquely determined by its restriction $\Delta|_S : S \to T$.

Suppose now that the canonical mapping $\text{Der}_R(T,T) \to \text{Der}_R(S,T)$ is bijective. Then each $d \in \text{Der}_R(S)$ has a unique extension Δ, which is a derivation of T/R with $\Delta|_S = d$; moreover, any derivation of T/S is trivial.

§ 4. Description of the Universal Extension in Special Cases

In this section we wish to describe the universal extension of a differential algebra in certain cases more closely
than it was possible in the general case. These descriptions
may help to determine explicitely the structure of the universal extension of a differential algebra, or of the universal differential algebra of an algebra.

Let R/R_o be an algebra and (Ω,δ) a differential algebra
of R/R_o.

1) Colimits

Let $S = \varinjlim S_\lambda$ be the colimit of a diagram D in $\mathcal{A}_R$ with
vertices S_λ. By the functorial property of the universal
extension there is a corresponding diagram D' in $\mathcal{D}_{R_o}$ (over the
diagram scheme that underlies D) having vertices Ω_{S_λ}. By 2.11
$\varinjlim \Omega_{S_\lambda}$ exists in $\mathcal{D}_{R_o}$ and is a differential algebra of S/R_o.
Let $i_\lambda : S_\lambda \to S$ and $j_\lambda : \Omega_{S_\lambda} \to \varinjlim \Omega_{S_\lambda}$ be the canonical
homomorphisms into the colimit. Then the composition
$\Omega \to \Omega_{S_\lambda} \xrightarrow{\lambda} \varinjlim \Omega_{S_\lambda}$ is independent of λ, because it coincides
in degree O with the structure morphism $\rho : R \to \varinjlim S_\lambda = S$.
Denote this composition by φ_S.

4.1. Proposition. $(\varinjlim \Omega_{S_\lambda}, \varphi_S)$ is the universal S-extension
of Ω. In other words:

$$\Omega_{\varinjlim S_\lambda} = \varinjlim \Omega_{S_\lambda} .$$

Proof. Let Ω^* be a differential algebra of S/R_o and $\Omega \to \Omega^*$
a ρ-homomorphism. By the universal property of Ω_{S_λ} there is

an i_λ-homomorphism $\psi_\lambda : \Omega_{S_\lambda} \to \Omega^*$. If $u : \Omega_{S_\lambda} \to \Omega_{S_{\lambda'}}$ is a morphism in the diagram D', then

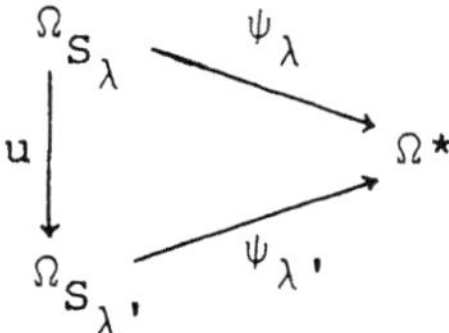

is commutative, since it is so in degree 0.

By the universal property of the colimit there is an S-homomorphism $\varinjlim \Omega_{S_\lambda} \to \Omega^*$, hence $\varinjlim \Omega_{S_\lambda} = \Omega_S$.

Since the tensor product is a special colimit, we have

4.2. Corollary. For algebras R/R_o and S/R_o

$$\Omega_{R\otimes_{R_o} S/R_o} = \Omega_{R/R_o} \otimes_{R_o} \Omega_{S/R_o} .$$

Applying 3.19 to this formula we obtain the following base change formula.

4.3. Corollary. $\quad \Omega_{R\otimes_{R_o} S/R} = R \otimes_{R_o} \Omega_{S/R_o} .$

Here $R \otimes_{R_o} \Omega_{S/R_o}$ has to be considered as an $R \otimes_{R_o} S$-algebra in the natural manner, and the differentiation d is given by the formula

$$d(r \otimes \omega) = r \otimes d_S\omega \qquad (r \in R, \omega \in \Omega_{S/R_o})$$

where d_S is the differentiation of Ω_{S/R_o}.

In fact, by 3.19 we have $\Omega_{R\otimes_{R_o} S/R} = \Omega_{R\otimes_{R_o} S/R_o}/I$, where I is the ideal generated by $\{d(r\otimes 1) \mid r \in R\}$. Under the isomorphism of 4.2 this ideal corresponds to $\bigoplus_{i>0} (\Omega^i_{R/R_o} \otimes_{R_o} \Omega_{S/R_o})$. The claim follows.

Given an R_o-derivation $\delta : R \to R\delta R$ we can apply the above formulas to the differential algebra $\Omega := R \ltimes R\delta R$ associated with δ and restrict Ω_S to elements of degree 1. Taking into account the way the colimit was constructed in the proof of 2.11 in various situations, we are led to some formulas for differential modules.

4.4. Formulas.

a) Let $\{S_\lambda\}_{\lambda \in \Lambda}$ be a direct system of R-algebras. Then

$$\Omega^1_{\varinjlim S_\lambda / \delta} \cong \varinjlim \Omega^1_{S_\lambda / \delta}.$$

Here the colimit on the right is taken in the category of R_o-modules. It can be made an S-module in the natural manner. The universal S-extension $d : S \to \varinjlim \Omega^1_{S_\lambda / \delta}$ of δ is the colimit of the universal S_λ-extensions $d_\lambda : S_\lambda \to \Omega^1_{S_\lambda / \delta}$ of δ.

For the proof compare with the construction of the direct limit of differential algebras in the proof of 2.11.

b) In the situation of 4.2

$$\Omega^1_{R \otimes_{R_o} S / R_o} \cong S \otimes_{R_o} \Omega^1_{R / R_o} \oplus R \otimes_{R_o} \Omega^1_{S / R_o}.$$

Both modules on the right are $R \otimes_{R_o} S$-modules and the universal derivation d is given by

$$d(r \otimes s) = s \otimes d_R r + r \otimes d_S s \quad (r \in R, s \in S).$$

For the proof consider the construction of the tensor product of two differential algebras in 2.11.

c) In the situation of 4.3

$$\Omega^1_{R \otimes_{R_o} S / R} \cong R \otimes_{R_o} \Omega^1_{S / R_o}$$

and the universal derivation d is given by

$$d(r \otimes s) = r \otimes d_S s \qquad (r \in R, s \in S).$$

As an application of 4.4a) let us show

4.5. Proposition.

For any algebra R/R_o there is a (unique) maximal subalgebra R'/R_o with $\Omega^1_{R'/R_o} = 0$.

For the proof consider the family of all subalgebras R_λ/R_o of R/R_o with $\Omega^1_{R_\lambda/R_o} = 0$. They form a direct system (with respect to the inclusion mappings) because of the following lemma.

4.6. Lemma.

Let R_1/R_o and R_2/R_o be subalgebras of an algebra R/R_o with $R = R_o[R_1, R_2]$. If $\Omega^1_{R_i/R_o} = 0$ for $i = 1, 2$, then $\Omega^1_{R/R_o} = 0$.

In fact, each $r \in R$ is of the form $r = \sum_{i=1}^{n} r_i^{(1)} r_i^{(2)}$ with $n \in \mathbb{N}$, $r_i^{(j)} \in R_j$ $(j=1,2)$. Since

$$dr = \sum_{i=1}^{n} (r_i^{(1)} dr_i^{(2)} + r_i^{(2)} dr_i^{(1)})$$

we have $\Omega^1_{R/R_o} = \langle \operatorname{im} \alpha_1, \operatorname{im} \alpha_2 \rangle$, where $\alpha_j : \Omega^1_{R_j/R_o} \to \Omega^1_{R/R_o}$ is the functorial mapping $(j=1,2)$. The claim of the lemma follows.

Now let R' be the union of all the subrings R_λ of R. Then $R' = \varinjlim R_\lambda$ and 4.4a) implies $\Omega^1_{R'/R_o} = \varinjlim \Omega^1_{R_\lambda/R_o} = 0$.

2) Limits

It can be shown that in general the universal extension of a differential algebra is <u>not</u> compatible with limits (see exercise 1). However, it does commute with finite products. Let $S = S_1 \times \ldots \times S_m$ be a direct product of R-algebras S_i with structure homomorphisms $\rho_i : R \to S_i$ $(i=1,\ldots,m)$. The structure of S as an R-algebra is given by $\rho : R \to S_1 \times \ldots \times S_m$ $(\rho(r) = (\rho_1(r),\ldots,\rho_m(r))$ for $r \in R)$. According to 2.11

$$\Omega_{S_1} \times \ldots \times \Omega_{S_m}$$

is a differential algebra of S/R_0 with the differentiation d given by

$$d(\omega_1,\ldots,\omega_m) = (d_1\omega_1,\ldots,d_m\omega_m) \qquad (\omega_i \in \Omega_{S_i})$$

where d_i is the differentiation of Ω_{S_i}. The canonical homomorphisms $\varphi_i : \Omega \to \Omega_{S_i}$ induce a ρ-homomorphism

$$\varphi: \Omega \to \Omega_{S_1} \times \ldots \times \Omega_{S_m} \qquad (\varphi(\omega) = (\varphi_1(\omega),\ldots,\varphi_m(\omega))).$$

<u>4.7. Proposition.</u> $(\Omega_{S_1} \times \ldots \times \Omega_{S_m}, \varphi)$ is the universal S-extension of Ω. In other words:

$$\Omega_{S_1 \times \ldots \times S_m} \cong \Omega_{S_1} \times \ldots \times \Omega_{S_m}.$$

Proof. Let (Ω^*, d^*) be a differential algebra of S/R_0 and $\psi : \Omega \to \Omega^*$ a ρ-homomorphism. Since $S = S_1 \times \ldots \times S_m$, there is a decomposition $1 = e_1 + \ldots + e_m$, $e_i^2 = e_i$ $(i=1,\ldots,m)$, $e_i e_j = 0$ $(i \neq j)$ with $e_i = (0,\ldots,1,\ldots,0)$.

From this we get a decomposition

$$\Omega^* = \Omega_1^* \times \ldots \times \Omega_m^* \quad \text{with} \quad \Omega_i^* := e_i \Omega^* \quad (i=1,\ldots,m)$$

where Ω_i^* is a graded algebra over $S_i = e_i S$ with the grading given by $(\Omega_i^*)^n := e_i(\Omega^*)^n$ $(i=1,\ldots,m)$. From $d^* e_1 + \ldots + d^* e_m = d^*(1) = 0$ and $2e_i d^* e_i = d^* e_i$ we obtain $2e_1 d^* e_1 + \ldots + 2e_m d^* e_m = 0$ and, after multiplication with e_i,

$$0 = 2e_i^2 d^* e_i = d^* e_i.$$

Therefore

$$d^*(e_i \omega) = e_i d^* \omega \in \Omega_i^* \qquad (\omega \in \Omega^*)$$

which shows that $d^*(\Omega_i^*) \subset \Omega_i^*$ $(i=1,\ldots,m)$. It is clear that

Ω_i^* is a differential algebra of S_i/R_o with the differentia-
tion $d_i^* := d|_{\Omega_i^*}$, and that the canonical projection $p_i : \Omega^* \to \Omega_i^*$
is a homomorphism of differential algebras. Moreover, Ω^* is
the direct product of the Ω_i^* in the category $\mathcal{D}_{R_o}$:

$$d^*(\omega_1,\ldots,\omega_m) = (d_1^*\omega_1,\ldots,d_m^*\omega_m) \qquad (\omega_i \in \Omega_i^*).$$

$\psi_i := p_i \circ \psi$ is a ρ_i-homomorphism from Ω to Ω_i^*, hence
there is an S_i-homomorphism $h_i : \Omega_{S_i} \to \Omega_i^*$ $(i=1,\ldots,m)$. Then
$h : \Omega_{S_1} \times \cdots \times \Omega_{S_m} \to \Omega^*$ $(h(\omega_1,\ldots,\omega_m) = (h_1(\omega_1),\ldots,h_m(\omega_m))$
is an S-homomorphism.

4.8. Corollary. Let $\delta : R \to M$ be a derivation of R/R_o
into an R-module M with $M = R\delta R$. Then

$$\Omega^1_{S_1 \times \ldots \times S_m/\delta} \cong \Omega^1_{S_1/\delta} \times \cdots \times \Omega^1_{S_m/\delta} \; .$$

Here the product has to be considered as an $S_1 \times \ldots \times S_m$-
module in the natural manner, and the universal extension
$d : S_1 \times \ldots \times S_m \to \Omega^1_{S_1 \times \ldots \times S_m/\delta}$ of δ is given by

$$d(s_1,\ldots,s_m) = (d_1 s_1,\ldots,d_m s_m)$$

where $d_i : S_i \to \Omega^1_{S_i/\delta}$ is the universal S_i-extension of δ
$(i=1,\ldots,m)$.

3) Polynomial algebras

Let S/R_o be an algebra. Then $\Omega \otimes_{R_o} \Omega_{S/R_o}$ is a differential algebra of $R \otimes_{R_o} S/R_o$ and

$$\varphi : \Omega \to \Omega \otimes_{R_o} \Omega_{S/R_o} \qquad (\omega \mapsto \omega \otimes 1)$$

a ρ-homomorphism, where $\rho : R \to R \otimes_{R_o} S$ is the mapping
$r \mapsto r \otimes 1$ $(r \in R)$.

4.9. Proposition. $(\Omega \otimes_{R_o} \Omega_{S/R_o}, \varphi)$ is the universal $R \otimes_{R_o} S$-extension of Ω. In other words:

$$\Omega_{R \otimes_{R_o} S} \cong \Omega \otimes_{R_o} \Omega_{S/R_o}.$$

Proof. Let Ω^* be a differential algebra of $R \otimes_{R_o} S/R_o$ and $\psi : \Omega \to \Omega^*$ a ρ-homomorphism. By the universal property of Ω_{S/R_o}, there is also a σ-homomorphism $\Omega_{S/R_o} \to \Omega^*$, where $\sigma : S \to R \otimes_{R_o} S$ is the mapping with $\sigma(s) = 1 \otimes s$ $(s \in S)$. By the universal property of the tensor product, there is an $R \otimes_{R_o} S$-homomorphism $\Omega \otimes_{R_o} \Omega_{S/R_o} \to \Omega^*$, as we wanted to show.

A polynomial algebra $R[\{X_\lambda\}_{\lambda \in \Lambda}]$ with a family of indeterminates $\{X_\lambda\}_{\lambda \in \Lambda}$ can be written

$$R[\{X_\lambda\}_{\lambda \in \Lambda}] = R \otimes_{R_o} S \quad \text{with} \quad S := R_o[\{X_\lambda\}_{\lambda \in \Lambda}].$$

Therefore we obtain

4.10. Corollary. $\Omega_{R[\{X_\lambda\}_{\lambda \in \Lambda}]} = \Omega \otimes_{R_o} \Omega_{R_o[\{X_\lambda\}]/R_o}.$

By 3.10a) and 2.5d) the structure of $\Omega_{R_o[X_1,\ldots,X_n]/R_o}$ is well-known. In fact $\Omega^p_{R_o[X_1,\ldots,X_n]/R_o}$ is the free $R_o[X_1,\ldots,X_n]$-module on the basis $\{dX_{i_1}\ldots dX_{i_p} \mid 1 \le i_1 < \ldots < i_p \le n\}$ for all $p \in \mathbb{N}$. Thus

$$\Omega^\rho_{R[X_1,\ldots,X_n]} = \bigoplus_{q+p=\rho} \Omega^q \otimes_{R_o} \Omega^p_{R_o[X_1,\ldots,X_n]/R_o} =$$

$$\bigoplus_{q+p=\rho} \ \bigoplus_{1 \le i_1 < \ldots < i_p \le n} (\Omega^q \otimes_{R_o} R_o[X_1,\ldots,X_n]) dX_{i_1}\ldots dX_{i_p}.$$

Since an arbitrary polynomial algebra is the union (direct limit) of the polynomial algebras in finitely many of its variables, $\Omega_{R_o[\{X_\lambda\}]/R_o}$ is by 4.1 the colimit and, in fact, the union of the differential algebras $\Omega_{R_o}[X_{\lambda_1},\ldots,X_{\lambda_n}]/R_o$, where $\{\lambda_1,\ldots,\lambda_n\}$ is ranging over all finite subsets of Λ. In particular,

$$\Omega^1_{R_o}[\{X_\lambda\}_{\lambda\in\Lambda}]/R_o = \bigoplus_{\mu\in\Lambda} R_o[\{X_\lambda\}]dX_\mu$$

where the universal derivation d is given by

$$df = \sum_{\lambda\in\Lambda} \frac{\partial f}{\partial X_\lambda}\, dX_\lambda \quad\text{for all } f \in R_o[\{X_\lambda\}].$$

If we specialize to the case that Ω is the differential algebra associated with an R_o-derivation $\delta : R \to M$, $M = R\delta R$, we obtain:

4.11. Formulas.

a) Under the assumptions of 4.9

$$\Omega^1_{R\otimes_{R_o} S}/\delta = S \otimes_{R_o} M \oplus R \otimes_{R_o} \Omega^1_{S/R_o}.$$

The universal extension $d : R \otimes_{R_o} S \to \Omega^1_{R\otimes_{R_o} S}/\delta$ of δ is given by

$$d(r\otimes s) = s\otimes\delta r + r\otimes d_S s \quad (r \in R,\ s \in S)$$

where $d_S : S \to \Omega^1_{S/R_o}$ is the universal derivation of S/R_o.

b) $\Omega^1_{R[\{X_\lambda\}_{\lambda\in\Lambda}]/\delta} = R[\{X_\lambda\}] \otimes_R M \oplus \bigoplus_{\mu\in\Lambda} R[\{X_\lambda\}]dX_\mu.$

The universal extension $d : R[\{X_\lambda\}] \to \Omega^1_{R[\{X_\lambda\}]/\delta}$ of δ is given by the formula

$$df = \delta f + \sum_{\lambda\in\Lambda} \frac{\partial f}{\partial X_\lambda}\, dX_\lambda \quad (f \in R[\{X_\lambda\}]),$$

where δf is defined as follows:

Let $f = \Sigma r_{\nu_1\ldots\nu_n} X_{\lambda_1}^{\nu_1}\ldots X_{\lambda_n}^{\nu_n}$ with $r_{\nu_1\ldots\nu_n} \in R$.

Then

$$\delta f := \Sigma X_{\lambda_1}^{\nu_1}\ldots X_{\lambda_n}^{\nu_n} \otimes \delta r_{\nu_1\ldots\nu_n}.$$

We say that this expression, which will frequently occur later, is obtained by "differentiation of the coefficients of f with respect to δ", leaving the variables untouched.

Proof. a) follows immediately from 4.9. To obtain b) one has to make the identifications

$$R[\{X_\lambda\}] \otimes_R M = R_o[\{X_\lambda\}] \otimes_{R_o} M$$

and

$$f = \Sigma X_{\lambda_1}^{\nu_1} \ldots X_{\lambda_n}^{\nu_n} \otimes r_{\nu_1 \ldots \nu_n} \in R_o[\{X_\lambda\}] \otimes_{R_o} R.$$

4) Residue class algebras

Let S/R be an algebra, I an ideal of S, and $\varepsilon : S \to S/I$ the canonical map. The ideal J of Ω_S generated by I and $dI := \{dx \mid x \in I\}$ is homogeneous and differentially closed by 3.13. Thus by 2.9, since $J \cap S = I$, Ω_S/J is a differential algebra of S/I. Its differentiation d* is induced by the differentiation d of Ω_S:

$$d^*(\omega+J) = d\omega + J \qquad \text{for } \omega \in \Omega_S.$$

Let $\varphi : \Omega \to \Omega_S/J$ be the composition of $\Omega \to \Omega_S$ with the canonical epimorphism $\Omega_S \to \Omega_S/J$.

4.12. Proposition. $(\Omega_S/J, \varphi)$ is the universal S/I-extension of Ω. In other words:

$$\Omega_{S/I} \cong \Omega_S/(I, dI).$$

Proof. Let (Ω^*, d^*) be a differential algebra of S/I and $\psi : \Omega \to \Omega^*$ a ρ-homomorphism, where ρ is the composition

$R \to S \overset{\varepsilon}{\to} S/I$. By the universal property of Ω_S there is an ε-homomorphism $h : \Omega_S \to \Omega*$. Then $I \subset \ker h$ and, since $\ker h$ is differentially closed, also $dI \subset \ker h$. Thus h induces an S/I-homomorphism $\Omega_S/(I,dI) \to \Omega*$.

4.12 is frequently used in the following manner: Each algebra S/R has a <u>presentation in terms of generators and relations</u>

$$(1) \qquad S = R[\{X_\lambda\}_{\lambda \in \Lambda}]/I \ , \quad I = (\{f_\nu\}_{\nu \in N}).$$

Then Ω_S can be "computed" by the formula in the next corollary.

<u>4.13. Corollary.</u>

$$\Omega_S \cong \Omega_{R[\{X_\lambda\}]}/(\{f_\nu, df_\nu\}_{\nu \in N}).$$

Here one has simply to observe that (I,dI) is already generated by the polynomials f_ν and their differentials df_ν, since the f_ν generate I: For $g = \Sigma g_\nu f_\nu \in I$ ($g_\nu \in R[\{X_\lambda\}]$) we have $dg = \Sigma g_\nu df_\nu + \Sigma f_\nu dg_\nu \in (\{f_\nu, df_\nu\}_{\nu \in N})$.

<u>4.14. Example.</u>   Suppose S/R is generated by one element x. Then $S = R[X]/I$. We also assume that I is generated by one polynomial f. Then

$$\Omega_S \cong \Omega_{R[X]}/(f,df).$$

By 4.10 we have an isomorphism of $R[X] \otimes_R \Omega$-modules

$$\Omega_{R[X]} \cong R[X] \otimes_R \Omega \oplus (R[X] \otimes_R \Omega)dX$$

where f is identified with $f \otimes 1 \in R[X] \otimes_R \Omega$ and df with $\delta f + (f'(X) \otimes 1)dX$ ($\delta f \in R[X] \otimes_R \Omega$ as in 4.11b)). Thus

$$\Omega_S = \Omega_{R[X]}/(f)/(f,df)/(f) = S \otimes_R \Omega \oplus (S \otimes_R \Omega)dX/(\delta f(x) + (f'(x) \otimes 1)dX),$$

$\delta f(x)$ denoting the image of δf in $S \otimes_R \Omega$ (X replaced by x).

In particular, if $f'(x)$ happens to be a unit in S, the canonical mapping $S \otimes_R \Omega \rightarrow \Omega_S$ induced by $\Omega \rightarrow \Omega_S$ is bijective and we have an isomorphism of Ω-modules

$$4.15 \qquad\qquad \Omega_S \cong S \otimes_R \Omega.$$

This formula may be applied to finite separably algebraic field extensions S/R (due to the theorem of the primitive element). Taking direct limits on both sides of 4.15 one sees that the formula also holds for arbitrary separable algebraic field extensions.

Applying 4.12 in the usual manner to a derivation $\delta : R \rightarrow M$ of R with $M = R\delta R$ we obtain :

4.16. Formula.   In the situation of 4.12

$$\Omega^1_{S/I/\delta} \cong \Omega^1_{S/\delta}/SdI.$$

The universal S/I-extension d' of δ is given by $d'(s+I) = ds + SdI$, where d is the universal S-extension of δ.

By 4.12 we have $\Omega^1_{S/I/\delta} = \Omega^1_{S/\delta}/I\Omega^1_{S/\delta} + SdI$, and therefore it remains to be shown that $I\Omega^1_{S/\delta} \subset SdI$. But for $s \in S$ and $x \in I$ we have $xds = d(xs)-sdx \in SdI$.

Formula 4.16 is one of the most important basic tools in the theory of differential modules. It can be reformulated in the following way:

4.17. Proposition.   There is an exact sequence of S/I-modules

$$I/I^2 \xrightarrow{\alpha} \Omega^1_{S/\delta}/I\Omega^1_{S/\delta} \xrightarrow{\beta} \Omega^1_{S/I/\delta} \rightarrow 0.$$

Here α is the mapping with $\alpha(x+I^2) = dx + I\Omega^1_{S/\delta}$ for all

$x \in I$ and β is induced by the functorial map $\Omega^1_{S/\delta} \to \Omega^1_{S/I/\delta}$.

Proof. α is welldefined: If $x+I^2 = x'+I^2$ $(x,x' \in I)$, then $x-x' \in I^2$ and $dx-dx' \in dI^2 \subset IdI \subset I\Omega^1_{S/\delta}$. α is S/I-linear: For $s_1,s_2 \in S$ and $x_1,x_2 \in I$ we have $\alpha(s_1x_1+s_2x_2+I^2) = d(s_1x_1+s_2x_2) + I\Omega^1_{S/\delta} = s_1dx_1+s_2dx_2+I\Omega^1_{S/\delta} = (s_1+I)\cdot\alpha(x_1+I^2)+(s_2+I)\cdot\alpha(x_2+I^2)$.

The image of α is the submodule $SdI + I\Omega^1_{S/\delta}/I\Omega^1_{S/\delta}$ of $\Omega^1_{S/\delta}/I\Omega^1_{S/\delta}$. The exactness of the sequence follows now from the formula 4.16.

<u>4.18. Corollary.</u> For each S/I-module N there is an exact sequence

$$0 \to \mathrm{Der}_\delta(S/I,N) \xrightarrow{\ \beta^*\ } \mathrm{Der}_\delta(S,N) \xrightarrow{\ \alpha^*\ } \mathrm{Hom}_{S/I}(I/I^2,N)$$

For $d' \in \mathrm{Der}_\delta(S/I,N)$ we have $\beta^*(d') = d' \circ \varepsilon$, where $\varepsilon : S \to S/I$ is the canonical epimorphism, and for $d \in \mathrm{Der}_\delta(S,N)$ the linear map $\alpha^*(d)$ is induced by d: $\alpha^*(d)(x+I^2) = dx$ for $x \in I$.

<u>Proof.</u> Apply $\mathrm{Hom}_{S/I}(-,N)$ to the exact sequences

$$0 \to \mathrm{im}(\alpha) \to \Omega^1_{S/\delta}/I\Omega^1_{S/\delta} \to \Omega^1_{S/I/\delta} \to 0$$

$$I/I^2 \to \mathrm{im}(\alpha) \to 0.$$

Then, since $\mathrm{Hom}_{S/I}(\mathrm{im}(\alpha),N) \to \mathrm{Hom}_{S/I}(I/I^2,N)$ is injective, we get an exact sequence

(2) $\quad 0 \to \mathrm{Hom}_{S/I}(\Omega^1_{S/I/\delta},N) \to \mathrm{Hom}_{S/I}(\Omega^1_{S/\delta}/I\Omega^1_{S/\delta},N) \to \mathrm{Hom}_{S/I}(I/I^2,N).$

By 3.21a) $\mathrm{Hom}_{S/I}(\Omega^1_{S/I/\delta},N) \cong \mathrm{Der}_\delta(S/I,N)$ and

$$\mathrm{Hom}_{S/I}(\Omega^1_{S/\delta}/I\Omega^1_{S/\delta},N) \cong \mathrm{Hom}_S(\Omega^1_{S/\delta},N) \cong \mathrm{Der}_\delta(S,N).$$

It is easy to check that the mappings in (2) are identified with the mappings $\alpha*$ and $\beta*$ described in the corollary.

It should be mentioned that the sequences in 4.17 and 4.18 can be extended to "long" exact sequences in the theory of cotangent functors (see the remarks at the end of §3).

Suppose now S/R is given by generators and relations as in (1). Then from 4.17 we obtain the following description of $\Omega^1_{S/\delta}$ in terms of generators and relations:

<u>4.19. Proposition.</u>   Let x_λ denote the image of X_λ in S. Then

$$\Omega^1_{S/\delta} = (S \otimes_R M \oplus \bigoplus_{\lambda \in \Lambda} SdX_\lambda)/U$$

where U is the submodule of $S \otimes_R M \oplus \bigoplus_{\lambda \in \Lambda} SdX_\lambda$ generated by the elements

$$\delta f_\nu(\{x_\lambda\}) + \sum_{\lambda \in \Lambda} \frac{\partial f_\nu}{\partial x_\lambda} dX_\lambda \qquad (\nu \in N).$$

Here $\delta f_\nu(\{x_\lambda\})$ is the image of δf_ν in $S \otimes_R M$ and $\frac{\partial f_\nu}{\partial x_\lambda}$ that of $\frac{\partial f_\nu}{\partial X_\lambda}$ in S.

The universal extension $d : S \to \Omega^1_{S/\delta}$ of δ can be described as follows: For $s \in S$ choose $f \in R[\{X_\lambda\}]$ with the image s in S. Then ds is the residue class mod U of

$$\delta f(\{x_\lambda\}) + \sum_{\lambda \in \Lambda} \frac{\partial f}{\partial x_\lambda} dX_\lambda \in S \otimes_R M \oplus \bigoplus_{\lambda \in \Lambda} SdX_\lambda.$$

Proof. Apply the exact sequence of 4.17 with S replaced by $R[\{X_\lambda\}]$ and S/I replaced by S. By 4.11b)

$$\Omega^1_{R[\{X_\lambda\}]/\delta}/I\Omega^1_{R[\{X_\lambda\}]/\delta} = S \otimes_R M \oplus \bigoplus_{\lambda \in \Lambda} SdX_\lambda$$

and one easily sees that the image of the mapping α of 4.17 is just U. The description of d comes from the corresponding description of d in 4.11b).

4.20. Example. For $S = R[x] = R[X]/(f)$ we have $\Omega^1_{S/\delta} = (S \otimes_R M \oplus SdX)/U$, where U is generated by $\delta f(x) + f'(x)dX$.

5) Algebras of fractions

Let $N \subset R$ be multiplicatively closed, and let R_N and Ω_N denote the algebras of fractions with respect to N. Let $\rho : R \to R_N$ ($\rho(r) = \frac{r}{1}$) be the canonical homomorphism and regard R_N and Ω_N as R_O-algebras via ρ. With $\Omega^n_N := \{\frac{\omega}{\nu} \mid \omega \in \Omega^n, \nu \in N\}$ we have $\Omega_N = \bigoplus_{n \in \mathbb{N}} \Omega^n_N$, hence Ω_N is a graded R_N-algebra. For $\frac{\omega}{\nu} \in \Omega_N$ ($\omega \in \Omega$, $\nu \in N$) we define

$$d_N(\tfrac{\omega}{\nu}) := \frac{\nu \cdot \delta\omega - \delta\nu \cdot \omega}{\nu^2} .$$

A simple calculation shows that the expression on the right is independent of the special representation of $\frac{\omega}{\nu}$ as a fraction. Moreover, one easily checks that (Ω_N, d_N) is a differential algebra of R_N/R_O, called the <u>fraction algebra</u> of (Ω, δ) with respect to N. The canonical map $\varphi : \Omega \to \Omega_N$ ($\omega \mapsto \frac{\omega}{1}$) is a ρ-homomorphism.

4.21. Proposition. (Ω_N, φ) is the universal R_N-extension of Ω. In other words:

$$\Omega_{R_N} \cong \Omega_N \cong R_N \otimes_R \Omega.$$

Proof. Let (Ω^*, d^*) be a differential algebra of R_N/R_O and $\psi : \Omega \to \Omega^*$ a ρ-homomorphism. By the universal property of the fraction algebra Ω_N there is a homomorphism of R_N-algebras $h : \Omega_N \to \Omega^*$ with $\psi = h \circ \varphi$, that is, $h(\frac{\omega}{\nu}) = \frac{1}{\nu} \cdot \psi(\omega)$ for all $\frac{\omega}{\nu} \in \Omega_N$ ($\omega \in \Omega$, $\nu \in N$). As ψ is homogeneous, so is h, and $h^O = \mathrm{id}_{R_N}$. That h is also compatible with differentiation is shown by the following calculation:

$$d*(h(\tfrac{\omega}{\nu})) = d*(\tfrac{1}{\nu}\cdot\psi(\omega)) = -\tfrac{1}{\nu^2}\,d*(\tfrac{\nu}{1})\cdot\psi(\omega) + \tfrac{1}{\nu}\,d*(\psi(\omega)) =$$

$$-\tfrac{1}{\nu^2}\,\psi(\delta\nu\cdot\omega) + \tfrac{1}{\nu}\,\psi(\delta\omega) = \tfrac{1}{\nu^2}\,\psi(\nu\cdot\delta\omega-\delta\nu\cdot\omega) = h(d_N(\tfrac{\omega}{\nu})).$$

4.22. Corollary. Let S/R be an algebra and $N \subset S$ multiplicatively closed.

a) $\Omega_{S_N} \cong (\Omega_S)_N.$

b) For each derivation $\delta : R \to M$ of R/R_o with $M = R\delta R$ we have

$$\Omega^1_{S_N/\delta} \cong (\Omega^1_{S/\delta})_N \cong S_N \otimes_S \Omega^1_{S/\delta} .$$

If d is the universal extension of δ to S and d_N the universal extension of δ to S_N, then

$$(3) \qquad\qquad d_N(\tfrac{s}{\nu}) = \frac{\nu ds - s d\nu}{\nu^2} \qquad (s \in S,\ \nu \in N).$$

Proof. a) follows from the transitive law of universal extensions and 4.21.

b) is the usual specialization of a) to 1-forms.

4.23. Corollary. Suppose $\Omega^1_{S/\delta}$ is a finitely presented S-module and P an arbitrary S-module. Then there is a canonical isomorphism

$$\mathrm{Der}_\delta(S,P)_N \xrightarrow{\sim} \mathrm{Der}_\delta(S_N,P_N)$$

that maps $d \in \mathrm{Der}_\delta(S,P)$ to the derivation d_N given by formula (3).

In fact, $\mathrm{Der}_\delta(S_N, S_N \otimes_S P) = \mathrm{Hom}_{S_N}(\Omega^1_{S_N/\delta}, S_N \otimes_S P)$

$= S_N \otimes_S \mathrm{Hom}_S(\Omega^1_{S/\delta}, P) = \mathrm{Der}_\delta(S,P)_N.$

We close this paragraph with examples, in which some of

the rules of this section are applied.

4.24. Examples.   a) Let $L = K(\{X_\lambda\}_{\lambda \in \Lambda}) = Q(K[\{X_\lambda\}_{\lambda \in \Lambda}])$ be the rational function field of a set $\{X_\lambda\}$ of indeterminates over a field K. If $\delta : K \to K\delta K$ is a derivation, then

$$\Omega^1_{L/\delta} = L \otimes_{K[\{X_\lambda\}]} \Omega^1_{K[\{X_\lambda\}]/\delta} =$$

$$L \otimes_{K[\{X_\lambda\}]} (K[\{X_\lambda\}] \otimes_K K\delta K \oplus \bigoplus_{\mu \in \Lambda} K[\{X_\lambda\}]dX_\mu) =$$

$$L \otimes_K K\delta K \oplus \bigoplus_{\lambda \in \Lambda} LdX_\lambda \quad \text{by 4.22b).}$$

If d is the universal L-extension of δ, then for $\frac{f}{g} \in L$

$$d(\frac{f}{g}) = \frac{1}{g^2}[g \cdot (\delta f + \Sigma \frac{\partial f}{\partial X_\lambda} dX_\lambda) - f \cdot (\delta g + \Sigma \frac{\partial g}{\partial X_\lambda} dX_\lambda)].$$

b) Let $R = \bigoplus_{i \in \mathbb{N}} R^i$ be a positively graded noetherian ring with $\mathbb{Q} \subset R^O$, and let $M := \bigoplus_{i > O} R^i$. The Euler derivation $\delta : R \to R$ (with $\delta x_i = i x_i$ for $x_i \in R^i$, see 1.5) is a surjection from R onto M. By the universal property of the module of differentials there is a canonical epimorphism

$$h : \Omega^1_{R/R^O} \to M \qquad (dx_i \mapsto \delta x_i)$$

and a canonical injection

$$\text{Hom}_R(M,R) \hookrightarrow \text{Hom}_R(\Omega^1_{R/R^O},R) = \text{Der}_{R^O}(R).$$

Suppose M does not entirely consist of zero divisors and let $K := Q(R)$. Put

$$M^{-1} := \{y \in K \mid yM \subset R\}.$$

Then the mapping

$$\alpha : M^{-1} \to \text{Hom}_R(M,R)$$

with $\alpha(y)(m) = ym$ (for $y \in M^{-1}, m \in M$) is an isomorphism. Hence we have a canonical injection

$$\alpha : M^{-1} \to \text{Der}_{R^O}(R) \qquad (\alpha(y) = y\delta).$$

In some cases α is a bijection: Suppose Ω^1_{K/R_o} is a free K-module of rank 1 (see next paragraph for situations in which this is the case). Then h induces an isomorphism

$$\Omega^1_{K/R^o} = K \otimes_R \Omega^1_{R/R^o} \to K \otimes_R M \cong K$$

and ker h is the same as the kernel of $\Omega^1_{R/R^o} \to K \otimes_R \Omega^1_{R/R^o}$, which is the torsion T of Ω^1_{R/R^o}. We conclude that $\mathrm{Hom}_R(M,R) \to \mathrm{Hom}_R(\Omega^1_{R/R^o},R) = \mathrm{Hom}_R(\Omega^1_{R/R^o}/T,R)$ is an iso-morphism, and therefore

$$M^{-1} \cong \mathrm{Der}_{R^o}(R) \qquad (y \mapsto y\delta).$$

Exercises

1) Let $R := K[t]$ be the polynomial algebra in a variable t over a field K. For $S := K[t]/(t^2)$ let $\alpha : R \to S$ be the canonical epimorphism and $\beta : R \to S$ the K-homomorphism with $\beta(t) = O$.

a) Show that $T := K[t^2,t^3] \subset K[t]$ is the equalizer of α and β (and hence $T[dT]$ is the equalizer of the induced maps $\alpha^*,\beta^*: \Omega_{R/K} \to \Omega_{S/K}$ in the category $\mathcal{D}_K$).

b) Show that the kernel of the K-homomorphism $\varphi : K[X,Y] \to K[t^2,t^3]$ with $\varphi(X) = t^2$, $\varphi(Y) = t^3$ is generated by X^3-Y^2.

c) Use φ to compute $\Omega_{T/K}$, and show that the canonical T-homomorphism $\Omega_{T/K} \to T[dT]$ is <u>not</u> bijective. (This statement is true for any field K, but if char.K = 2 or 3, it is particularly easy to see).

2) For a field K let $R = K[X_1,\ldots,X_n]/(f)$, where f is a non-constant polynomial (R is called a <u>hypersurface algebra</u> over K). Let x_i be the image of X_i in R. For $\mathfrak{z} \in \mathrm{Spec}(R)$ the following statements are equivalent:

   a) $\Omega^1_{R_{\mathfrak{z}}/K}$ is a free $R_{\mathfrak{z}}$-module of rank n-1.

   b) $(\dfrac{\partial f}{\partial x_1},\ldots,\dfrac{\partial f}{\partial x_n}) \not\subset \mathfrak{z}$.

3) Let (Ω,δ) be a differential algebra of an algebra R/R_o with $\mathbb{Q} \subset R_o$. Consider $R[X] \otimes_R \Omega = R_o[X] \otimes_{R_o} \Omega$ as a complex with differentiation $\mathrm{id}_{R_o[X]} \otimes \delta$. The mapping $\alpha : R_o[X] \otimes_{R_o} \Omega \to \Omega_{R[X]}$ that identifies $R_o[X] \otimes_{R_o} \Omega$ with $(R_o[X] \otimes_{R_o} \Omega)dX$ (multiply by dX) is a complex homomorphism and coker α is a complex, which is up to a shift in degree isomorphic to $R_o[X] \otimes_{R_o} \Omega$. Conclude:

   a) If $H^p_{DR}(\Omega) = 0$ for $p > 0$, then $H^p_{DR}(\Omega_{R[X]}) = 0$ for p > 0.

   b) $H^p_{DR}(\Omega_{R_o[X_1,\ldots,X_n]/R_o}) = 0$ for p > 0.

4) Let K be a field of characteristic 0 and $L := K(X)$. Is $H^1_{DR}(\Omega_{L/K}) = 0$?

5) Let K be a field and A a subalgebra of the polynomial algebra K[X]. Let $T(\Omega^1_{A/K})$ denote the torsion of $\Omega^1_{A/K}$. Show that $\Omega^1_{A/K}/T(\Omega^1_{A/K})$ is isomorphic to $\sum\limits_{f \in A} Af'$, the A-submodule of K[X] generated by the derivatives f' of the $f \in A$.

§ 5. Differential Modules of Field Extensions.

The purpose of this section is to investigate, which informations are contained in the differential modules of field extensions. Here we obtain the first applications of the theory of differential modules: Some results about separability and the structure of field extensions.

A field extension L/K is called finitely generated, or an <u>algebraic function field</u>, if elements $x_1,\ldots,x_n \in L$ exist with $L = K(x_1,\ldots,x_n)$. A transcendence basis $\{x_\lambda\}_{\lambda \in \Lambda}$ of a field extension L/K is called <u>separating</u>, if $L/K(\{x_\lambda\}_{\lambda \in \Lambda})$ is separable (algebraic). An algebraic function field is called <u>separable</u>, if it has a separating transcendence basis. The definition of separability for an arbitrary field extension will be given later.

Consider first an algebra S/R and a derivation $\delta : R \to M$ of R into an R-module M with $M = R\delta R$. By 3.24 there is a canonical exact sequence

$$(1) \qquad S \otimes_R R\delta R \overset{\alpha}{\to} \Omega^1_{S/\delta} \overset{\beta}{\to} \Omega^1_{S/R} \to O.$$

For many questions one has to study the kernel of α, which we denote by $T(S/\delta)$. If S'/R is another algebra and $\sigma : S \to S'$ an R-homomorphism, we have a canonical commutative diagram with exact rows

$$
\begin{array}{ccccccccc}
O \to T(S/\delta) & \to & S \otimes_R R\delta R & \overset{\alpha}{\longrightarrow} & \Omega^1_{S/\delta} & \overset{\beta}{\longrightarrow} & \Omega^1_{S/R} & \to & O \\
& & \downarrow{\sigma \otimes id} & & \downarrow & & \downarrow & & \\
O \to T(S'/\delta) & \to & S' \otimes_R R\delta R & \overset{\alpha'}{\longrightarrow} & \Omega^1_{S'/\delta} & \overset{\beta'}{\longrightarrow} & \Omega^1_{S'/R} & \to & O
\end{array}
$$

and consequently $\sigma \otimes id$ induces an S-linear map

$$T(\sigma/\delta) : T(S/\delta) \to T(S'/\delta)$$

which makes T a functor of S. If R is a field, and σ is injective, then $\sigma \otimes id$ and $T(\sigma/\delta)$ are also injective, hence $T(S/\delta)$ can be identified with its image in $T(S'/\delta)$.

The following proposition allows the reduction of some questions about arbitrary field extensions to algebraic function fields.

<u>5.1. Proposition.</u> Let L/K be a field extension and $\delta : K \to K\delta K$ a derivation. Then

$$T(L/\delta) = \bigcup_Z T(Z/\delta),$$

where Z runs over all intermediate fields of L/K, for which Z/K is finitely generated.

Proof. Let $\{x_\lambda\}_{\lambda \in \Lambda}$ be a system of generators of the field extension L/K, $\{X_\lambda\}_{\lambda \in \Lambda}$ a family of indeterminates and $\varphi : K[\{X_\lambda\}] \to L$ the K-homomorphism with $\varphi(X_\lambda) = x_\lambda$ ($\lambda \in \Lambda$). Let $S := \text{im } \varphi$, $I := \ker \varphi$, and let $\{F_\nu\}_{\nu \in N}$ be a system of generators of I. By 4.19 we have an exact sequence

$$O \to U \to S \otimes_K K\delta K \oplus \bigoplus_{\lambda \in \Lambda} SdX_\lambda \to \Omega^1_{S/\delta} \to O$$

where U is generated by the elements

$$(2) \qquad \delta F_\nu(\{x_\lambda\}) + \sum_\lambda \frac{\partial F_\nu}{\partial x_\lambda} dX_\lambda \qquad (\nu \in N).$$

Since L is the quotient field of S and $\Omega^1_{L/\delta} = L \otimes_S \Omega^1_{S/\delta}$, we also have an exact sequence

$$O \to U' \to L \otimes_K K\delta K \oplus \bigoplus_{\lambda \in \Lambda} LdX_\lambda \to \Omega^1_{L/\delta} \to O$$

where U' is the L-vector space generated by the elements (2). Obviously $T(L/\delta) = U' \cap L \otimes_K K\delta K$. Hence each $z \in T(L/\delta)$ can be written

$$(3) \qquad z = \sum_{i=1}^n \alpha_i \otimes m_i \qquad (\alpha_i \in L, \; m_i \in K\delta K)$$

74

and

$$(3') \quad z = \sum_{\nu \in N} \beta_\nu \cdot (\delta F_\nu(\{x_\lambda\}) + \sum_\lambda \frac{\partial F_\nu}{\partial x_\lambda} dX_\lambda) \qquad (\beta_\nu \in L).$$

In relation (3') only finitely many β_ν are different from zero, say $\beta_{\nu_1}, \ldots, \beta_{\nu_r}$. Let $X_{\lambda_1}, \ldots, X_{\lambda_s}$ be the indeterminates which actually occur in F_{ν_j} ($j=1,\ldots,r$). Put $Z := K(\alpha_1, \ldots, \alpha_n, \beta_{\nu_1}, \ldots, \beta_{\nu_r}, x_{\lambda_1}, \ldots, x_{\lambda_s})$. The relations (3) and (3') then also hold in $Z \otimes_K K\delta K \oplus \bigoplus_{i=1}^{s} ZdX_{\lambda_i}$, hence $z \in T(Z/\delta)$.

5.2. Corollary.   If L/K is separably algebraic, then $T(L/\delta) = 0$, and we have a canonical isomorphism

$$L \otimes_K K\delta K \overset{\sim}{\to} \Omega^1_{L/\delta}.$$

Proof. Using (1) the second statement follows from the first, since $\Omega^1_{L/K} = 0$ (1.22a). In order to prove the first statement, we may assume L/K is finite (5.1). It was shown in 4.15 that $L \otimes_K K\delta K \cong \Omega^1_{L/\delta}$ in this case, hence $T(L/\delta) = 0$.

5.3. Corollary.   Suppose L/K has a separating transcendence basis $\{x_\lambda\}_{\lambda \in \Lambda}$. Then

a) $\Omega^1_{L/K} = \bigoplus_{\lambda \in \Lambda} Ldx_\lambda$.

b) $\dim_L \Omega^1_{L/K} < \infty$ if and only if $\mathrm{Trdeg}(L/K) < \infty$.

c) $\Omega^1_{L/K} = 0$ if and only if L/K is algebraic.

Proof. $Z := K(\{x_\lambda\}_{\lambda \in \Lambda})$ is isomorphic over K to the quotient field of $K[\{X_\lambda\}_{\lambda \in \Lambda}]$. By 4.24a) we have $\Omega^1_{Z/K} = \bigoplus_{\lambda \in \Lambda} Zdx_\lambda$. Since L/Z is separably algebraic, there is an isomorphism $L \otimes_Z \Omega^1_{Z/K} \overset{\sim}{\to} \Omega^1_{L/K}$ by 5.2, which proves a). b) and c) are then clear.

In particular, we see that for a separable algebraic function field L/K

$$\dim_L \Omega^1_{L/K} = \operatorname{Trdeg}(L/K)$$

moreover, for any separating transcendence basis $\{x_1,\ldots,x_t\}$ of L/K the differentials $\{dx_1,\ldots,dx_t\}$ form a basis of $\Omega^1_{L/K}$. This result will be strengthened later.

5.4. Corollary. Suppose K has characteristic 0. Then a family $\{x_\lambda\}_{\lambda \in \Lambda}$ of elements of L is a transcendence basis of L/K if and only if $\{dx_\lambda\}_{\lambda \in \Lambda}$ is a basis of $\Omega^1_{L/K}$.

Proof. It remains to show that $\{x_\lambda\}_{\lambda \in \Lambda}$ is a transcendence basis of L/K, if $\{dx_\lambda\}_{\lambda \in \Lambda}$ is a basis of $\Omega^1_{L/K}$. If $\{x_\lambda\}_{\lambda \in \Lambda}$ were algebraically dependent over K, a finite subfamily $\{x_{\lambda_1},\ldots,x_{\lambda_m}\}$ were algebraically dependent over K. With $Z := K(x_{\lambda_1},\ldots,x_{\lambda_m})$ we had $\dim_Z \Omega^1_{Z/K} < m$ by 5.3a), since $\operatorname{Trdeg}(Z/K) < m$. But then the differentials dx_{λ_i} of the x_{λ_i} in $\Omega^1_{Z/K}$ were linearly dependent over Z, hence also their images in $\Omega^1_{L/K}$, a contradiction.

5.5. Examples.

a) Let K be a field of characteristic 0, $R := K[\![X]\!]$ and $L := Q(R) = K((X))$. It is known that $\operatorname{Trdeg}(L/K) = \infty$, hence $\dim_L \Omega^1_{L/K} = \infty$. Consequently $\Omega^1_{R/K}$ cannot be finitely generated. We see that the natural K-derivation $d : R \to RdX$, $f \mapsto f'dX$ is <u>not</u> universal.

b) Let k be a field of charakteristic $p > 0$, t an indeterminate and $K := k(t)$. The polynomial

$$f := X^p + tY^p + t \in K[X,Y]$$

is irreducible, and $R = K[X,Y]/(f)$ is a domain. $L := Q(R)$ has transcendence degree 1 over K.

Let x,y be the images of X and Y in R. Then by 4.19

$$\Omega^1_{R/K} = RdX \oplus RdY/\langle\frac{\partial f}{\partial x} dX + \frac{\partial f}{\partial y} dY\rangle = Rdx \oplus Rdy$$

because $\frac{\partial f}{\partial X} = \frac{\partial f}{\partial Y} = 0$. $\Omega^1_{L/K} = Ldx \oplus Ldy$ is a vector space of dimension 2, hence L/K <u>cannot</u> be separable.

For a field K of characteristic zero, all that can be said about $\Omega^1_{L/K}$ is contained already in 5.4. From now on we assume that char.K =: p > O. Beside the notion of separating transcendence basis in this case the notion of p-basis is also of importance. Suppose S/R is an algebra, where S is of prime characteristic p. A system $\{x_\lambda\}_{\lambda\in\Lambda}$ of generators of the algebra $S/R[S^p]$ is called a <u>p-generating set of S/R</u>. For such systems the elements

$$(4) \qquad x_{\lambda_1}^{\nu_1} \cdot \ldots \cdot x_{\lambda_n}^{\nu_n} \qquad (n \in \mathbb{N},\ 0 \leq \nu_i \leq p-1)$$

form a system of generators of S as an $R[S^p]$-module. $\{x_\lambda\}_{\lambda\in\Lambda}$ is called a <u>p-basis of S/R</u>, if the elements (4) form a basis of $S/R[S^p]$. This is equivalent with the fact that the kernel of the $R[S^p]$-homomorphism

$$\varphi : R[S^p][\{X_\lambda\}_{\lambda\in\Lambda}] \to S \qquad (X_\lambda \mapsto x_\lambda)$$

is generated by the polynomials $X_\lambda^p - x_\lambda^p$ $(\lambda \in \Lambda)$.

<u>5.6. Proposition.</u> Let $\{x_\lambda\}_{\lambda\in\Lambda}$ be a p-generating set of S/R. Then $\{dx_\lambda\}_{\lambda\in\Lambda}$ is a system of generators of $\Omega^1_{S/R}$. $\{x_\lambda\}_{\lambda\in\Lambda}$ is a p-basis of S/R if and only if $\{dx_\lambda\}_{\lambda\in\Lambda}$ is a basis of $\Omega^1_{S/R}$.

Proof. For any derivation d of S/R we have $d(R[S^p]) = 0$, hence $\Omega^1_{S/R} = \Omega^1_{S/R[S^p]}$. The first assertion of 5.6 is now clear.

If $\{x_\lambda\}_{\lambda \in \Lambda}$ is a p-basis of S/R, using

$$S = R[S^p][\{x_\lambda\}]/(\{x_\lambda^p - x_\lambda^p\})$$

and 4.19 we obtain $\Omega^1_{S/R} = \bigoplus_{\lambda \in \Lambda} S dX_\lambda = \bigoplus_{\lambda \in \Lambda} S dx_\lambda$, since for $f_\lambda := X_\lambda^p - x_\lambda^p$ we have $\frac{\partial f_\lambda}{\partial x_\mu} = 0$ for all $\lambda, \mu \in \Lambda$.

Conversely, assume $\{dx_\lambda\}_{\lambda \in \Lambda}$ is a basis of $\Omega^1_{S/R}$ and f an element of the kernel I of the map φ defined above. Since $X_\lambda^p - x_\lambda^p \in I$ we may assume $\deg_{X_\lambda} f \leq p-1$ for all $\lambda \in \Lambda$. Since $f(\{x_\lambda\}) = 0$, we also have $\sum_\lambda \frac{\partial f}{\partial x_\lambda} \cdot dx_\lambda = 0$, hence $\frac{\partial f}{\partial x_\lambda} = 0$ for all $\lambda \in \Lambda$, since $\{dx_\lambda\}$ is linearly independent over S. By induction we see that all partial derivatives of f are in I. In $S[\{X_\lambda\}_{\lambda \in \Lambda}]$ the polynomial f can be written

$$f = \sum_{0 \leq \nu_i \leq p-1} a_{\nu_1 \ldots \nu_n} (X_{\lambda_1} - x_{\lambda_1})^{\nu_1} \ldots (X_{\lambda_n} - x_{\lambda_n})^{\nu_n}$$

with $a_{\nu_1 \ldots \nu_n} \in S$ and certain $\lambda_1, \ldots, \lambda_n \in \Lambda$. Then

$$\nu_1! \cdot \ldots \cdot \nu_n! \cdot a_{\nu_1 \ldots \nu_n} = \frac{\partial^{\nu_1 + \ldots + \nu_n} f}{(\partial x_{\lambda_1})^{\nu_1} \ldots (\partial x_{\lambda_n})^{\nu_n}} = 0 \text{ for all}$$

$(\nu_1, \ldots, \nu_n)$. Since $\nu_i \leq p-1$, this yields $a_{\nu_1 \ldots \nu_n} = 0$ and f = 0. Hence I is generated by $\{X_\lambda^p - x_\lambda^p\}_{\lambda \in \Lambda}$, and $\{x_\lambda\}$ is a p-basis of S/R.

The proposition implies that the cardinality of a p-basis of S/R is independent of the chosen p-basis, since the corresponding fact is true for bases of free modules. This cardinality is called the p-degree of S/R and is denoted p-deg(S/R). It is known and easy to see that any field extension L/K has a p-basis and hence a p-degree.

For field extensions, 5.6 can be strengthened in the following way

5.7. Proposition. Let L/K be a field extension and $\{x_\lambda\}_{\lambda \in \Lambda}$ a family of elements of L.

a) $\{x_\lambda\}_{\lambda \in \Lambda}$ is a p-basis of L/K if and only if $\{dx_\lambda\}_{\lambda \in \Lambda}$ is a basis of $\Omega^1_{L/K}$.

b) $\dim_L \Omega^1_{L/K} = \text{p-deg}(L/K)$.

c) $\Omega^1_{L/K} = 0$ if and only if $L = K[L^p]$.

Proof. b) and c) follow from 5.6. For a) we have only to show that if $\{dx_\lambda\}$ is a basis of $\Omega^1_{L/K}$, then $\{x_\lambda\}_{\lambda \in \Lambda}$ is a p-basis of L/K. With $L' := K[L^p][\{x_\lambda\}_{\lambda \in \Lambda}]$ we have

$$\Omega^1_{L/L'} = \Omega^1_{L/K}/\langle\{dx_\lambda\}\rangle = 0$$

hence $L = L'[L^p]$ by c). But since $L^p \subset L'$, this implies that $L = L'$ and $\{x_\lambda\}$ is a p-generating set of L/K. By 5.6 it is then even a p-basis.

5.8. Corollary. If L/K has a separating transcendence basis $\{x_\lambda\}$, then $\{x_\lambda\}$ is also a p-basis and $\text{p-deg}(L/K) = \text{Trdeg}(L/K)$.

This follows from 5.3 and 5.7a), and is our first result, in which differentials are only used in the proof.

For algebraic function fields one has more precise statements.

5.9. Proposition. For an algebraic function field L/K the following statements are equivalent:

a) $\Omega^1_{L/K} = 0$.

b) L/K is separably algebraic.

Proof. b) $\to$ a) was already shown in 1.22a).

a) $\to$ b). If $\Omega^1_{L/K} = 0$, then $L = K[L^p] = K[L^{p^2}] = \ldots = K[L^{p^e}]$ for all $e > 0$ by 5.7c). Let $\{x_1, \ldots, x_t\}$ be a transcendence basis of L/K, L' the set of all elements of L that are separable over $K(x_1, \ldots, x_t)$. Then L/L' is finite and purely inseparable: There are elements $y_1, \ldots, y_r \in L$ with $y_i^{p^e} \in L'$ for some $e \in \mathbb{N}$ $(i=1, \ldots, r)$ such that $L = L'[y_1, \ldots, y_r]$. But $L^{p^e} = L'^{p^e}[y_1^{p^e}, \ldots, y_r^{p^e}] \subset L'$, hence $L = K[L^{p^e}] = L'$. Thus $L/K(x_1, \ldots, x_t)$ is separably algebraic. By 5.3 we have $t = \mathrm{Trdeg}(L/K) = \dim_L \Omega^1_{L/K} = 0$, hence L/K is separably algebraic.

5.10. Theorem.   Let L/K be an algebraic function field. Then

a) $p\text{-deg}(L/K) \geq \mathrm{Trdeg}(L/K)$.

b) From each p-basis of L/K a transcendence basis of L/K can be selected.

c) $p\text{-deg}(L/K) = \mathrm{Trdeg}(L/K)$ if and only if L/K is separable.

d) If L/K is separable, for elements $x_1, \ldots, x_t \in L$ the following statements are equivalent:

α) $\{x_1, \ldots, x_t\}$ is a p-basis of L/K.

β) $\{x_1, \ldots, x_t\}$ is a separating transcendence basis of L/K.

γ) $\{dx_1, \ldots, dx_t\}$ is a basis of $\Omega^1_{L/K}$.

Proof. Let $\{x_1, \ldots, x_\tau\}$ be a p-basis of L/K and $L' := K(x_1, \ldots, x_\tau)$. Then $\Omega^1_{L/L'} = \Omega^1_{L/K}/\langle dx_1, \ldots, dx_\tau \rangle = 0$, hence L/L' is separably algebraic by 5.9 and $\mathrm{Trdeg}(L/K) = \mathrm{Trdeg}(L'/K)$. Since $x_1, \ldots, x_\tau$ generate L'/K, a transcendence basis can be selected from them, which is then also a transcendence basis of L/K. This proves b) and a). If p-deg(L/K)= Trdeg(L/K), $\{x_1,\ldots,x_\tau\}$ has already to be a transcendence basis of L/K, hence L/K is separable and c) is true (5.8).

To show d), suppose now that L/K is separable. We have just seen that $\alpha) \rightarrow \beta)$. $\beta) \rightarrow \gamma)$ follows from 5.3 and $\gamma) \rightarrow \alpha)$ from 5.7a). Thus the proof is complete.

<u>5.11. Corollary.</u>   Let L/K be an algebraic function field, $t := \mathrm{Trdeg}(L/K)$, $\tau := \mathrm{p\text{-}deg}(L/K)$.

a) If L/K is separable, then L/K can be generated by $t+1$ elements.

b) If $\tau > t$, then L/K can be generated by τ (and not by fewer) elements.

Proof. a) is a consequence of the theorem of the primitive element.

b) Let $\{x_1,\ldots,x_\tau\}$ be a p-basis of L/K. Select a transcendence basis from it, say $\{x_{\tau-t+1},\ldots,x_\tau\}$. Let $L' := K(x_{\tau-t+1},\ldots,x_\tau)$ and L" the field of all elements of L, which are separable over L'. We have $L^{p^e} \subset L"$ for some $e \in \mathbb{N}$. Then $L = K[L^p][x_1,\ldots,x_\tau] = L'[L^p][x_1,\ldots,x_{\tau-t}] =$ $L"[L^p][x_1,\ldots,x_{\tau-t}] = L"[L^{p^e}][x_1,\ldots,x_{\tau-t}] = L"[x_1,\ldots,x_{\tau-t}]$.
By the theorem of the primitive element (as stated in van der Waerden [vdW],p.141) $L"[x_1]$ can be generated over L' by one element x. We obtain $L = K(x_{\tau-t+1},\ldots,x_\tau)[x,x_2,\ldots,x_{\tau-t}]$ $= K(x,x_2,\ldots,x_\tau)$ and L/K has τ generators. Since $\dim_L \Omega^1_{L/K} = \tau$ it is clear that $\tau-1$ elements will not do.

We let K now be again an arbitrary field, $\delta : K \to K\delta K$ a derivation and L/K a field extension. We wish to study the vector space $T(L/\delta)$ more closely.

For an intermediate field Z of L/K, let δ_Z denote the universal Z-extension of δ and $d_{Z/K}$ the universal derivation of Z/K. By the definition of T we have a commutative diagram with exact rows and columns

$$
\begin{array}{ccccccccc}
 & & & & 0 & & 0 & & \\
 & & & & \downarrow & & \downarrow & & \\
 & & & & T(L/\delta_Z) & & T(L/d_{Z/K}) & & \\
 & & & & \downarrow & & \downarrow & & \\
0 \to L \otimes_Z T(Z/\delta) & \to & L \otimes_Z (Z \otimes_K K\delta K) & \to & L \otimes_Z \Omega^1_{Z/\delta} & \to & L \otimes_Z \Omega^1_{Z/K} & \to & 0 \\
 & & \| \wr & & \downarrow & & \downarrow & & \\
0 \longrightarrow T(L/\delta) & \longrightarrow & L \otimes_K K\delta K & \longrightarrow & \Omega^1_{L/\delta} & \longrightarrow & \Omega^1_{L/K} & \to & 0 \\
 & & & & \downarrow & & \downarrow & & \\
 & & & & \Omega^1_{L/Z} & = & \Omega^1_{L/Z} & & \\
 & & & & \downarrow & & \downarrow & & \\
 & & & & 0 & & 0 & &
\end{array}
$$

which induces the following commutative diagram with exact rows and columns

$$
\begin{array}{ccccccc}
 & 0 & & 0 & & 0 & \\
 & \downarrow & & \downarrow & & \downarrow & \\
T(L/\delta)/L \otimes_Z T(Z/\delta) & \longrightarrow & T(L/\delta_Z) & \longrightarrow & T(L/d_{Z/K}) & & \\
 \downarrow & & \downarrow & & \downarrow & & \\
0 \to L \otimes_Z (Z \otimes_K K\delta K)/L \otimes_Z T(Z/\delta) & \to & L \otimes_Z \Omega^1_{Z/\delta} & \to & L \otimes_Z \Omega^1_{Z/K} & \to & 0 \\
 \downarrow & & \downarrow & & \downarrow & & \\
0 \longrightarrow L \otimes_K K\delta K/T(L/\delta) & \longrightarrow & \Omega^1_{L/\delta} & \longrightarrow & \Omega^1_{L/K} & \to & 0 \\
 \downarrow & & \downarrow & & \downarrow & & \\
 0 & & \Omega^1_{L/Z} & = & \Omega^1_{L/Z} & & \\
 & & \downarrow & & \downarrow & & \\
 & & 0 & & 0 & &
\end{array}
$$

By the snake lemma we get an exact sequence

$$0 \to L \otimes_Z T(Z/\delta) \to T(L/\delta) \to T(L/\delta_Z) \to T(L/d_{Z/K}) \to 0$$

which, when connected with the sequence in the right hand column of the last diagram, gives us a six-term exact sequence

(5) $\quad O \to L \otimes_Z T(Z/\delta) \to T(L/\delta) \to T(L/\delta_Z) \to L \otimes_Z \Omega^1_{Z/K}$

$\quad \to \Omega^1_{L/K} \to \Omega^1_{L/Z} \to O.$

We shall see soon, how (5) can be used for induction arguments.

5.12. <u>Theorem.</u> a) If L/K has a separating transcendence basis, then $T(L/\delta) = O$ and hence the canonical sequence

$$O \to L \otimes_K K\delta K \to \Omega^1_{L/\delta} \to \Omega^1_{L/K} \to O$$

is exact.

b) (Cartier's inequality) If char.$K =: p > O$ and L/K is finitely generated, then

$$\dim_L T(L/\delta) \le p\text{-deg}(L/K) - \text{Trdeg}(L/K)$$

with equality in case δ is the universal derivation of K (over $\mathbf{Z}$).

Proof. a) Let $\{x_\lambda\}$ be a separating transcendence basis of L/K, $Z := K(\{x_\lambda\})$ and δ_Z the universal Z-extension of δ. Then $T(L/\delta_Z) = O$ by 5.2, since L/Z is separably algebraic. Moreover, by 4.24a)

$$\Omega^1_{Z/\delta} = Z \otimes_K K\delta K \oplus \bigoplus_\lambda Zdx_\lambda$$

and consequently $T(Z/\delta) = O$. The exact sequence (5) shows that also $T(L/\delta) = O$.

b) Consider at first the special case that L/K is an inseparable extension of degree p. Then there is an $x \in L \smallsetminus K$ with $\xi := x^p \in K$ and $L = K[x] = K[X]/(X^p-\xi)$. By 4.19 we have

$$\Omega^1_{L/\delta} = L \otimes_K K\delta K \oplus LdX/U$$

where U is the vector space spanned by $1 \otimes \delta\xi$ in $L \otimes_K K\delta K$.

Therefore $T(L/\delta) = L \cdot (1 \otimes \delta\xi)$ and

$$\dim_L T(L/\delta) \leq 1 = p\text{-deg}(L/K) - \text{Trdeg}(L/K).$$

If δ is the universal derivation of K, then $K\delta K = \Omega^1_{K/K^p}$. Since $x \notin K$, we have $\xi \notin K^p$, and ξ is part of a p-basis of K/K^p. Therefore $\delta\xi \neq 0$ by 5.6, and we obtain in this case $\dim_L T(L/\delta) = 1$.

In the general case, choose a transcendence basis $\{x_1, \ldots, x_t\}$ of L/K. Let Z be the field of all elements of L that are separable over $K(x_1, \ldots, x_t)$. There is a sequence $\{y_1, \ldots, y_r\}$ of elements $y_i \in L$ such that $L = Z[y_1, \ldots, y_r]$ and $y_i^p \in Z[y_1, \ldots, y_{i-1}]$, $y_i \notin Z[y_1, \ldots, y_{i-1}]$ for $i=1, \ldots, r$.

For Z/K the statement in b) is already verified by a). The general case will follow by induction once the following statement has been proved:

(*) Let Z be an intermediate field of L/K such that $L = Z[y]$ with $y^p \in Z$, $y \notin Z$. If b) holds for $T(Z/\delta)$, then it also holds for $T(L/\delta)$.

In order to prove (*) we apply (5) to the situation at hand. Observe that Z/K is finitely generated, since L/K is. By assumption we have

$$\dim_L (L \otimes_Z T(Z/\delta)) \leq p\text{-deg}(Z/K) - \text{Trdeg}(Z/K)$$

and by the special case considered above $\dim_Z T(L/\delta_Z) \leq 1$. In both inequalities we have equality, if δ is the universal derivation of K (hence δ_Z is the universal derivation of Z). Moreover, by 5.6

$$\dim_L (L \otimes_Z \Omega^1_{Z/K}) = p\text{-deg}(Z/K)$$
$$\dim_L \Omega^1_{L/K} = p\text{-deg}(L/K) \text{ and } \dim_L \Omega^1_{L/Z} = 1.$$

From (5) we see now that $\dim_L T(L/\delta) =$

$\dim_L (L \otimes_Z T(Z/\delta)) + \dim_L T(L/\delta_Z) - \dim_L (L \otimes_Z \Omega^1_{Z/K}) + \dim_L \Omega^1_{L/K} - \dim_L \Omega^1_{L/Z}$

$\leq$ p-deg(Z/K)-Trdeg$(Z/K)+1-($p-deg$(Z/K))+$p-deg$(L/K)-1$

$=$ p-deg(L/K) - Trdeg(L/K)

and we have equality, if δ is the universal derivation of K.

The theorem allows several field theoretic applications.

<u>5.13. Corollary.</u>   Let K_o be a field of characteristic $p > 0$, and let K/K_o and L/K be algebraic function fields. Then
p-deg$(L/K)+$p-deg$(K/K_o) \geq$p-deg$(L/K_o) \geq$Trdeg$(L/K)+$p-deg(K/K_o).
If L/K is separable or $K_o = K^p$, then

$$\text{p-deg}(L/K_o) = \text{Trdeg}(L/K) + \text{p-deg}(K/K_o).$$

Proof.   Let δ be the universal derivation of K/K_o. In the exact sequence

(6) $O \to T(L/\delta) \to L \otimes_K \Omega^1_{K/K_o} \to \Omega^1_{L/K_o} \to \Omega^1_{L/K} \to O$

all vector spaces are finite dimensional. Applying 5.12b) we obtain
p-deg$(L/K_o) = \dim_L \Omega^1_{L/K_o} =$
$\dim_L \Omega^1_{L/K} + \dim_K \Omega^1_{K/K_o} - \dim_L T(L/\delta) \geq$
p-deg$(L/K) + $p-deg$(K/K_o) - ($p-deg$(L/K) - $Trdeg$(L/K)) =$
Trdeg$(L/K) + $p-deg$(K/K_o)$.

When L/K is separable or $K_o = K^p$ (hence δ is the universal derivation of K), then we have equality in 5.12b) and consequently here as well. The estimate
p-deg$(L/K_o) \leq$ p-deg$(L/K) + $p-deg$(K/K_o)$ follows from (6),

since $\dim_L T(L/\delta) \geq 0$.

5.14. <u>Corollary.</u> Let L/K be a separable algebraic function field and Z an intermediate field. Then Z/K is separable as well.

Proof. It is known that Z/K is also an algebraic function field. Let δ be the universal derivation of K/$\mathbb{Z}$. Then $T(L/\delta) = 0$ by 5.12a),and from (5) we see that $T(Z/\delta) = 0$, hence p-deg(Z/K) = Trdeg(Z/K) by 5.12b), which implies that Z/K is separable by 5.10c).

An arbitrary field extension L/K is defined to be <u>separable</u>, if each intermediate field Z of L/K, which is finitely generated over K, is separable. By the corollary this definition generalizes the definition of separability for algebraic function fields. If L/K is separable, then <u>any</u> intermediate field Z of L/K is separable over K. Of course, in characteristic zero all field extensions are separable.

5.15. <u>Corollary.</u> For a field extension L/K the following statements are equivalent:

a) L/K is separable.

b) $T(L/\delta) = 0$ for each derivation $\delta : K \to K\delta K$.

c) $T(L/d) = 0$ for the universal derivation d of K over $\mathbb{Z}$.

Proof. By 5.1 it is enough to consider finitely generated field extensions,where char.K > 0. a) $\to$ b) follows from 5.12a). Conversely,from $T(L/d) = 0$ we obtain by 5.12b) that p-deg(L/K) = Trdeg(L/K), hence L/K is separable.

5.16. <u>Example.</u> If a field extension L/K has a separating transcendence basis, then it is separable. This follows from 5.12a) and 5.15. The converse of the statement in 5.16 is not true in general (see exercise 4b)).

5.17. Corollary.   If Z/K and L/Z are separable field extensions, so is L/K.

Proof.   Let $\delta : K \to K\delta K$ be a derivation and δ_Z its universal Z-extension. By (5) we have an exact sequence

$$O \to L \otimes_Z T(Z/\delta) \to T(L/\delta) \to T(L/\delta_Z).$$

By 5.15, $T(Z/\delta) = T(L/\delta_Z) = O$, hence $T(L/\delta) = O$, and therefore L/K is separable, again by 5.15.

5.18. Proposition   (F.K.Schmidt). Over a perfect field K each field extension L/K is separable.

Proof.   It is enough to consider a field K of characteristic $p > O$ and an algebraic function field L/K. Choose an intermediate field Z of L/K such that Z/K is separable and L/Z is a purely inseparable algebraic extension. We have $L^{p^e} \subset Z$ for some $e \in \mathbb{N}$. By 5.14, with Z/K also L^{p^e} is separable over $K = K^{p^e}$. Since L/K and L^{p^e}/K^{p^e} are isomorphic by Frobenius, the field extension L/K is separable too.

If in the inequality of 5.12b) the equality sign holds, we can expect more precise results than in the general case. For the rest of this section we shall assume once again that K is a field of characteristic $p > O$.

5.19. Definition.   Let L/K be an algebraic function field.
a) A derivation $\delta : K \to K\delta K$ is called <u>admissible</u> for L/K,

if $\dim_K K\delta K < \infty$ and $\dim_L T(L/\delta) = p\text{-deg}(L/K) - \text{Trdeg}(L/K)$.

b) A subfield K_o of K is called <u>admissible</u> for L/K, if $K^p \subset K_o \subset K$, $[K:K_o] < \infty$ and if the universal derivation d_{K/K_o} is admissible for L/K.

<u>5.20. Remarks.</u>

a) Let $\delta : K \to K\delta K$ be a derivation with $\dim_K K\delta K < \infty$. Then δ is admissible for L/K if and only if

$$\dim_L \Omega^1_{L/\delta} = \dim_K K\delta K + \text{Trdeg}(L/K).$$

A subfield $K_o \subset K$ with $K^p \subset K_o$, $[K:K_o] < \infty$ is admissible for L/K if and only if

$$\dim_L \Omega^1_{L/K_o} = p\text{-deg}(K/K_o) + \text{Trdeg}(L/K).$$

b) The trivial derivation of K is admissible for L/K if and only if L/K is separable.

c) If $[K:K^p] < \infty$, then $K_o = K^p$ is an admissible field for each algebraic function field L/K.

   a) follows from the definition of $T(L/\delta)$, and b) is then a consequence of 5.10c). c) follows from the last part of 5.12b).

<u>5.21. Proposition</u>   (Existence of admissible derivations). Let L/K be an algebraic function field and $K' \subset K$ a sub-field with $K^p \subset K'$, $[K:K'] < \infty$. Then there is a subfield $K_o \subset K'$ which is admissible for L/K.

Proof.   Let $\{x_\lambda\}_{\lambda \in \Lambda}$ be a p-basis of K'/K^p and $\{y_1,\ldots,y_r\}$ a p-basis of K/K'. Then $\{x_\lambda\} \cup \{y_1,\ldots,y_r\}$ is a p-basis of K/K^p. Let d denote the universal derivation of K/K^p. Then

$$\Omega^1_{K/K^p} = \bigoplus_{\lambda \in \Lambda} K dx_\lambda \oplus \bigoplus_{i=1}^{r} K dy_i \,.$$

By 5.12b), $T(L/d) \subset L \otimes_K \Omega^1_{K/K^p}$ is an L-vector space of dimension $p\text{-deg}(L/K) - \mathrm{Trdeg}(L/K)$. Hence there exists a finite subset Λ' of Λ such that

$$T(L/d) \subset \bigoplus_{\lambda \in \Lambda'} L \cdot (1 \otimes dx_\lambda) \oplus \bigoplus_{i=1}^{r} L \cdot (1 \otimes dy_i) \,.$$

Put $K_o := K^p[\{x_\lambda\}_{\lambda \in \Lambda \smallsetminus \Lambda'}]$, and let δ be the universal derivation of K/K_o. Obviously $K^p \subset K_o \subset K'$ and $[K:K_o] < \infty$. Moreover, one has a commutative diagram with exact rows and columns

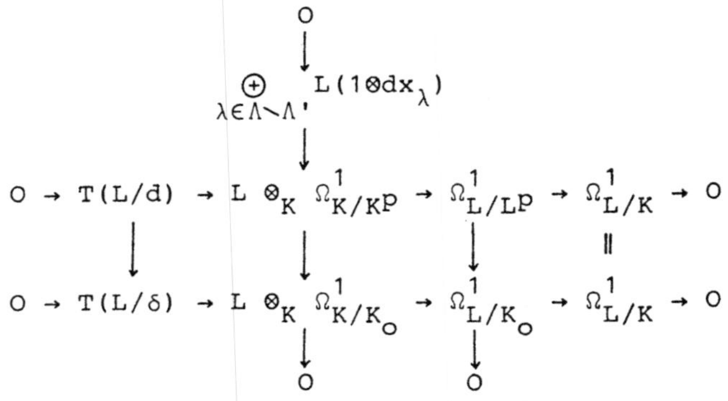

$T(L/d) \to T(L/\delta)$ is an injection, since

$T(L/d) \cap \bigoplus_{\lambda \in \Lambda \smallsetminus \Lambda'} L(1 \otimes dx_\lambda) = 0$. Hence $\dim_L T(L/\delta) \geq \dim_L T(L/d) = p\text{-deg}(L/K) - \mathrm{Trdeg}(L/K)$ and by 5.12b) we have equality. Therefore δ (and K_o) are admissible for L/K.

From the proof it is clear that one can find simultaneously an admissible $K_o \subset K'$ for finitely many algebraic function fields L/K.

In order to give a further characterization of admissible fields we recall the definition of linear disjointness: Let L/K be a field extension and Z_i ($i=1,2$) two intermediate fields of L/K. Z_1 and Z_2 are called <u>linearly disjoint over K</u>, if the canonical homomorphism

$$Z_1 \otimes_K Z_2 \to [Z_1 \cdot Z_2] \quad (a \otimes b \to ab)$$

is an isomorphism. Here $[Z_1 \cdot Z_2]$ denotes the subring of L

generated by $Z_1 \cup Z_2$. The condition is equivalent with the

following: Elements of Z_1 that are linearly independent

over K are also linearly independent over Z_2.

<u>5.22. Proposition.</u>   Let L/K be an algebraic function field

and $K_0 \subset K$ a subfield with $K^p \subset K_0$, $[K:K_0] < \infty$. Then the

following conditions are equivalent:

a) K_0 is admissible for L/K.

b) K_0 and L^p are linearly disjoint over K^p.

c) Any p-basis of K_0/K^p is also a p-basis of $K_0[L^p]$ over L^p.

Proof.   That b) and c) are equivalent follows easily from

the definition of linear disjointness. In order to show

a) $\leftrightarrow$ b) let $\{x_1,\ldots,x_t\}$ be a transcendence basis of L/K and

$Z := K(x_1,\ldots,x_t)$. Then K_0 and $Z^p = K^p(x_1^p,\ldots,x_t^p)$ are linear-

ly disjoint over K^p, which means that $K_0[Z^p] \cong K_0 \otimes_{K^p} Z^p$.

The following diagram illustrates the situation:

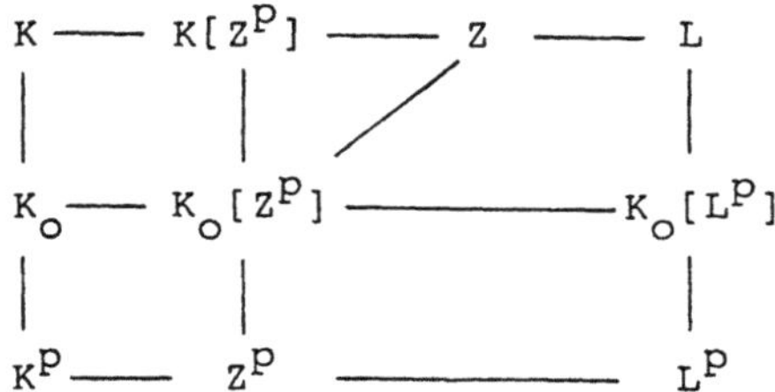

Since $K_0 \otimes_{K^p} L^p = (K_0 \otimes_{K^p} Z^p) \otimes_{Z^p} L^p = K_0[Z^p] \otimes_{Z^p} L^p$, we see

that K_0 and L^p are linearly disjoint over K^p if and only

if $K_0[Z^p]$ and L^p are linearly disjoint over Z^p. This last condition is equivalent with $[K_o[L^p]:K_o[Z^p]] = [L^p:Z^p] = [L:Z]$, hence also with the condition $[L:K_o[L^p]] = [Z:K_o[Z^p]]$. This equation holds if and only if

$$p\text{-deg}(L/K_o) = p\text{-deg}(Z/K_o) = \text{Trdeg}(Z/K) + p\text{-deg}(K/K_o) = \text{Trdeg}(L/K) + p\text{-deg}(K/K_o).$$

By 5.20a) the last equation holds if and only if K_o is admissible for L/K.

5.23. Corollary. If K_o is admissible for L/K, so is K_o' for each subfield $K_o' \subset K_o$ with $K^p \subset K_o'$, $[K_o:K_o'] < \infty$.

5.24. Corollary (Separability criterion of MacLane).
Let L/K be an arbitrary field extension. Then the following conditions are equivalent:
a) L/K is separable.
b) K and L^p are linearly disjoint over K^p.

Proof. One can reduce immediately to the case of a finitely generated extension L/K. Then the claim follows from 5.20b) and 5.22.

The reader should keep in mind that the statements about the dimension of differential modules of field extensions are also true for the corresponding derivation modules, since these are the duals of the differential modules.

Exercises

1) a) Let K be a field of characteristic 0. For polynomials $F_1,\ldots,F_n \in K[X_1,\ldots,X_n]$ the following statements are equivalent:

α) $\{F_1,\ldots,F_n\}$ is algebraically dependent over K.

β) $\det\left(\dfrac{\partial F_i}{\partial X_j}\right) = 0$

b) In case n=2 let F_i be homogeneous polynomials of degree n_i (i=1,2) and $d := \gcd(n_1,n_2)$. Then α) and β) are also equivalent with

γ) There exists an $\alpha \in K^\times$ such that $F_1^{\frac{n_2}{d}} = \alpha F_2^{\frac{n_1}{d}}$.

c) Monomials $X_1^{\alpha_{1j}} \cdot \ldots \cdot X_n^{\alpha_{nj}} \in K(X_1,\ldots,X_n)$ (j=1,...,t) with $\alpha_{ij} \in \mathbb{Z}$ are algebraically independent over K if and only if the vectors $(\alpha_{1j},\ldots,\alpha_{nj}) \in \mathbb{Q}^n$ (j=1,...,t) are linearly independent over $\mathbb{Q}$.

2) Let K be a field of characteristic p > 0.

a) K is perfect if and only if $\Omega^1_{K/K^p} = 0$.

b) If K is perfect and $K_0 \subset K$ a subfield such that K/K_0 is finitely generated, then K/K_0 is finite and separable.

3) Let L/K be a separable field extension, where K has characteristic p > 0. If $\{x_\lambda\}_{\lambda \in \Lambda}$ is a p-basis of L/K, then $\{x_\lambda\}_{\lambda \in \Lambda}$ is algebraically independent over K.

4) a) Give an example of a field extension L/K such that $\Omega^1_{L/K} = 0$ and L/K is <u>not</u> separable.

b) Give an example of a separable field extension L/K, which has no separating transcendence basis (Hint: Theorem of F.K.Schmidt).

92

5) Use MacLane's criterion (5.24) to show that for any field K the field $L = K((X_1,\ldots,X_n)) := Q(K[\![X_1,\ldots,X_n]\!])$ is separable over K.

6) Let S/R be a ring extension, where R has prime characteristic p and $S^p \subset R$. Suppose S/R has a p-basis $\{x_1,\ldots,x_n\}$.

a) $Z(\Omega_{S/R}) = B(\Omega_{S/R}) \oplus R[x_1^{p-1}dx_1,\ldots,x_n^{p-1}dx_n]$ (Use induction on n).

b) Consider R as an S-module via the homomorphism $F : S \to R$, $F(s) = s^p$ for all $s \in S$. Show that there is a ring-homomorphism $\gamma : Z(\Omega_{S/R}) \to R \otimes_S \Omega_{S/R}$ with $\gamma(B(\Omega_{S/R})) = 0$, $\gamma(r) = r\otimes 1$ for $r \in R$ and $\gamma(x_i^{p-1}dx_i) = 1\otimes dx_i$ $(i=1,\ldots,n)$.

c) Show that the set of all $s \in S$ with $\gamma(s^{p-1}ds) = 1\otimes ds$ is all of S. Conclude that γ is independent of the choice of a p-basis of S/R.

d) Show that γ induces an isomorphism

$$C : H_{DR}(\Omega_{S/R}) \to R \otimes_S \Omega_{S/R}.$$

In particular, if $R = S^p$ and S is reduced, we have $H_{DR}(\Omega_{S/R}) \cong \Omega_{S/R}$. C is called <u>Cartier operator</u>. For more details, see [C].

§ 6. Differential Modules of Local Rings

The structure of the differential module of a local ring
is, of course, much more complicated than that of a field.
We wish to determine first its minimal number of generators.

Let $(R,\mathfrak{m})$ be a local ring with residue field $K := R/\mathfrak{m}$
and $p := \mathrm{char}.K$. Let d denote the universal derivation of
$R/\mathbb{Z}$. By 4.17 there is a canonical exact sequence of K-vector
spaces

$$(1) \qquad \mathfrak{m}/\mathfrak{m}^2 \xrightarrow{\alpha} \Omega^1_{R/\mathbb{Z}}/\mathfrak{m}\Omega^1_{R/\mathbb{Z}} \xrightarrow{\beta} \Omega^1_{K/\mathbb{Z}} \to 0$$

where $\alpha(x+\mathfrak{m}^2) = dx+\mathfrak{m}\Omega^1_{R/\mathbb{Z}}$ for all $x \in \mathfrak{m}$. We have $pR \subset \mathfrak{m}$, and
for each $r \in R$ we obtain $d(pr) = pdr \in \mathfrak{m}\Omega^1_{R/\mathbb{Z}}$. Therefore α
induces a K-linear map

$$\alpha' \;:\; \mathfrak{m}/\mathfrak{m}^2+pR \to \Omega^1_{R/\mathbb{Z}}/\mathfrak{m}\Omega^1_{R/\mathbb{Z}}$$

with $\alpha'(x+\mathfrak{m}^2+pR) = dx+\mathfrak{m}\Omega^1_{R/\mathbb{Z}}$.

6.1. <u>Theorem.</u>   The canonical sequence

$$(2) \qquad 0 \to \mathfrak{m}/\mathfrak{m}^2+pR \xrightarrow{\alpha'} \Omega^1_{R/\mathbb{Z}}/\mathfrak{m}\Omega^1_{R/\mathbb{Z}} \xrightarrow{\beta} \Omega^1_{K/\mathbb{Z}} \to 0$$

is exact.

Proof. It remains to be shown that α' is injective. For
$\overline{R} := R/\mathfrak{m}^2+pR$, $\overline{\mathfrak{m}} := \mathfrak{m}/\mathfrak{m}^2+pR$ we have $\Omega^1_{\overline{R}/\mathbb{Z}} = \Omega^1_{R/\mathbb{Z}}/Rd(\mathfrak{m}^2+pR)$ by
4.16 and, since $d(\mathfrak{m}^2+pR) = d\mathfrak{m}^2+pdR \subset \mathfrak{m}\Omega^1_{R/\mathbb{Z}}$, we obtain
$\Omega^1_{R/\mathbb{Z}}/\mathfrak{m}\Omega^1_{R/\mathbb{Z}} = \Omega^1_{R/\mathbb{Z}}/Rd(\mathfrak{m}^2+pR) / \mathfrak{m}\Omega^1_{R/\mathbb{Z}}/Rd(\mathfrak{m}^2+pR) = \Omega^1_{\overline{R}/\mathbb{Z}}/\overline{\mathfrak{m}}\,\Omega^1_{\overline{R}/\mathbb{Z}}$.
α' can be identified with the corresponding map

$$\overline{\alpha} \;:\; \overline{\mathfrak{m}} \to \Omega^1_{\overline{R}/\mathbb{Z}}/\overline{\mathfrak{m}}\Omega^1_{\overline{R}/\mathbb{Z}} \qquad (\overline{\alpha}(\overline{x}) = \overline{d}\overline{x} + \overline{\mathfrak{m}}\Omega^1_{\overline{R}/\mathbb{Z}})$$

where $\overline{d}$ denotes the universal derivation of $\overline{R}/\mathbb{Z}$. It is
enough to show, that $\overline{\alpha}$ is injective. In other words: We

have to prove the theorem only in the case that $\mathfrak{m}^2 + pR = 0$ in the original local ring R.

Then R is a complete and separated local ring having the same characteristic as its residue field K. By Cohen's structure theorem (Matsumura $[M_1]$, Thm.60), which is needed here only in a mild form ($\mathfrak{m}^2 = 0$), there is a field $K' \subset R$, which is mapped isomorphically onto K under the canonical epimorphism $R \to K$.

Let $\{x_\lambda\}_{\lambda \in \Lambda}$ be a basis of the K-vector space $\mathfrak{m}$. Since $\mathfrak{m}^2 = 0$, we have

$$R = K' \oplus \bigoplus_{\lambda \in \Lambda} K'x_\lambda \ , \quad x_\lambda x_\mu = 0 \ \text{for} \ \lambda,\mu \in \Lambda.$$

Therefore the kernel I of the canonical K'-homomorphism

$$K'[\{X_\lambda\}_{\lambda \in \Lambda}] \to R \qquad (X_\lambda \mapsto x_\lambda)$$

is generated by all products $X_\lambda X_\mu$ ($\lambda,\mu \in \Lambda$). By 4.19 we have

$$\Omega^1_{R/\mathbb{Z}} \cong (R \otimes_{K'} \Omega^1_{K'/\mathbb{Z}} \oplus \bigoplus_{\lambda \in \Lambda} RdX_\lambda)/U$$

where U is the submodule of $\bigoplus_{\lambda \in \Lambda} RdX_\lambda$ generated by the elements $x_\lambda dX_\mu + x_\mu dX_\lambda$ ($\lambda,\mu \in \Lambda$). Therefore

$$\Omega^1_{R/\mathbb{Z}}/\mathfrak{m}\Omega^1_{R/\mathbb{Z}} \cong \Omega^1_{K/\mathbb{Z}} \oplus \bigoplus_{\lambda \in \Lambda} KdX_\lambda$$

and α is identified with the mapping given by $\sum_{i=1}^{n} \alpha_i x_{\lambda_i} \mapsto \sum_{i=1}^{n} \alpha_i dX_{\lambda_i}$ ($\alpha_i \in K$). Clearly this mapping is injective.

<u>6.2. Corollary.</u>   Suppose that both $\mathfrak{m}$ and $\Omega^1_{R/\mathbb{Z}}$ are finitely generated. Then

$$\mu(\Omega^1_{R/\mathbb{Z}}) = \text{edim } R/pR + \dim_K \Omega^1_{K/\mathbb{Z}} \ .$$

Proof. By Nakayama's lemma $\mu(\Omega^1_{R/\mathbb{Z}}) = \dim_K(\Omega^1_{R/\mathbb{Z}}/m\Omega^1_{R/\mathbb{Z}}) =$
$\dim_K(m/m^2+pR) + \dim_K\Omega^1_{K/\mathbb{Z}}$. The ring $\tilde{R} := R/pR$ is local with
maximal ideal $\tilde{m} := m/pR$ and residue field K. Since
$m/m^2+pR = m/pR/m^2+pR/pR = \tilde{m}/\tilde{m}^2$, we see that $\dim_K(m/m^2+pR) =$
$\dim_K \tilde{m}/\tilde{m}^2 = \text{edim } \tilde{R}$.

The term $\dim_K\Omega^1_{K/\mathbb{Z}}$ is known by the results of § 5:

$$\dim_K \Omega^1_{K/\mathbb{Z}} = \begin{cases} \text{Trdeg}(K/\mathbb{Q}) & , \text{ if } p = 0 \\[2em] \text{p-deg}(K/K^p) & , \text{ if } p > 0. \end{cases}$$

In fact from 6.2 and Nakayama's lemma we get the following
more precise statement.

<u>6.3. Corollary.</u>   Under the assumptions of 6.2 let
$x_1,\ldots,x_m \in m$ be elements whose residue classes in m/pR
form a minimal system of generators of that ideal in R/pR,
and let $y_1,\ldots,y_t \in R$ be elements whose residue classes $\overline{y_i}$
in K form a transcendence basis of $K/\mathbb{Q}$, in case $p = 0$, and
a p-basis of K/K^p, in case $p > 0$. Then
$\{dx_1,\ldots,dx_m,dy_1,\ldots,dy_t\}$ is a minimal system of generators
of $\Omega^1_{R/\mathbb{Z}}$.

Now let (S,n) be a local ring with residue field L. Let
R be an arbitrary ring and $\rho : R \to S$ a ring homomorphism.
We want to study $\Omega^1_{S/R}$. $\mathfrak{p} := \rho^{-1}(n)$ is a prime ideal of R,
and ρ induces a ring homomorphism $\rho' : R_{\mathfrak{p}} \to S$ with
$\rho'(\mathfrak{p}R_{\mathfrak{p}}) \subset n$. By 1.9f) we have $\Omega^1_{S/R} = \Omega^1_{S/R_{\mathfrak{p}}}$, and therefore
we may also assume that R is a local ring with maximal
ideal m and $\rho(m) \subset n$ in other words: S/R is a local algebra.
$K := R/m$ may be regarded as a subfield of L.

<u>6.4. Theorem.</u>   Under these assumptions let δ_K denote the universal derivation of $K/\mathbb{Z}$. Then there is an exact sequence of L-vector spaces

$$T' \to T(L/\delta_K) \to \mathfrak{m}/\mathfrak{m}^2 + \mathfrak{m}S \to \Omega^1_{S/R}/\mathfrak{m}\Omega^1_{S/R} \to \Omega^1_{L/K} \to O,$$

where T' is the kernel of the canonical homomorphism

$$L \otimes_K \Omega^1_{R/\mathbb{Z}}/\mathfrak{m}\Omega^1_{R/\mathbb{Z}} \to \Omega^1_{S/\mathbb{Z}}/\mathfrak{m}\Omega^1_{S/\mathbb{Z}} \ .$$

Proof.   Let $p := \operatorname{char}.K$. In the commutative diagram

$$
\begin{array}{ccccccccc}
 & & & & O & & O & & \\
 & & & & \downarrow & & \downarrow & & \\
 & & & & T' & & T(L/\delta_K) & & \\
 & & & & \downarrow & & \downarrow & & \\
O \to & L \otimes_K \mathfrak{m}/\mathfrak{m}^2 + pR & \to & L \otimes_K \Omega^1_{R/\mathbb{Z}}/\mathfrak{m}\Omega^1_{R/\mathbb{Z}} & \to & L \otimes_K \Omega^1_{K/\mathbb{Z}} & \to & O \\
 & \downarrow & & \downarrow & & \downarrow & & \\
O \longrightarrow & \mathfrak{m}/\mathfrak{m}^2 + pS & \longrightarrow & \Omega^1_{S/\mathbb{Z}}/\mathfrak{m}\Omega^1_{S/\mathbb{Z}} & \longrightarrow & \Omega^1_{L/\mathbb{Z}} & \to & O \\
 & \downarrow & & \downarrow & & \downarrow & & \\
 & \mathfrak{m}/\mathfrak{m}^2 + \mathfrak{m}S & \longrightarrow & \Omega^1_{S/R}/\mathfrak{m}\Omega^1_{S/R} & \longrightarrow & \Omega^1_{L/K} & \to & O \\
 & \downarrow & & \downarrow & & \downarrow & & \\
 & O & & O & & O & & \\
\end{array}
$$

the first two rows are exact by 6.1. Clearly the left hand column is exact. The exactness of the last two columns follows from the definition of T' and $T(L/\delta_K)$. The snake lemma gives us now the exact sequence of the theorem.

<u>6.5. Corollary.</u>   a) If L/K is separable, then the sequence

$$O \to \mathfrak{m}/\mathfrak{m}^2 + \mathfrak{m}S \to \Omega^1_{S/R}/\mathfrak{m}\Omega^1_{S/R} \to \Omega^1_{L/K} \to O$$

is exact.

b) If, moreover, S is essentially of finite type over R, then

$$\mu(\Omega^1_{S/R}) = \operatorname{edim} S/\mathfrak{m}S + \operatorname{Trdeg}(L/K).$$

a) follows from 6.4, since $T(L/\delta_K) = 0$, if L/K is separable (5.15). Under the assumptions of b) $\Omega^1_{S/R}$ is finitely generated and L/K an algebraic function field. The formula in b) follows now by Nakayama's lemma from a), since $\dim_L \Omega^1_{L/K} = \mathrm{Trdeg}(L/K)$ by 5.3.

Next we consider the case that L/K is not separable.

6.6. Lemma.   Let $(R,\mathfrak{m})$ be a local ring and $p > 0$ an integer. Then each subring $R' \subset R$ with $R^p \subset R'$ is also local, and $\mathfrak{m}' := \mathfrak{m} \cap R'$ is its maximal ideal.

Let us show that the elements of $R' \smallsetminus \mathfrak{m}'$ are units in R'. $x \in R' \smallsetminus \mathfrak{m}'$ is certainly a unit in R, so $xy = 1$ for some $y \in R$. But then $xx^{p-1}y^p = 1$ and $x^{p-1}y^p \in R'$, hence x is also a unit of R'.

6.7. Theorem.   Suppose that under the conditions of 6.4 K is a field of characteristic $p > 0$ and let $\mathfrak{m}' := \mathfrak{m} \cap R[S^p]$. Then:

a) There is a canonical exact sequence

$$0 \to \mathfrak{m}/\mathfrak{m}^2 + \mathfrak{m}'S \to \Omega^1_{S/R}/\mathfrak{m}\Omega^1_{S/R} \to \Omega^1_{L/K} \to 0.$$

b) If S/R is essentially of finite type, then

$$\mu(\Omega^1_{S/R}) = \mathrm{edim}\, S/\mathfrak{m}'S + p\text{-}\deg(L/K).$$

Proof.   By the lemma $R' := R[S^p]$ is a local ring with maximal ideal $\mathfrak{m}'$. Clearly $K[L^p] =: K'$ is the residue field of R'. For the universal derivation d of S/R we have $d(R[S^p]) \subset p\Omega^1_{S/R} \subset \mathfrak{m}\Omega^1_{S/R}$ and, in particular, $d\mathfrak{m}' \subset \mathfrak{m}\Omega^1_{S/R}$. Hence the canonical map $\mathfrak{m}/\mathfrak{m}^2 \to \Omega^1_{S/R}/\mathfrak{m}\Omega^1_{S/R}$ induces a map

$\mathfrak{n}/\mathfrak{n}^2 + \mathfrak{m}'S \to \Omega^1_{S/R}/\mathfrak{m}\Omega^1_{S/R}$ with the same image. Moreover, $\Omega^1_{S/R}/\mathfrak{m}\Omega^1_{S/R} = \Omega^1_{S/R'}/\mathfrak{m}\Omega^1_{S/R'}$ and $\Omega^1_{L/K} = \Omega^1_{L/K[L^p]} = \Omega^1_{L/K'}$.

By 6.4 we have an exact sequence

$$T' \to T(L/\delta_{K'}) \to \mathfrak{n}/\mathfrak{n}^2 + \mathfrak{m}'S \to \Omega^1_{S/R'}/\mathfrak{m}\Omega^1_{S/R'} \to \Omega^1_{L/K'} \to 0$$

where $\delta_{K'}$ is the universal derivation of $K'/\mathbb{Z}$ and T' the kernel of the canonical map $L \otimes_{K'} \Omega^1_{R'/\mathbb{Z}}/\mathfrak{m}'\Omega^1_{R'/\mathbb{Z}} \to \Omega^1_{S/\mathbb{Z}}/\mathfrak{m}\Omega^1_{S/\mathbb{Z}}$. We shall show that $T' \to T(L/\delta_{K'})$ is surjective, which then gives us the exact sequence in a). b) follows immediately from a) and Nakayama's lemma.

Let $\{x_\lambda\}_{\lambda \in \Lambda}$ be a p-basis of L/K', $\{y_\lambda\}_{\lambda \in \Lambda}$ a system of representatives of $\{x_\lambda\}$ in S and $\delta_{R'}$ the universal derivation of $R'/\mathbb{Z}$. Then by 4.19

$$\Omega^1_{L/\mathbb{Z}} = L \otimes_{K'} \Omega^1_{K'/\mathbb{Z}}/<\{\delta_{K'}(x_\lambda^p)\}_{\lambda \in \Lambda}> \oplus \bigoplus_{\lambda \in \Lambda} L dx_\lambda$$

and $T(L/\delta_{K'})$ is the subspace of $L \otimes_{K'} \Omega^1_{K'/\mathbb{Z}}$ spanned by $\{1 \otimes \delta_{K'}(x_\lambda^p)\}_{\lambda \in \Lambda}$. On the other hand, $y_\lambda^p \in R'$ and $\delta_{R'}(y_\lambda^p)$ is contained in the kernel of the canonical map $\Omega^1_{R'/\mathbb{Z}} \to \Omega^1_{S/\mathbb{Z}}/\mathfrak{m}\Omega^1_{S/\mathbb{Z}}$. If η_λ denotes the image of $\delta_{R'}(y_\lambda^p)$ in $\Omega^1_{R'/\mathbb{Z}}/\mathfrak{m}'\Omega^1_{R'/\mathbb{Z}}$, then $1 \otimes \eta_\lambda \in T'$ and $1 \otimes \delta_{K'}(x_\lambda^p)$ is the image of $1 \otimes \eta_\lambda$ in $T(L/\delta_{K'})$, q.e.d.

The previous results allow us to give some vanishing criteria for the differential module of an algebra.

6.8. Proposition.   Let S/R be an algebra which is essentially of finite type. Then the following conditions are equivalent:

a) $\Omega^1_{S/R} = 0$.

b) For all $\mathfrak{p} \in \mathrm{Spec}(S)$ we have (with $\mathfrak{q} := \mathfrak{p} \cap R$)

$$\mathfrak{p}S_\mathfrak{p} = \mathfrak{q}S_\mathfrak{p}$$

and $k(\mathfrak{p})/k(\mathfrak{q})$ is separably algebraic.

b') The conditions of b) hold for all $\mathfrak{p} \in \mathrm{Max}(S)$.

Proof. By the local-global principle we may assume that S/R is a local algebra. Let $\mathfrak{n}$ be the maximal ideal of S, $\mathfrak{m}$ the maximal ideal of R, $L := S/\mathfrak{n}$, $K := R/\mathfrak{m}$.

a) $\to$ b'). If $\Omega^1_{S/R} = 0$, then $\Omega^1_{L/K} = 0$ as well. Since L/K is finitely generated, L/K is separably algebraic by 5.9. Applying 6.5b) we see that edim $S/\mathfrak{m}S = 0$, in other words $\mathfrak{n} = \mathfrak{m}S$.

b') $\to$ a). Under the assumptions of b') we have $\Omega^1_{L/K} = 0$ and $\mathfrak{n}/\mathfrak{n}^2 + \mathfrak{m}S = 0$, hence $\Omega^1_{S/R}/\mathfrak{n}\,\Omega^1_{S/R} = 0$ by 6.5a), thus $\Omega^1_{S/R} = 0$ by Nakayama's lemma, since $\Omega^1_{S/R}$ is finitely generated.

6.9. <u>Corollary.</u>   Let K be a field and A/K an algebra which is essentially of finite type. Then the following are equivalent:

a) $\Omega^1_{A/K} = 0$.

b) $A \cong L_1 \times \ldots \times L_m$, where L_i/K $(i=1,\ldots,m)$ is a finite separable field extension.

Proof. b) $\to$ a). By 4.8

$$\Omega^1_{L_1 \times \ldots \times L_m/K} \cong \Omega^1_{L_1/K} \times \ldots \times \Omega^1_{L_m/K}$$

and by 5.9 we have $\Omega^1_{L_i/K} = 0$, hence $\Omega^1_{A/K} = 0$.

a) $\to$ b). By 6.8, $\mathfrak{p}A_\mathfrak{p} = 0$ for each $\mathfrak{p} \in \mathrm{Max}(A)$, consequently $\mathrm{Max}(A) = \mathrm{Min}(A)$. Since A is noetherian, $\mathrm{Min}(A)$ is a finite set, say $\{\mathfrak{p}_1, \ldots, \mathfrak{p}_m\}$. From $\mathfrak{p}_i A_{\mathfrak{p}_i} = 0$ $(i=1,\ldots,m)$ we obtain $\mathfrak{p}_1 \cap \ldots \cap \mathfrak{p}_m = 0$, and by the Chinese remainder theorem

$$A \cong A/\mathfrak{p}_1 \times \ldots \times A/\mathfrak{p}_m.$$

The $A/\mathfrak{p}_i$ are fields and we have $\Omega^1_{A/\mathfrak{p}_i/K} = 0$, therefore $A/\mathfrak{p}_i$ is finite and separable over K $(i=1,\ldots,b)$ by 5.9.

Remember that for an algebra S/R a prime $\mathfrak{p} \in \mathrm{Spec}(S)$ is called <u>unramified over R</u> (or S/R <u>unramified at $\mathfrak{p}$</u>), if $\mathfrak{p}S_\mathfrak{p} = \mathfrak{q}S_\mathfrak{p}$, where $\mathfrak{q} := \mathfrak{p} \cap R$, and $k(\mathfrak{p})/k(\mathfrak{q})$ is separably algebraic. Otherwise S/R is called <u>ramified at $\mathfrak{p}$</u>. S/R is called <u>étale at $\mathfrak{p}$</u>, if S/R is unramified at $\mathfrak{p}$ and $S_\mathfrak{p}/R_\mathfrak{q}$ is flat. S/R is called unramified (étale), if S/R is unramified (étale) at all $\mathfrak{p} \in \mathrm{Spec}(S)$.

In the situation of 6.8, S/R is unramified if and only if $\Omega^1_{S/R} = 0$.

The set
$$V_{S/R} := \{\mathfrak{p} \in \mathrm{Spec}(S) \mid S/R \text{ is ramified at } \mathfrak{p}\}$$
is called the <u>ramification locus</u> of S/R. The <u>étale locus</u> of S/R is the set of all $\mathfrak{p} \in \mathrm{Spec}(S)$, at which S/R is étale.

<u>6.10. Corollary.</u>   Under the assumptions of 6.8 we have
$$V_{S/R} = \mathrm{Supp}(\Omega^1_{S/R}).$$

In particular, $V_{S/R}$ is a closed subset of $\mathrm{Spec}(S)$. If R is noetherian, the étale locus of S/R is open in $\mathrm{Spec}(S)$.

By 6.8, $\mathfrak{p} \in \mathrm{Spec}(S)$ is unramified over R if and only if $\Omega^1_{S_\mathfrak{p}/R_\mathfrak{q}} = (\Omega^1_{S/R})_\mathfrak{p} = 0$, hence the formula of 6.10 holds. The étale locus is the intersection of $\mathrm{Spec}(S) \setminus V_{S/R}$ with the set of all $\mathfrak{p} \in \mathrm{Spec}(S)$, at which S/R is flat. If R is noetherian, this set is open in $\mathrm{Spec}(S)$ (Matsumura $[M_1]$,Thm.53).

Since $\mathrm{Supp}(\Omega^1_{S/R})$ is the set of all $\mathfrak{p} \in \mathrm{Spec}(S)$ with $\mathfrak{p} \supset \mathrm{Ann}(\Omega^1_{S/R})$, we may write under the assumptions of 6.8

$$V_{S/R} = \{ \mathfrak{p} \in \mathrm{Spec}(S) \mid \mathfrak{p} \supset \mathrm{Ann}(\Omega^1_{S/R}) \}.$$

6.8 allows easy proofs of some statements about unramified extensions.

<u>6.11. Corollary.</u>   Let $\mathfrak{q} \subset \mathfrak{p}$ be elements of $\mathrm{Spec}(S)$. If $\mathfrak{p}$ is unramified over R, so is $\mathfrak{q}$.

If $(\Omega^1_{S/R})_{\mathfrak{p}} = 0$, then $(\Omega^1_{S/R})_{\mathfrak{q}} = 0$.

<u>6.12. Corollary.</u>   Assume that the algebra T/S is essentially of finite type.

a) If T/R is unramified, so is T/S.

b) If T/S and S/R are unramified, so is T/R.

Use the exact sequence

$$T \otimes_S \Omega^1_{S/R} \to \Omega^1_{T/R} \to \Omega^1_{T/S} \to 0$$

and 6.8.

<u>6.13. Corollary.</u>   a) Let R'/R be an arbitrary algebra. If S/R is unramified, so is $R' \otimes_R S/R'$. The converse is true, if R'/R is faithfully flat.

b) Let S/R and T/R be algebras that are essentially of finite type. If S/R and T/R are unramified, so is $S \otimes_R T/R$. The converse is true if S/R and T/R are faithfully flat.

For a) use the formula $\Omega^1_{R' \otimes_R S/R'} = R' \otimes_R \Omega^1_{S/R}$ and for b) the formula $\Omega^1_{S \otimes_R T/R} = T \otimes_R \Omega^1_{S/R} \oplus S \otimes_R \Omega^1_{T/R}$ (see 4.4).

If S/R is an algebra, then $\mathfrak{q} \in \mathrm{Spec}(R)$ is call <u>unramified</u> in S, if all $\mathfrak{p} \in \mathrm{Spec}(S)$ lying over $\mathfrak{q}$ are unramified over R, otherwise $\mathfrak{q}$ is called ramified in S. Let $V_{S/R}$ denote the set of all $\mathfrak{q} \in \mathrm{Spec}(R)$ that are ramified in S.

6.14. Corollary. Under the assumptions of 6.8 suppose that $\mathrm{Spec}(S) \to \mathrm{Spec}(R)$ is a closed map. Then $v_{S/R}$ is a closed subset of $\mathrm{Spec}(R)$.

In fact, $v_{S/R}$ is the image of the closed set $V_{S/R}$ in $\mathrm{Spec}(R)$.

6.15. Examples. a) Let $K/\mathbb{Q}$ be a finite field extension (i.e. K an algebraic number field) and let R be the integral closure of $\mathbb{Z}$ in K (the ring of integers of K). Then there are only finitely many prime numbers p such that (p) is ramified in R. In fact, $v_{R/\mathbb{Z}}$ is a closed subset of $\mathrm{Spec}(\mathbb{Z})$ by 6.14. $v_{R/\mathbb{Z}}$ cannot be all of $\mathrm{Spec}(\mathbb{Z})$, since $\Omega^1_{K/\mathbb{Q}} = K \otimes_R \Omega^1_{R/\mathbb{Z}} = \mathbb{Q} \otimes_{\mathbb{Z}} \Omega^1_{R/\mathbb{Z}} = 0$, hence the annihilator of $\Omega^1_{R/\mathbb{Z}}$ as a $\mathbb{Z}$-module is $\neq 0$. $v_{R/\mathbb{Z}}$ is therefore a finite set.

b) Let S/R be a finite ring extension, where R and S are Dedekind domains. Assume that $L := Q(S)$ is separable over $K := Q(R)$. As in a) we have $\Omega^1_{L/K} = L \otimes_S \Omega^1_{S/R} = 0$, and hence $\Omega^1_{S/R}$ is a torsion module. $V_{S/R}$ is a finite set, and so is $v_{S/R}$.

For each $\mathfrak{q} \in \mathrm{Spec}(R)$ the extension ideal $\mathfrak{q}S$ has a unique representation

$$\mathfrak{q}S = \mathfrak{p}_1^{e_1} \cdot \ldots \cdot \mathfrak{p}_r^{e_r}$$

as a power product of primes $\mathfrak{p}_i \in \mathrm{Spec}(S)$. e_i is called the __ramification index__ of $\mathfrak{p}_i$ over R. $\mathfrak{p}_i$ is ramified over R if and only if $e_i > 1$ or $k(\mathfrak{p}_i)/k(\mathfrak{q})$ is inseparable.

By the Chinese remainder theorem we have

$$\Omega^1_{S/R} = \prod_{\mathfrak{p} \in V_{S/R}} \Omega^1_{S_\mathfrak{p}/R_\mathfrak{q}} \qquad (\mathfrak{q} := \mathfrak{p} \cap R).$$

Since $S_\mathfrak{p}$ is a discrete valuation ring, hence a principal ideal domain, we have $\mu_\mathfrak{p}(\Omega^1_{S_\mathfrak{p}/R_\mathfrak{q}}) \le 1$, if $k(\mathfrak{p})/k(\mathfrak{q})$ is separable, and $\mu_\mathfrak{p}(\Omega^1_{S_\mathfrak{p}/R_\mathfrak{q}}) \le 1 + p\text{-deg}(k(\mathfrak{p})/k(\mathfrak{q}))$ otherwise (6.7b)).

Let $e_\mathfrak{p}$ be the ramification index of $\mathfrak{p}$ over R and assume that $k(\mathfrak{p})/k(\mathfrak{q})$ is separable. Then

$$\Omega^1_{S_\mathfrak{p}/R_\mathfrak{q}} \cong S_\mathfrak{p}/\mathfrak{p}^{\varepsilon_\mathfrak{p}} S_\mathfrak{p}$$

for some $\varepsilon_\mathfrak{p} \in \mathbb{N}$. We wish to compare the integers $e_\mathfrak{p}$ and $\varepsilon_\mathfrak{p}$. With $\bar{S}_\mathfrak{p} := S_\mathfrak{p}/\mathfrak{q}S_\mathfrak{p} = S_\mathfrak{p}/\mathfrak{p}^{e_\mathfrak{p}}S_\mathfrak{p}$ we have

$$\Omega^1_{\bar{S}_\mathfrak{p}/k(\mathfrak{q})} \cong \Omega^1_{S_\mathfrak{p}/R_\mathfrak{q}}/\mathfrak{q}\Omega^1_{S_\mathfrak{p}/R_\mathfrak{q}} \cong S_\mathfrak{p}/\mathfrak{p}^{\text{Min}(e_\mathfrak{p},\varepsilon_\mathfrak{p})} S_\mathfrak{p}$$

By Cohen's structure theorem

$$\bar{S}_\mathfrak{p} = k(\mathfrak{p})[t]/(t^{e_\mathfrak{p}})$$

and therefore $\Omega^1_{\bar{S}_\mathfrak{p}/k(\mathfrak{q})} = \bar{S}_\mathfrak{p}dt/<e_\mathfrak{p}\tau^{e_\mathfrak{p}-1}dt> \cong \bar{S}_\mathfrak{p}/(e_\mathfrak{p}\tau^{e_\mathfrak{p}-1})$, where τ is the image of t in $\bar{S}_\mathfrak{p}$ (a generator of the maximal ideal of $\bar{S}_\mathfrak{p}$).

Comparing the two expressions for $\Omega^1_{\bar{S}_\mathfrak{p}/k(\mathfrak{q})}$ we obtain:
a) If $e_\mathfrak{p} \not\equiv 0 \bmod p$, where $p := \text{char.}k(\mathfrak{q})$, then $\varepsilon_\mathfrak{p} = e_\mathfrak{p}-1$ and hence

$$\Omega^1_{S_\mathfrak{p}/R_\mathfrak{q}} \cong S_\mathfrak{p}/\mathfrak{p}^{e_\mathfrak{p}-1} S_\mathfrak{p}.$$

b) If $e_\mathfrak{p} \equiv 0 \bmod p$, then $\varepsilon_\mathfrak{p} \ge e_\mathfrak{p}$.

$\mathfrak{p}$ is called <u>tamely ramified</u> over R, if $e_\mathfrak{p} > 1$, $e_\mathfrak{p} \not\equiv 0 \bmod p$ and $k(\mathfrak{p})/k(\mathfrak{q})$ is separable.

The example shows that $\Omega^1_{S/R}$ not only "knows", which primes of S are ramified over R, but sometimes "discovers" the ramification index of tamely ramified primes too. For more information about modules of differentials of discrete valuation rings, see [BK].

The following is an important structure theorem for un-ramified extensions, a generalization of the theorem of the primitive element for finite separable field extensions.

6.16. Theorem.   Let S/R be a local algebra, where $(R,\mathcal{M})$ is noetherian and S/R essentially finite.

a) If S/R is unramified, then there is a prime $\mathcal{N} \in \mathrm{Spec}(R[X])$ with $\mathcal{N} \cap R = \mathcal{M}$, a monic polynomial $f \in R[X]$ with $f' \notin \mathcal{N}$, and an R-epimorphism $\alpha : R[X]_{\mathcal{N}}/(f) \to S$, which is an iso-morphism, if S/R is étale.

b) If in addition R and S have the same residue field, there is an α as in a), with $\mathcal{N} = (\mathcal{M},X)$ and $f(0) \in \mathcal{M}$.

Proof.   That S/R is essentially finite means that $S = T_{\mathfrak{p}}$, where T/R is a finite subalgebra of S/R and $\mathfrak{p} \in \mathrm{Spec}(T)$. If S/R is unramified, then $L := T_{\mathfrak{p}}/\mathfrak{p}T_{\mathfrak{p}} = T_{\mathfrak{p}}/\mathcal{M}T_{\mathfrak{p}}$ is a finite and separable extension field of $K := R/\mathcal{M}$. Therefore we can find a $\xi \in L\smallsetminus\{0\}$ with $L = K[\xi]$.

Let $\mathfrak{p}_1,\ldots,\mathfrak{p}_r$ be the maximal ideals of the semilocal ring T (all lying over $\mathcal{M}$). By the Chinese remainder theorem

$$(3) \qquad\qquad T/\mathcal{M}T = T_1 \times \ldots \times T_r \ .$$

with $T_i := T_{\mathfrak{p}_i}/\mathcal{M}T_{\mathfrak{p}_i}$ $(i=1,\ldots,r)$. Assume $\mathfrak{p}_1 = \mathfrak{p}$. Then $T_1 = L$ and we can choose an $x \in T$ with image ξ in T_1 and image 0 in T_i for $i=2,\ldots,r$.

Write $T' := R[x]$, $\mathfrak{p}' := \mathfrak{p} \cap T'$. Since x is contained in all maximal ideals of T but $\mathfrak{p}$, we conclude that $\mathfrak{p}$ is the only prime of T lying over $\mathfrak{p}'$, hence $S = T_{\mathfrak{p}} = T_{\mathfrak{p}'}$ and $T_{\mathfrak{p}'}/T'_{\mathfrak{p}'}$ is finite. But $T_{\mathfrak{p}'}/T'_{\mathfrak{p}'}$ is also unramified and both rings have the same residue field L. By Nakayama $S = T'_{\mathfrak{p}'}$.

After replacing T by T' we may assume $T = R[x]$. Then $T/\mathcal{m}T = K[X]/(\bar{f})$ with a monic polynomial $\bar{f}$, and the decomposition (3) corresponds to the factorization of $\bar{f}$ into irreducible polynomials. Since L is one of the factors of $T/\mathcal{m}T$, the minimal polynomial $\bar{g}$ of ξ over K is one of the factors of $\bar{f}$, and it is not a multiple factor. Therefore $\bar{f} = \bar{g} \cdot \bar{\varphi}$ with $\bar{\varphi} \in K[X]$, and $\bar{g}$ does not divide $\bar{\varphi}$.

Let n be the degree of $\bar{f}$. By Nakayama $T = R+Rx+\ldots+Rx^{n-1}$, and there is a monic polynomial $f \in R[X]$ with $f(x) = 0$ and image $\bar{f}$ in $K[X]$. Let $\mathcal{u}$ be the inverse image of $\not{\mathcal{p}}$ under the R-epimorphism

$$\alpha_0 : R[X] \to R[x] \qquad (X \mapsto x).$$

α_0 induces an R-epimorphism $\alpha : R[X]_{\mathcal{u}}/(f) \to S$ and a K-epimorphism $\bar{\alpha} : K[X]_{\bar{\mathcal{u}}}/(\bar{f}) \to L$, where $\bar{\mathcal{u}}$ is the image of $\mathcal{u}$ in $K[X]$. Clearly $\bar{\alpha}$ is an isomorphism.

Assume $f' \in \mathcal{u}$. Then $\bar{f}'(\xi) = 0$. But $\bar{f}'(\xi) = \bar{g}'(\xi) \cdot \bar{\varphi}(\xi)$, since $\bar{g}(\xi) = 0$. We have $\bar{\varphi}(\xi) \neq 0$, since $\bar{g}$ does not divide $\bar{\varphi}$, and $\bar{g}'(\xi) \neq 0$, because ξ is separable over K. Thus we arrived at a contradiction, so $f' \notin \mathcal{u}$.

Suppose now that S/R is étale and let I be the kernel of α. The exact sequence

$$0 \to I \to R[X]_{\mathcal{u}}/(f) \xrightarrow{\alpha} S \to 0$$

remains exact after tensorization with K over R, because S is a flat R-module. We obtain an exact sequence

$$0 \to I/\mathcal{m}I \to K[X]_{\bar{\mathcal{u}}}/(f) \xrightarrow{\bar{\alpha}} L \to 0$$

and, as we already know, $\bar{\alpha}$ is an isomorphism. We conclude with Nakayama that $I = 0$.

In case L = K there is a $\rho \in R$ with image ξ in L. The R-isomorphism $R[Y] \to R[X]$ $(Y \mapsto X-\rho)$ induces an R-epimorphism

$$\tilde{\alpha} \; : \; R[Y]_{\tilde{\mathit{n}}}/(\tilde{f}) \to S$$

where $\tilde{\mathit{n}}$ is the inverse image of n in $R[Y]$ and $\tilde{f}$ the polynomial with $\tilde{f}(Y) = f(Y+\rho)$. Again $\tilde{\alpha}$ is an isomorphism, if S/R is étale. Since $x-\rho \in \not{\mathit{p}}$ we have $Y \in \tilde{\mathit{n}}$, and thus $\tilde{\mathit{n}} = (\mathit{m}, Y)$. Moreover, $\tilde{f}(0) = f(\rho) \in \mathit{m}$, since $\overline{f}(\xi) = 0$, and $\tilde{f}' \notin \tilde{\mathit{n}}$, since $f' \notin \mathit{n}$. The proof of the theorem is now complete.

6.17. <u>Remark.</u>   Let S/R be a local algebra as in 6.16, and assume S/R is essentially <u>of finite type</u> and unramified. Since $S/\mathit{m}S$ is finite over R/m, one can conclude from Zariski's Main Theorem (B.16) and B.12 that S/R is essentially finite. Therefore theorem 6.16 also holds under the weaker assumption that S/R be essentially of finite type.

An interesting unsolved problem is the <u>Jacobian problem</u> <u>about polynomial rings</u>, which we wish to describe here: Let K be a domain with $\mathbb{Z} \subset K$ and $f_1, \ldots, f_n \in K[X_1, \ldots, X_n]$ polynomials, such that the functional determinant $\det(\frac{\partial f_i}{\partial X_j})$ is a unit of K. Does it follow that

$$K[f_1, \ldots, f_n] = K[X_1, \ldots, X_n]?$$

Put $K[f_1, \ldots, f_n] =: K[f]$, $K[X_1, \ldots, X_n] =: K[X]$. The assumption about the functional determinant is equivalent with the condition that the canonical map

$$(4) \qquad K[X] \otimes_{K[f]} \Omega^1_{K[f]/K} \to \Omega^1_{K[X]/K}$$

is bijective, since the image of this map is the submodule of $\Omega^1_{K[X]/K} = \bigoplus_{i=1}^{n} K[X]dX_i$ generated by $\{\sum \frac{\partial f_j}{\partial X_i} dX_i\}_{j=1,\ldots,n}$.

It follows then that $\Omega^1_{K[X]/K[f]} = 0$, and 6.8 implies that $K[X]/K[f]$ is unramified. Especially, if K is a field, then $K(X)/K(f)$ is an algebraic extension, hence $\{f_1,\ldots,f_n\}$ a transcendence basis of $K(X)/K$. Since $K[X]$ and $K[f]$ are regular rings, $K[X]/K[f]$ is flat (Matsumura $[M_1]$, Thm. 51), thus $K[X]/K[f]$ is even étale.

If (4) is an isomorphism, then

$$\mathrm{Der}_K(K[X],K[X]) \to \mathrm{Der}_K(K[f],K[X])$$

is an isomorphism too, which implies that each element of $\mathrm{Der}_K(K[f])$ is the restriction to $K[f]$ of a unique element of $\mathrm{Der}_K(K[X])$.

Positive solutions of the Jacobian problem are known e.g., if in addition to (4) one of the following conditions is satisfied:

a) There is a field L containing K with $L(f) = L(X)$.

b) $K[X]$ is finite over $K[f]$.

c) K is a field and $K(X)/K(f)$ a Galois extension.

For easy proofs in the spirit of these lectures see Angermüller [Ang]. A general discussion of the Jacobian problem and the work related to it, is given by Bass, Connell and Wright [BCW]. For extensive treatments of the ramification theory of commutative algebras the reader is referred to Iversen [I] and Scheja-Storch $[SS_1]$.

We now turn to special problems in characteristic $p > 0$. As an application of 6.8 we get the following criterion for the existence of a p-basis.

<u>6.18. Proposition.</u>   Let R be a local ring of prime characteristic p, $R_0 \to R$ a ring homomorphism such that

$R/R_o[R^p]$ is finite. Then the following conditions are equivalent:

a) R/R_o has a p-basis.

b) Ω^1_{R/R_o} is a free R-module.

Proof. a) $\to$ b) was shown in 5.6.

b) $\to$ a). Ω^1_{R/R_o} has a basis of the form $\{dx_1,\ldots,dx_\tau\}$ with $x_1,\ldots,x_\tau \in R$; this follows from $\Omega^1_{R/R_o} = RdR$ and Nakayama's lemma. Let $R' := R_o[R^p][x_1,\ldots,x_\tau]$. By 6.6 this is a local ring and, since $\Omega^1_{R/R'} = 0$, the algebra R/R' is unramified. The residue class field of R is separable over that of R' but also purely inseparable, since $R^p \subset R'$, hence the two residue class fields are equal.

Let $\mathfrak{m}$ be the maximal ideal of R, $\mathfrak{m}'$ that of R'. Then $\mathfrak{m} = \mathfrak{m}'R$ and $R/\mathfrak{m}'R = R'/\mathfrak{m}'$, consequently $R = R'+\mathfrak{m}'R$. Since R/R' is finite, Nakayama's lemma implies that $R = R'$ and hence $\{x_1,\ldots,x_\tau\}$ is a p-basis of R/R_o by 5.6.

In §5 we have discussed the notion of an admissible derivation for algebraic function fields. We now generalize this notion to algebras S/R that are essentially of finite type. The reasons for introducing this notion are similar to those mentioned in the field theoretic case and will become clearer in the next section.

6.19. <u>Definition.</u> A derivation $\delta : R \to R\delta R$ of R is called <u>admissible for $\mathfrak{p} \in \mathrm{Spec}(S)$</u>, if $R_\mathfrak{q}\delta R_\mathfrak{q}$ ($\mathfrak{q} := \mathfrak{p} \cap R$) is a free $R_\mathfrak{q}$-module of finite rank, say r, and $\mu_\mathfrak{p}(\Omega^1_{S/\delta}) = \mathrm{edim}\, S_\mathfrak{p}/\mathfrak{q}S_\mathfrak{p} + \mathrm{Trdeg}(k(\mathfrak{p})/k(\mathfrak{q})) + r$. A differential algebra (Ω,d) of R is <u>admissible for $\mathfrak{p} \in \mathrm{Spec}(S)$</u>, if

$\Omega_{\mathfrak{p}}$ is an exterior differential algebra and d : R → Ω^1 an admissible derivation for $\mathfrak{p}$.

6.20. Examples. a) Let L/K be an algebraic function field. A derivation δ : K → KδK is admissible for L/K in the sense of 5.19 if and only if δ is admissible for the zero ideal of L.

b) Suppose in the situation of 6.19 that $k(\mathfrak{p})/k(\mathfrak{q})$ is separable. Then by 6.5b) the trivial derivation of R is admissible for $\mathfrak{p}$. Thus the existence of admissible derivations is only in question, if inseparable residue field extensions occur.

In the rest of this section we prove in some cases the existence of admissible derivations (differential algebras).

6.21. Proposition. Suppose R is a reduced noetherian ring of prime characteristic p and $\Omega^1_{R/\mathbb{Z}}$ is finitely generated. Then $\Omega_{R/\mathbb{Z}}$ is admissible for each $\mathfrak{p} \in \mathrm{Spec}(S)$ with $\mathfrak{p} \cap R \in \mathrm{Min}(R)$.

Proof. For such $\mathfrak{p}$ let $\mathfrak{q} := \mathfrak{p} \cap R$. Then $R_{\mathfrak{q}}$ is a field, since R is reduced, $\Omega_{R_{\mathfrak{q}}/\mathbb{Z}}$ is an exterior algebra and clearly $\Omega^1_{R_{\mathfrak{q}}/\mathbb{Z}}$ a free $R_{\mathfrak{q}}$-module of finite rank r. By 6.2 we have

$$\mu_{\mathfrak{p}}(\Omega^1_{S/\mathbb{Z}}) = \mathrm{edim}\ S_{\mathfrak{p}} + \dim_{k(\mathfrak{p})} \Omega^1_{k(\mathfrak{p})/\mathbb{Z}'}$$
$$r = \mu_{\mathfrak{q}}(\Omega^1_{R/\mathbb{Z}}) = \mathrm{edim}\ R_{\mathfrak{q}} + \dim_{k(\mathfrak{q})} \Omega^1_{k(\mathfrak{q})/\mathbb{Z}}$$

and therefore

$$\mu_{\mathfrak{p}}(\Omega^1_{S/\mathbb{Z}}) = \mathrm{edim}\ S_{\mathfrak{p}}/\mathfrak{q}S_{\mathfrak{p}} + \dim_{k(\mathfrak{p})} \Omega^1_{k(\mathfrak{p})/\mathbb{Z}} - \dim_{k(\mathfrak{q})} \Omega^1_{k(\mathfrak{q})/\mathbb{Z}} + r$$
$$= \mathrm{edim}\ S_{\mathfrak{p}}/\mathfrak{q}S_{\mathfrak{p}} + \mathrm{Trdeg}(k(\mathfrak{p})/k(\mathfrak{q})) + r, \text{ since by 5.18 } k(\mathfrak{p})$$

and $k(\mathfrak{q})$ are separable over their prime fields.

For a noetherian local ring R of prime characteristic p
Brezuleanu and Radu [BR_1] have shown that $\Omega^1_{R/\mathbb{Z}}$ is finitely
generated if and only if R/R^p is finite. Their proof is
beyond the scope of these lectures.

<u>6.22. Proposition.</u>   Let R be a reduced noetherian ring of
prime characteristic p and S/R an algebra that is essentially
of finite type. Assume R has a p-basis over R^p. Then there
is a subring $R_0 \subset R$ with $R^p \subset R_0$ such that R/R_0 is finite
and Ω_{R/R_0} is a free R-module which is admissible for each
$\mathfrak{p} \in \mathrm{Spec}(S)$ with $\mathfrak{p} \cap R \in \mathrm{Min}(R)$.

Proof.   Let $\{x_\lambda\}_{\lambda \in \Lambda}$ be a p-basis of R/R^p, hence
$\Omega^1_{R/R^p} = \bigoplus_{\lambda \in \Lambda} R\delta x_\lambda$. The algebra S/R has a presentation

$$S = R[X_1, \ldots, X_n]_N / (F_1, \ldots, F_m)$$

with a multiplicatively closed set $N \subset R[X_1, \ldots, X_n]$ and
$F_1, \ldots, F_m \in R[X_1, \ldots, X_n]$. By 4.19

$$\Omega^1_{S/\mathbb{Z}} = \bigoplus_{\lambda \in \Lambda} S \cdot (1 \otimes \delta x_\lambda) \oplus \bigoplus_{i=1}^{n} S dX_i / U ,$$

where U is generated by

$$\delta F_j(x_1, \ldots, x_n) + \sum_{i=1}^{n} \frac{\partial F_j}{\partial x_j} dX_j \quad (j=1, \ldots, m)$$

x_i being the image of X_i in S.

We choose a finite subset $\Lambda_0 \subset \Lambda$ such that
$$U \subset \bigoplus_{\lambda \in \Lambda_0} S \cdot (1 \otimes \delta x_\lambda) \oplus \bigoplus_{i=1}^{n} S dX_i \quad \text{and put} \quad R_0 := R^p[\{x_\lambda\}_{\lambda \in \Lambda \setminus \Lambda_0}].$$
Then $\{x_\lambda\}_{\lambda \in \Lambda_0}$ is a p-basis of R/R_0 and Ω^1_{R/R_0} is free. We
shall show that R_0 satisfies the requirements of the propo-
sition.

Take a prime $\mathfrak{p} \in \mathrm{Spec}(S)$ with $\mathfrak{g} := \mathfrak{p} \cap R \in \mathrm{Min}(R)$ and
put $\mathfrak{M} := \mathfrak{p} S_\mathfrak{p}$, $L := k(\mathfrak{p})$, $K := k(\mathfrak{g}) = R_\mathfrak{g}$. By 6.1 we have an

exact sequence

$$0 \to \mathfrak{m}/\mathfrak{m}^2 \to \Omega^1_{S_{\mathfrak{p}}/\mathbb{Z}}/\mathfrak{m}\Omega^1_{S_{\mathfrak{p}}/\mathbb{Z}} \to \Omega^1_{L/\mathbb{Z}} \to 0.$$

Let d denote the universal derivation of $K/\mathbb{Z}$. By the definition of $T(L/d)$ there is an exact sequence

$$0 \to T(L/d) \to L \otimes_K \Omega^1_{K/\mathbb{Z}} \to \Omega^1_{L/\mathbb{Z}} \to \Omega^1_{L/K} \to 0.$$

By the choice of Λ_o

$$\Omega^1_{S/\mathbb{Z}} = \bigoplus_{\lambda \in \Lambda \smallsetminus \Lambda_o} S(1 \otimes \delta x_\lambda) \oplus E$$

with

$$E := \bigoplus_{\lambda \in \Lambda_o} S(1 \otimes \delta x_\lambda) \oplus \bigoplus_{i=1}^{n} S dX_i/U.$$

It is clear that $E \cong \Omega^1_{S/R_o}$.

Since $\mathfrak{m}/\mathfrak{m}^2$ and $T(L/d)$ are finite dimensional vector spaces over L, we may assume, after enlarging Λ_o by finitely many elements, if necessary, that the image of $\mathfrak{m}/\mathfrak{m}^2$ in $\Omega^1_{S_{\mathfrak{p}}/\mathbb{Z}}/\mathfrak{m}\Omega^1_{S_{\mathfrak{p}}/\mathbb{Z}}$ is contained in $E_{\mathfrak{p}}/\mathfrak{m}E_{\mathfrak{p}}$ and the image of $T(L/d)$ in $L \otimes_K \Omega^1_{K/\mathbb{Z}}$ is contained in $\sum_{\lambda \in \Lambda_o} L(1 \otimes d\bar{x}_\lambda)$, where $\bar{x}_\lambda$ denotes the image of x_λ in K. (The last sum is direct since the $d\bar{x}_\lambda$ $(\lambda \in \Lambda)$ form a basis of Ω^1_{K/K^p}).

Since we have to enlarge Λ_o in a manner that may depend on $\mathfrak{p}$, we write $\Lambda_o^{\mathfrak{p}}$ for the new subset of Λ and put $R_o^{\mathfrak{p}} := R^p[\{x_\lambda\}_{\lambda \in \Lambda \smallsetminus \Lambda_o^{\mathfrak{p}}}]$.

By the choice of $\Lambda_o^{\mathfrak{p}}$ the sequence

$$(5) \quad 0 \to \mathfrak{m}/\mathfrak{m}^2 \to \Omega^1_{S_{\mathfrak{p}}/R_o^{\mathfrak{p}}}/\mathfrak{m}\Omega^1_{S_{\mathfrak{p}}/R_o^{\mathfrak{p}}} \to \Omega^1_{L/R_o^{\mathfrak{p}}} \to 0$$

is exact and from the diagram

$$\begin{array}{ccccccccc}
& & & & O & & & & \\
& & & & \downarrow & & & & \\
& & & & N & & & & \\
& & & & \downarrow & & & & \\
0 \to & \mathfrak{m}/\mathfrak{m}^2 & \to & \Omega^1_{S_\mathfrak{p}/\mathbb{Z}}/\mathfrak{m}\,\Omega^1_{S_\mathfrak{p}/\mathbb{Z}} & \to & \Omega^1_{L/\mathbb{Z}} & \to & O \\
& \| & & \downarrow & & \downarrow & & \\
0 \to & \mathfrak{m}/\mathfrak{m}^2 & \to & \Omega^1_{S_\mathfrak{p}/R_0^\mathfrak{p}}/\mathfrak{m}\,\Omega^1_{S_\mathfrak{p}/R_0^\mathfrak{p}} & \to & \Omega^1_{L/R_0^\mathfrak{p}} & \to & O \\
& & & \downarrow & & \downarrow & & \\
& & & O & & O & &
\end{array}$$

with $N := \bigoplus_{\lambda\in\Lambda\setminus\Lambda_0^\mathfrak{p}} S_\mathfrak{p}(1\otimes\delta x_\lambda)\,/\,\mathfrak{m}\bigoplus_{\lambda\in\Lambda\setminus\Lambda_0^\mathfrak{p}} S_\mathfrak{p}(1\otimes\delta x_\lambda)$ we see that there is also an exact sequence

$$0 \to \bigoplus_{\lambda\in\Lambda\setminus\Lambda_0^\mathfrak{p}} L\,d\bar{x}_\lambda \to \Omega^1_{L/\mathbb{Z}} \to \Omega^1_{L/R_0^\mathfrak{p}} \to O\ .$$

Finally, the diagram

$$\begin{array}{ccccccc}
& O & & O & & & \\
& \downarrow & & \downarrow & & & \\
& \bigoplus\limits_{\lambda\in\Lambda\setminus\Lambda_0^\mathfrak{p}} L(1\otimes d\bar{x}_\lambda) & = & \bigoplus\limits_{\lambda\in\Lambda\setminus\Lambda_0^\mathfrak{p}} L\,d\bar{x}_\lambda & & & \\
& \downarrow & & \downarrow & & & \\
0 \to T(L/d) \to & L\otimes_K \Omega^1_{K/\mathbb{Z}} & \longrightarrow & \Omega^1_{L/\mathbb{Z}} & \to \Omega^1_{L/K} \to O \\
& \downarrow & & \downarrow & \| & \\
& L\otimes_K \Omega^1_{K/R_0^\mathfrak{p}} & \longrightarrow & \Omega^1_{L/R_0^\mathfrak{p}} & \to \Omega^1_{L/K} \to O \\
& \downarrow & & \downarrow & & \\
& O & & O & &
\end{array}$$

shows that there is an exact sequence

(6) $\quad 0 \to T(L/d) \to L\otimes_K \Omega^1_{K/R_0^\mathfrak{p}} \to \Omega^1_{L/R_0^\mathfrak{p}} \to \Omega^1_{L/K} \to O.$

From (5),(6), and 5.12b) we conclude that

$\mu_\mathfrak{p}(\Omega^1_{S/R_0^\mathfrak{p}}) = \operatorname{edim} S_\mathfrak{p} + \dim_L \Omega^1_{L/R_0^\mathfrak{p}} =$

$\operatorname{edim} S_\mathfrak{p} + \dim_L \Omega^1_{L/K} - \dim_L T(L/d) + \dim_K \Omega^1_{K/R_0^\mathfrak{p}} =$

$\operatorname{edim} S_\mathfrak{p} + \operatorname{Trdeg}(k(\mathfrak{p})/k(\mathfrak{q})) + \dim_{R_\mathfrak{q}} \Omega^1_{R_\mathfrak{q}/R_0^\mathfrak{p}}\ .$

Since $\Omega^1_{S/R_0^\mathfrak{p}} \cong \Omega^1_{S/R_0} \oplus \bigoplus_{\lambda\in\Lambda_0^\mathfrak{p}\setminus\Lambda_0} S(1\otimes\delta x_\lambda)$ and

$$\Omega^1_{R/R_{\mathfrak{p}_0}} \cong \Omega^1_{R/R_0} \oplus \bigoplus_{\lambda \in \Lambda_{\mathfrak{p}_0} \smallsetminus \Lambda_0} R\delta x_\lambda \quad \text{we finally obtain}$$

$$\mu_{\mathfrak{p}}(\Omega^1_{S/R_0}) = \text{edim } S_{\mathfrak{p}} + \text{Trdeg}(k(\mathfrak{p})/k(\mathfrak{q})) + \dim_{R_{\mathfrak{q}}} \Omega^1_{R_{\mathfrak{q}}/R_0}, \quad \text{q.e.d.}$$

Proposition 6.22 can be applied, if $R = K$ is a field of characteristic $p > 0$, which we now assume.

<u>6.23. Definition.</u>   Let A be an affine K-algebra. A subfield $K_0 \subset K$ is called <u>admissible for A</u>, if $K^p \subset K_0$, $[K:K_0] < \infty$, and Ω_{K/K_0} is admissible for all $\mathfrak{p} \in \text{Spec}(A)$.

<u>6.24. Corollary.</u>   Let $A_1,\ldots,A_m$ be affine K-algebras. Then there exists a subfield $K_0 \subset K$ that is admissible for $A_1,\ldots,A_m$.

For one affine K-algebra this is an immediate consequence of 6.22. But since we may enlarge the set Λ_0 in the proof of 6.22 by finitely many elements of Λ, we may choose it in such a way that the corresponding subfield K_0 is admissible for $A_1,\ldots,A_m$.

<u>6.25. Corollary.</u>   Let R and S be affine K-algebras. Suppose R is reduced, $R \subset S$ and K_0 is a subfield of K that is admissible for both R and S. Then Ω^1_{R/K_0} is admissible for all $\mathfrak{p} \in \text{Spec}(S)$ with $\mathfrak{p} \cap R \in \text{Min}(R)$.

In fact, with $\mathfrak{q} := \mathfrak{p} \cap R$ we have $\mu_{\mathfrak{p}}(\Omega^1_{S/K_0}) =$

$\text{edim } S_{\mathfrak{p}} + \text{Trdeg}(k(\mathfrak{p})/K) + \text{p-deg}(K/K_0) =$

$\text{edim } S_{\mathfrak{p}} + \text{Trdeg}(k(\mathfrak{p})/k(\mathfrak{q})) + \text{Trdeg}(k(\mathfrak{q})/K) + \text{p-deg}(K/K_0) =$

$\text{edim } S_{\mathfrak{p}} + \text{Trdeg}(k(\mathfrak{p})/k(\mathfrak{q})) + \dim_{R_{\mathfrak{q}}} \Omega^1_{R_{\mathfrak{q}}/K_0}.$

Exercises

1) For a squarefree integer a, form the quadratic number field $K := \mathbb{Q}(\sqrt{a})$. Let R be the integral closure of $\mathbb{Z}$ in K (the ring of integers of K).

   a) If $a \not\equiv 1 \bmod 4$, then $R = \mathbb{Z} \oplus \mathbb{Z}\sqrt{a}$, if $a \equiv 1 \bmod 4$, then $R = \mathbb{Z} \oplus \mathbb{Z} \cdot \dfrac{1-\sqrt{a}}{2}$.

   b) Determine the prime ideals (p) of $\mathbb{Z}$ that are ramified in R.

2) For a noetherian ring R of prime characteristic p the following conditions are equivalent:

   a) R is a direct product of finitely many perfect fields of characteristic p.

   b) $\Omega^1_{R/R^p} = 0$.

   c) $R = R^p$.

3) Let S/R be a local algebra, where S is a finite R-module.

   a) If $\mu(\Omega^1_{S/R}) \leq 1$, then $S = R[x]$ with some $x \in S$. (Hint: We may assume that R is a field (Nakayama). Using 5.11 one can construct an $x \in S$ with $\Omega^1_{S/R} = Sdx$ whose image in the residue field L of S is a primitive element of L/R).

   b) If S/R is étale, there is an R-isomorphism $S \cong R[X]/(f)$, where f is a monic polynomial and $f'(x)$ is a unit of S.

4) Under the hypothesis of exercise 3) assume that $\mu(\Omega^1_{S/R}) = r$. Then $S = R[x_1,\ldots,x_r]$ with suitable elements $x_1,\ldots,x_r \in S$. Conclude that $\mu(\Omega^1_{S/R}) \geq \operatorname{edim} S/\mathfrak{m}S$, where $\mathfrak{m}$ is the maximal ideal of R.

5) Let S/R be an algebra that is essentially of finite type and unramified. Then the kernel of the map $S \otimes_R S \to S$

(a $\otimes$ b $\mapsto$ ab) is generated by an idempotent element of
$S \otimes_R S$.

6) Let K be a field and R/K an algebra, which is essentially
of finite type.

a) If R/K is unramified, so is R'/K for any subalgebra
R'/K of R/K.

b) If R_1/K and R_2/K are subalgebras of R/K which are
essentially of finite type and unramified, so is
$[R_1 \cdot R_2]/K$.

c) A finite dimensional algebra R/K contains a unique
maximal unramified subalgebra.

7) An algebra R/R_o is called <u>formally unramified</u> (formally
smooth, formally étale), if for any algebra S/R_o and
any ideal I of S with $I^2 = 0$ the canonical mapping
$\mathrm{Hom}_{R_o}^{alg}(R,S) \to \mathrm{Hom}_{R_o}^{alg}(R,S/I)$ is injective (surjective,
bijective). Here $\mathrm{Hom}_{R_o}^{alg}$ denotes the set of all R_o-alge-
bra homomorphisms. Show that the following conditions
are equivalent:

a) R/R_o is formally unramified.
b) $\Omega^1_{R/R_o} = 0$
(Use §1, exercise 10 and 1.21).

8) a) Let R/R_o be an algebra and $\wp \in \mathrm{Spec}(R)$. The $k(\wp)$-
vector space $T_\wp(R/R_o) := \mathrm{Der}_{R_o}(R,k(\wp))$ is called the
<u>tangent space</u> of R/R_o at the point $\wp$. Show that

$$\dim_{k(\wp)} T_\wp(R/R_o) = \mu_\wp(\Omega^1_{R/R_o})$$

if Ω^1_{R/R_o} is finitely generated.
b) Let f : R $\to$ S be a homomorphism of R_o-algebras,
$\mathfrak{P} \in \mathrm{Spec}(S)$, $\wp := f^{-1}(\mathfrak{P})$. Show that there is a canonical

$k(\mathfrak{p})$-linear map

$$T_{\mathfrak{p}}(f) : T_{\mathfrak{p}}(S/R_o) \to T_{\mathfrak{q}}(R/R_o) \otimes_{k(\mathfrak{q})} k(\mathfrak{p})$$

(which corresponds to the functorial map $S \otimes_R \Omega^1_{R/R_o} \to \Omega^1_{S/R_o}$ and is called the <u>tangent map of f at the point</u> $\mathfrak{p}$).

Also show that there is an exact sequence

$$O \to T_{\mathfrak{p}}(S/R) \to T_{\mathfrak{p}}(S/R_o) \xrightarrow{T_{\mathfrak{p}}(f)} T_{\mathfrak{q}}(R/R_o) \otimes_{k(\mathfrak{q})} k(\mathfrak{p})$$

c) Under the assumptions of b) let R be noetherian, and suppose that S/R is essentially of finite type. Show that $T_{\mathfrak{p}}(f)$ is injective if and only if S/R is unramified at $\mathfrak{p}$.

d) Let I be an ideal of R, $f : R \to R/I$ the canonical epimorphism, and let $\mathfrak{q} \in \mathrm{Spec}(R)$ with $\mathfrak{q} \supset I$ be given. Show that there is an exact sequence of $k(\mathfrak{q})$-vector spaces (with $\bar{\mathfrak{q}} := \mathfrak{q}/I$)

$$O \to T_{\bar{\mathfrak{q}}}(R/I/R_o) \xrightarrow{T_{\bar{\mathfrak{q}}}(f)} T_{\mathfrak{q}}(R/R_o) \to \mathrm{Hom}_{R/I}(I_{\mathfrak{q}}/I^2_{\mathfrak{q}}, k(\mathfrak{q})).$$

§ 7. Differential Modules of Affine Algebras

The results of the last section will now be applied to affine algebras and their localizations. Let A be an affine algebra over a field K. We shall use the following facts about the Krull dimension (dim A) of A and the dimension of the localizations of A (see appendix B for generalizations):

If A is an integral domain with quotient field L, then

$$(1) \qquad \dim A = \mathrm{Trdeg}(L/K).$$

If for $\mathfrak{p} \in \mathrm{Spec}(A)$ the local ring $A_{\mathfrak{p}}$ is a domain with quotient field L, then

$$(2) \qquad \dim A_{\mathfrak{p}} = \mathrm{Trdeg}(L/K) - \mathrm{Trdeg}(k(\mathfrak{p})/K)$$
$$= \mathrm{Trdeg}(L/K) - \dim A/\mathfrak{p}.$$

A (or $A_{\mathfrak{p}}$) is called <u>equidimensional</u>, if for all minimal primes $\mathfrak{q}$ of A (of $A_{\mathfrak{p}}$) the dimension of $A/\mathfrak{q}$ (of $A_{\mathfrak{p}}/\mathfrak{q}$) is independent of $\mathfrak{q}$. If $\mathfrak{p}', \mathfrak{p} \in \mathrm{Spec}(A)$ with $\mathfrak{p}' \subset \mathfrak{p}$ are given, and $A_{\mathfrak{p}}$ is equidimensional, then

$$(3) \qquad \dim A_{\mathfrak{p}} = \dim A_{\mathfrak{p}'} + \dim A_{\mathfrak{p}}/\mathfrak{p}'A_{\mathfrak{p}}.$$

For a noetherian local ring R we call edim R − dim R the <u>regularity defect of R</u>.

<u>7.1. Proposition.</u>   For $\mathfrak{p}', \mathfrak{p} \in \mathrm{Spec}(A)$ with $\mathfrak{p}' \subset \mathfrak{p}$ let $A_{\mathfrak{p}}$ be equidimensional. Then

$$\mathrm{edim}\, A_{\mathfrak{p}} - \dim A_{\mathfrak{p}} \geq \mathrm{edim}\, A_{\mathfrak{p}'} - \dim A_{\mathfrak{p}'}.$$

Proof.   If char.K := p > 0, let $\delta : K \to K\delta K$ be an admissible derivation for A. If char.K = 0, put $\delta := 0$. Then by 6.5 and definition 6.19

$$\mu_{\mathfrak{p}}(\Omega^1_{A/\delta}) = \operatorname{edim} A_{\mathfrak{p}} + \operatorname{Trdeg}(k(\mathfrak{p})/K) + \dim_K K\delta K,$$

$$\mu_{\mathfrak{p}'}(\Omega^1_{A/\delta}) = \operatorname{edim} A_{\mathfrak{p}'} + \operatorname{Trdeg}(k(\mathfrak{p}')/K) + \dim_K K\delta K.$$

Since $\mu_{\mathfrak{p}} \geq \mu_{\mathfrak{p}'}$ we obtain (using (2) and (3))

$$\operatorname{edim} A_{\mathfrak{p}} - \operatorname{edim} A_{\mathfrak{p}'} \geq \operatorname{Trdeg}(k(\mathfrak{p}')/K) - \operatorname{Trdeg}(k(\mathfrak{p})/K)$$

$$= \dim A_{\mathfrak{p}}/\mathfrak{p}'A_{\mathfrak{p}} = \dim A_{\mathfrak{p}} - \dim A_{\mathfrak{p}'}.$$

The inequality of 7.1 holds (with a different proof,
of course) for arbitrary noetherian local rings (see [Le]).
It implies that, if $A_{\mathfrak{p}}$ is a regular local ring, so is $A_{\mathfrak{p}'}$.

7.2. Theorem (Regularity criterion in characteristic zero).
If char.$K = 0$, then for each $\mathfrak{p} \in \operatorname{Spec}(A)$ the following
conditions are equivalent:
a) $A_{\mathfrak{p}}$ is a regular local ring.
b) $\Omega^1_{A_{\mathfrak{p}}/K}$ is a free $A_{\mathfrak{p}}$-module.
Its rank is then $\dim A_{\mathfrak{p}} + \operatorname{Trdeg}(k(\mathfrak{p})/K)$.

Proof. a) $\to$ b). With $L := Q(A_{\mathfrak{p}})$ we have by 6.5b)
$$\mu_{\mathfrak{p}}(\Omega^1_{A/K}) = \dim A_{\mathfrak{p}} + \operatorname{Trdeg}(k(\mathfrak{p})/K) = \operatorname{Trdeg}(L/K) = \dim_L \Omega^1_{L/K} =$$
$\dim_L (L \otimes_{A_{\mathfrak{p}}} \Omega^1_{A_{\mathfrak{p}}/K})$. Then $\Omega^1_{A_{\mathfrak{p}}/K}$ is a free $A_{\mathfrak{p}}$-module by the
following lemma:

7.3. Lemma. Let M be a finitely generated module over an
integral domain R with quotient field L. If
$\mu(M) = \dim_L (L \otimes_R M)$, then M is a free R-module.

Let $\{m_1, \ldots, m_r\}$ a system of generators of M with
$r := \mu(M)$. Then $\{1 \otimes m_1, \ldots, 1 \otimes m_r\}$ generates $L \otimes_R M$ and is a
basis of that vector space. Hence $\{m_1, \ldots, m_r\}$ is linearly
independent over R.
b) $\to$ a). We first show that the ring $R := A_{\mathfrak{p}}$ is "generically

reduced", that is, $R_{\mathfrak{y}}$ is a field (or equivalently $\mathfrak{y}R_{\mathfrak{y}} = 0$) for each $\mathfrak{y} \in \text{Min}(R)$.

With $\mathfrak{m} := \mathfrak{y}R_{\mathfrak{y}}$ we have an exact sequence (6.5a))

$$0 \to \mathfrak{m}/\mathfrak{m}^2 \to \Omega^1_{R_{\mathfrak{y}}/K}/\mathfrak{m}\Omega^1_{R_{\mathfrak{y}}/K} \to \Omega^1_{k(\mathfrak{y})/K} \to 0.$$

For each $x \in \mathfrak{m} \smallsetminus \mathfrak{m}^2$ it follows from Nakayama's lemma that dx is part of a minimal system of generators of $\Omega^1_{R_{\mathfrak{y}}/K}$, hence part of a basis of $\Omega^1_{R_{\mathfrak{y}}/K} = R_{\mathfrak{y}} \otimes_R \Omega^1_{R/K}$.

Suppose that $\mathfrak{m} \neq (0)$, and choose $x \in \mathfrak{m} \smallsetminus \mathfrak{m}^2$.

Since $\mathfrak{m}$ is the only prime of $R_{\mathfrak{y}}$, we know that x is nilpotent: $x^{\rho} = 0$, $x^{\rho-1} \neq 0$ for some $\rho > 1$. Then $dx^{\rho} = \rho x^{\rho-1} dx = 0$, but $\rho x^{\rho-1} \neq 0$, since $\text{char}.K = 0$. Thus $dx = 0$, contradicting what was shown above. Hence $\mathfrak{m} = (0)$, and R is generically reduced.

Choose $\mathfrak{y} \in \text{Min}(R)$ with $\dim R/\mathfrak{y} = \dim R = \dim A_{\mathfrak{g}}$. Then $R_{\mathfrak{y}}$ is a field, and by (2) we have $\dim A_{\mathfrak{g}} = \dim R/\mathfrak{y} = \text{Trdeg}(R_{\mathfrak{y}}/K) - \text{Trdeg}(k(\mathfrak{g})/K)$. On the other hand by 6.5 $\text{edim } A_{\mathfrak{g}} + \text{Trdeg}(k(\mathfrak{g})/K) = \mu_{\mathfrak{g}}(\Omega^1_{A/K}) = \dim_{R_{\mathfrak{y}}}\Omega^1_{R_{\mathfrak{y}}/K} = \text{Trdeg}(R_{\mathfrak{y}}/K)$, hence $\text{edim } A_{\mathfrak{g}} = \dim A_{\mathfrak{g}}$, q.e.d.

<u>7.4. Corollary.</u> a) A is generically reduced (resp. reduced) if and only if $\Omega^1_{A_{\mathfrak{y}}/K}$ is a free $A_{\mathfrak{y}}$-module for each $\mathfrak{y} \in \text{Min}(A)$ (each $\mathfrak{y} \in \text{Ass}(A)$).

b) A is regular if and only if $\Omega^1_{A/K}$ is a projective A-module.

It is a much more difficult question under which conditions $\Omega^1_{A/K}$ is, for a regular affine K-algebra, a free A-module (see Exercise 2)).

<u>7.5. Theorem</u> (Regularity criterion for characteristic p > 0). Assume that $\text{char}.K =: p > 0$ and that A is generically reduced.

Let $\delta : K \to K\delta K$ be an admissible derivation for A. For each $\mathfrak{g} \in \mathrm{Spec}(A)$ the following conditions are equivalent:

a) $A_{\mathfrak{g}}$ is a regular local ring.

b) $\Omega^1_{A_{\mathfrak{g}}/\delta}$ is a free $A_{\mathfrak{g}}$-module.

c) $\Omega^1_{A_{\mathfrak{g}}/L}$ is a free $A_{\mathfrak{g}}$-module.

d) $A_{\mathfrak{g}}$ has a p-basis over $(A_{\mathfrak{g}})^p$.

Under these conditions

$$\mathrm{rank}\ \Omega^1_{A_{\mathfrak{g}}/\delta} = \dim A_{\mathfrak{g}} + \mathrm{Trdeg}(k(\mathfrak{g})/K) + \dim_K K\delta K.$$

Proof.   a) $\leftrightarrow$ b) can be shown as in the proof of 7.2. Here the proof is easier, since we already assume A to be generically reduced.

If $A_{\mathfrak{g}}$ is regular with quotient field L, then
$$\mu_{\mathfrak{g}}(\Omega^1_{A/\delta}) = \dim A_{\mathfrak{g}} + \mathrm{Trdeg}(k(\mathfrak{g})/K) + \dim_K K\delta K =$$
$\mathrm{Trdeg}(L/K) + \dim_K K\delta K = \dim_L \Omega^1_{L/\delta}$, and $\Omega^1_{A_{\mathfrak{g}}/\delta}$ is a free $A_{\mathfrak{g}}$-module by 7.3.

Conversely, if $\Omega^1_{A_{\mathfrak{g}}/\delta}$ is free, put $R := A_{\mathfrak{g}}$ and choose $\mathfrak{y} \in \mathrm{Min}(R)$ with $\dim R/\mathfrak{y} = \dim R$. Then $R_{\mathfrak{y}}$ is a field and from

$\mathrm{edim}\ A_{\mathfrak{g}} + \mathrm{Trdeg}(k(\mathfrak{g})/K) + \dim_K K\delta K = \mu_{\mathfrak{g}}(\Omega^1_{A/\delta}) =$
$\dim_{R_{\mathfrak{y}}} \Omega^1_{R_{\mathfrak{y}}/\delta} = \mathrm{Trdeg}(R_{\mathfrak{y}}/K) + \dim_K K\delta K$ we obtain
$\mathrm{edim}\ A_{\mathfrak{g}} = \mathrm{Trdeg}(R_{\mathfrak{y}}/K) - \mathrm{Trdeg}(k(\mathfrak{g})/K) = \dim R = \dim A_{\mathfrak{g}}$.

a) $\leftrightarrow$ c). The construction of admissible derivations in the proof of 6.22 shows that there is an admissible subfield $K_o \subset K$ for A such that

$$\Omega^1_{A/L} \cong \Omega^1_{A/K_o} \oplus A \otimes_{K_o} \Omega^1_{K_o/L}.$$

Here $A \otimes_{K_o} \Omega^1_{K_o/L}$ is a free A-module and therefore $\Omega^1_{A_{\mathfrak{g}}/L}$ is a free $A_{\mathfrak{g}}$-module if and only if $\Omega^1_{A_{\mathfrak{g}}/K_o}$ is one, which is equivalent with $A_\mathfrak{p}$ being regular (a) $\leftrightarrow$ b)).

It remains to be shown: If $A_\mathfrak{p}$ is regular, then $A_\mathfrak{p}$ has a p-basis over $(A_\mathfrak{p})^p$.

Choose K_o as above, put $L := Q(A_\mathfrak{p})$ and let $\{x_\lambda\}_{\lambda \in \Lambda}$ be a p-basis of K_o/K^p. Since K_o is linearly disjoint to L^p over K^p (by 5.22), $\{x_\lambda\}_{\lambda \in \Lambda}$ is also a p-basis of $K_o[(A_\mathfrak{p})^p]$ over $(A_\mathfrak{p})^p$. Since $\Omega^1_{A_\mathfrak{p}/K_o}$ is a free $A_\mathfrak{p}$-module, there is a p-basis $\{y_1, \ldots, y_t\}$ of $A_\mathfrak{p}$ over $K_o[(A_\mathfrak{p})^p]$ by 6.18. Then $\{x_\lambda\}_{\lambda \in \Lambda} \cup \{y_1, \ldots, y_t\}$ is a p-basis of $A_\mathfrak{p}$ over $(A_\mathfrak{p})^p$.

<u>7.6. Corollary.</u>   The following conditions are equivalent:

a) A is regular.

b) If $\delta : K \to K\delta K$ is an admissible derivation for A, then $\Omega^1_{A/\delta}$ is a projective A-module.

c) $\Omega^1_{A/\mathbf{L}}$ is a projective A-module.

As for c) observe that $\Omega^1_{A/\mathbf{L}}$ is a direct sum of a finite and a free A-module.

In 7.5 and 7.6 the assumption that A be generically reduced cannot be omitted: If K is a perfect field of characteristic $p > 0$ and $A = K[X]/(X^p)$, then $\Omega^1_{A/K} = AdX$ is free but A is not regular.

For a noetherian ring R

$$\text{Reg}(R) := \{\mathfrak{p} \in \text{Spec}(R) \mid R_\mathfrak{p} \text{ is regular}\}$$

is called the <u>regular locus</u> of R and

$\text{Sing}(R) := \text{Spec}(R) \smallsetminus \text{Reg}(R)$ the <u>singular locus</u> (<u>singular set</u> of $\text{Spec}(R)$).

<u>7.7. Proposition</u>   (Closedness of the singular locus).
For an affine algebra A over a field K the set $\text{Sing}(A)$ is

closed in Spec(A).

Proof. If char.K = 0, then Reg(A) is the set of all $\mathscr{g} \in$ Spec(A) for which $(\Omega^1_{A/K})_{\mathscr{g}}$ is free (7.2). This is well-known to be an open subset of Spec(A) (cf.D.12). If char.K =: p>0, the proof is more complicated, since 7.5 gives only a weaker statement. If A is reduced, we can argue as above. In the general case assume that Ass(A) = $\{\mathscr{g}_1, \ldots, \mathscr{g}_t\}$. If $\mathscr{g} \in$ Spec(A) contains two different elements of Ass(A), then $A_{\mathscr{g}}$ cannot be a domain, nor can it be regular. Assume $\mathscr{g}$ contains only one element of Ass(A), say $\mathscr{g}_1$, which is necessarily in Min(A). Then $\mathscr{g}_1 A_{\mathscr{g}} = 0$ if and only if $\mathscr{g}_1 A_{\mathscr{g}_1} = 0$. In this case $\mathscr{g} \in$ Reg(A) if and only if $\bar{\mathscr{g}} \in$ Reg($A/\mathscr{g}_1$), where $\bar{\mathscr{g}} := \mathscr{g}/\mathscr{g}_1$. We see that Sing(A) is the union of $V(\mathscr{g}_i) \cap V(\mathscr{g}_j)$ ($i \neq j$), of $V(\mathscr{g}_i)$ for $\mathscr{g}_i \in$ Ass(A) with $\mathscr{g}_i A_{\mathscr{g}_i} \neq 0$, and of $\varphi_i($Sing($A/\mathscr{g}_i$)) for $\mathscr{g}_i \in$ Ass(A) with $\mathscr{g}_i A_{\mathscr{g}_i} = 0$, where $\varphi_i :$ Spec($A/\mathscr{g}_i$) $\to$ Spec(A) is the canonical map. Since Sing($A/\mathscr{g}_i$) is closed in Spec($A/\mathscr{g}_i$), we conclude that Sing(A) is closed too.

It is known that for an arbitrary noetherian ring R the set Reg(R) need not be open in Spec(R).

7.2 and 7.5 are closely related to the <u>Jacobian criterion for regularity</u>, which we shall now derive.

Choose a presentation of A/K

(4) $\qquad A = K[X_1, \ldots, X_n]/(F_1, \ldots, F_m) = K[x_1, \ldots, x_n]$

where x_i is the residue of X_i in A. Take $\mathscr{g} \in$ Spec(A) and let ξ_i denote the image of x_i in $k(\mathscr{g})$. Moreover, $\frac{\partial F_k}{\partial x_i}$ is the image of $\frac{\partial F_k}{\partial X_i}$ in A and $\frac{\partial F_k}{\partial \xi_i}$ its image in $k(\mathscr{g})$. Put $K_0 := K$, if char.K = 0, and choose K_0 to be admissible for A, if

char.K > O. Let δ denote the universal derivation of K/K_O,
and let $\Omega^1_{K/K_O} = K\delta t_1 \oplus \ldots \oplus K\delta t_r$ $(t_1,\ldots,t_r \in K)$. From (4)
we then get the following presentation of the differential
module

$$(5) \qquad \Omega^1_{A_\gamma/K_O} = A_\gamma \otimes_K \Omega^1_{K/K_O} \oplus \bigoplus_{i=1}^n A_\gamma dX_i / U$$

where U is generated by

$$\delta F_k(x_1,\ldots,x_n) + \sum_{i=1}^n \frac{\partial F_k}{\partial x_i} dX_i \qquad (k=1,\ldots,m).$$

The first summand is of the form

$$\delta F_k(x_1,\ldots,x_n) = \sum_{j=1}^r \varphi_{kj} \cdot (1 \otimes \delta t_j) \text{ with } \varphi_{kj} \in A_\gamma.$$

We denote the image of φ_{kj} in $k(\gamma)$ by $\frac{\partial F_k}{\partial t_j}$ and call

$$J(\gamma) := \begin{pmatrix} \dfrac{\partial F_1}{\partial \xi_1}, \ldots, \dfrac{\partial F_1}{\partial \xi_n}, & \dfrac{\partial F_1}{\partial t_1}, \ldots, \dfrac{\partial F_1}{\partial t_r} \\ \vdots \qquad \vdots & \vdots \qquad \vdots \\ \dfrac{\partial F_m}{\partial \xi_1}, \ldots, \dfrac{\partial F_m}{\partial \xi_n}, & \dfrac{\partial F_m}{\partial t_1}, \ldots, \dfrac{\partial F_m}{\partial t_r} \end{pmatrix}$$

the <u>Jacobian matrix at the point</u> γ associated with the
presentation (4) (and the choice of K_O). Its coefficients
are elements of $k(\gamma)$. Observe that the $\dfrac{\partial F_k}{\partial t_j}$ do not occur,
if K is a perfect field.

<u>7.8. Theorem</u>   (Jacobian criterion for regularity).
Assume that A is equidimensional. Then
a) $\operatorname{rank}(J(\gamma)) \leq n - \dim A$.
b) $\gamma \in \operatorname{Reg}(A)$ if and only if $\operatorname{rank}(J(\gamma)) = n - \dim A$.

Proof.   We have $\mu_\gamma(\Omega^1_{A/K_O}) = \operatorname{edim} A_\gamma + \operatorname{Trdeg}(k(\gamma)/K) + r$.
On the other hand, from (5) we see that

$$\Omega^1_{A_{\mathscr{g}}/K_o} / {}_{\mathscr{g}}\Omega^1_{A_{\mathscr{g}}/K_o} \cong \bigoplus_{j=1}^{r} k(\mathscr{g}) \cdot \Delta_j \oplus \bigoplus_{i=1}^{n} k(\mathscr{g}) dX_i / \overline{U}, \text{ where } \Delta_j \text{ is the}$$

image of $1 \otimes \delta t_j$ (mod $\mathscr{g}$) and $\overline{U}$ is generated by

$$\sum_{j=1}^{r} \frac{\partial F_k}{\partial t_j} \Delta_j + \sum_{i=1}^{n} \frac{\partial F_k}{\partial \xi_i} dX_i \qquad (k=1,\ldots,m).$$

Therefore, we also have

$$\mu_{\mathscr{g}}(\Omega^1_{A/K_o}) = n+r - \mathrm{rank}(J(\mathscr{g}))$$

and we obtain

$$\begin{aligned}
\mathrm{rank}(J(\mathscr{g})) &= n+r - (\mathrm{edim}\, A_{\mathscr{g}} + \mathrm{Trdeg}(k(\mathscr{g})/K) + r) \\
&= n - (\mathrm{edim}\, A_{\mathscr{g}} - \dim A_{\mathscr{g}}) - (\dim A_{\mathscr{g}} + \mathrm{Trdeg}(k(\mathscr{g})/K)) \\
&= n - (\mathrm{edim}\, A_{\mathscr{g}} - \dim A_{\mathscr{g}}) - \dim A,
\end{aligned}$$

where we used in the last equality the fact that A is equi-dimensional (3). We see that in the general case $\mathrm{rank}(J(\mathscr{g})) \leq n - \dim A$ and that the equality sign holds if and only if $A_{\mathscr{g}}$ is regular.

The Jacobian criterion is particularly useful to decide regularity in explicit examples, while 7.2 and 7.5 are used in general considerations. The following (sufficient) criterion for regularity is also frequently helpful.

<u>7.9. Proposition.</u> Let R be a noetherian ring, S/R an algebra of finite type. Assume that for $\mathscr{p} \in \mathrm{Spec}(S)$, $\mathscr{g} := \mathscr{p} \cap R$ the ring $R_{\mathscr{g}}$ is regular and $\delta : R \to R\delta R$ is a derivation such that $(R\delta R)_{\mathscr{g}}$ is a free $R_{\mathscr{g}}$-module of rank r. If

$$\mu_{\mathscr{p}}(\Omega^1_{S/\delta}) \leq r + \dim S_{\mathscr{p}} - \dim R_{\mathscr{g}} + \mathrm{Trdeg}(k(\mathscr{p})/k(\mathscr{g}))$$

then $S_{\mathscr{p}}$ is regular.

Proof. Write $S = R[X_1,\ldots,X_n]/I$, $P := R[X_1,\ldots,X_n]$, and let $\mathscr{n}$ be the inverse image of $\mathscr{p}$ in P. Then $S_{\mathscr{p}} = P_{\mathscr{n}}/IP_{\mathscr{n}}$ and $P_{\mathscr{n}}$

are regular local rings, since $R_{\mathfrak{q}}$ is regular.

The exact sequence of $S_{\mathfrak{p}}$-modules

$$I_{\mathfrak{n}}/I_{\mathfrak{n}}^2 \to \Omega^1_{P_{\mathfrak{n}}/\delta}/I\Omega^1_{P_{\mathfrak{n}}/\delta} \to \Omega^1_{S_{\mathfrak{p}}/\delta} \to 0$$

induces, after tensorization with $k(\mathfrak{n}) = k(\mathfrak{p})$, the exact

sequence

$$I_{\mathfrak{n}}/\mathfrak{n}I_{\mathfrak{n}} \overset{\alpha}{\to} \Omega^1_{P_{\mathfrak{n}}/\delta}/\mathfrak{n}\Omega^1_{P_{\mathfrak{n}}/\delta} \overset{\beta}{\to} \Omega^1_{S_{\mathfrak{p}}/\delta}/\mathfrak{p}\Omega^1_{S_{\mathfrak{p}}/\delta} \to 0.$$

Since $\Omega^1_{P_{\mathfrak{n}}/\delta}$ is a free $P_{\mathfrak{n}}$-module of rank r+n and

$\mu_{\mathfrak{p}}(\Omega^1_{S/\delta}) \leq r + \dim S_{\mathfrak{p}} - \dim R_{\mathfrak{q}} + \mathrm{Trdeg}(k(\mathfrak{p})/k(\mathfrak{q}))$ by

assumption, we see that ker β has at least the dimension

$s := n + \dim R_{\mathfrak{q}} - \dim S_{\mathfrak{p}} - \mathrm{Trdeg}(k(\mathfrak{p})/k(\mathfrak{q}))$.

By B.1 we have $\dim P_{\mathfrak{n}} = n + \dim R_{\mathfrak{q}} - \mathrm{Trdeg}(k(\mathfrak{n})/k(\mathfrak{q}))$ and

hence $s = \dim P_{\mathfrak{n}} - \dim S_{\mathfrak{p}}$.

Let $a_1, \ldots, a_{\sigma} \in I$ be elements whose images in

$\Omega^1_{P_{\mathfrak{n}}/\delta}/\mathfrak{n}\Omega^1_{P_{\mathfrak{n}}/\delta}$ form a basis of ker β ($\sigma \geq s$), and let

$I^* := (a_1, \ldots, a_{\sigma}) \cdot P_{\mathfrak{n}}$. In the canonical commutative diagram

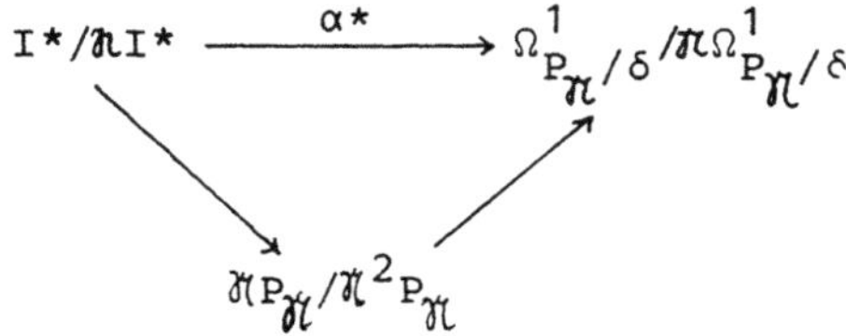

the mapping α^* (which is induced by α) is injective, hence

the images of $a_1, \ldots, a_{\sigma}$ in $\mathfrak{n}P_{\mathfrak{n}}/\mathfrak{n}^2P_{\mathfrak{n}}$ are linearly independent

over $k(\mathfrak{n})$. This implies that $a_1, \ldots, a_{\sigma}$ form part of a re-

gular system of parameters of $P_{\mathfrak{n}}$. $P_{\mathfrak{n}}/I^*$ is a regular local

ring of dimension $\dim P_{\mathfrak{n}} - \sigma \leq \dim P_{\mathfrak{n}} - s = \dim S_{\mathfrak{p}}$.

Since $S_{\mathfrak{p}}$ is a homomorphic image of $P_{\mathfrak{n}}/I^*$, this is only

possible, if $S_{\mathfrak{p}} = P_{\mathfrak{n}}/I^*$ and hence $S_{\mathfrak{p}}$ is regular.

7.10. <u>Corollary.</u>   If under the assumptions of 7.9 $R \subset S$ are domains with quotient fields K and L respectively, then $\Omega^1_{S_{\mathfrak{p}}/\delta}$ is a free $S_{\mathfrak{p}}$-module of rank r+Trdeg(L/K).

Proof. Since $R_{\mathfrak{q}}$ is universally catenary,we have by B.1
$$\dim S_{\mathfrak{p}} + \mathrm{Trdeg}(k(\mathfrak{p})/k(\mathfrak{q})) = \dim R_{\mathfrak{q}} + \mathrm{Trdeg}(L/K).$$
Using the exact sequence

$$0 \to T(L/\delta) \to L \otimes_K K\delta K \to \Omega^1_{L/\delta} \to \Omega^1_{L/K} \to 0$$

and 5.12,we see that $\mu_{\mathfrak{p}}(\Omega^1_{S/\delta}) \geq \dim_L \Omega^1_{L/\delta} \geq r + \mathrm{Trdeg}(L/K) =$ $r + \dim S_{\mathfrak{p}} - \dim R_{\mathfrak{q}} + \mathrm{Trdeg}(k(\mathfrak{p})/k(\mathfrak{q})) \geq \mu_{\mathfrak{p}}(\Omega^1_{S/\delta})$. We thus have equality,and by 7.3 $\Omega^1_{S_{\mathfrak{p}}/\delta}$ is free of rank r+Trdeg(L/K).

We now use the regularity criteria to study the behavior of regularity under base change.

7.11. <u>Lemma.</u>   Let L/K be an arbitrary field extension and K'/K a separable field extension. Then $K' \otimes_K L$ is a reduced ring.

Proof. We may assume that K'/K is finitely generated. Then K'/K has a separating transcendence basis $\{x_1,\ldots,x_t\}$. Since
$$K' \otimes_K L = K' \otimes_{K(x_1,\ldots,x_t)} (K(x_1,\ldots,x_t) \otimes_K L)$$
$$\subset K' \otimes_{K(x_1,\ldots,x_t)} L(x_1,\ldots,x_t)$$
it suffices to consider the case that K'/K is a finite separable field extension: $K' = K[X]/(f)$ with a separable polynomial $f \in K[X]$. But then $K' \otimes_K L \cong L[X]/(f)$. As a polynomial of L[X],the separable polynomial f has no multiple factors. By the Chinese remainder theorem $K' \otimes_K L$ is therefore the direct product of (separable) extension fields of L, hence reduced.

The lemma generalizes as follows.

7.12. **Proposition.**   Let A be an affine algebra over a field K and K'/K a separable field extension. For $\mathscr{G} \in \operatorname{Spec}(A)$ let $A_{\mathscr{G}}$ be regular. Then $K' \otimes_K A_{\mathscr{G}}$ is a regular ring too.

Proof.   If char.K = 0, we may use 7.2. $\Omega^1_{K' \otimes_K A_{\mathscr{G}}/K'} = K' \otimes_K \Omega^1_{A_{\mathscr{G}}/K}$ is a free module over $K' \otimes_K A_{\mathscr{G}}$, because $\Omega^1_{A_{\mathscr{G}}/K}$ is a free $A_{\mathscr{G}}$-module. Therefore $K' \otimes_K A_{\mathscr{G}}$ is regular.

Now let char.K =: p > 0. Since $K' \otimes_K A_{\mathscr{G}}$ is a reduced ring by 7.11, we may use 7.5. Let $\{x_\lambda\}_{\lambda \in \Lambda}$ be a p-basis of K/K^p. Since K'^p and K are linearly disjoint over K^p (5.24), $\{x_\lambda\}_{\lambda \in \Lambda}$ is also part of a p-basis of K'/K'^p. Moreover, $K' \otimes_K A_{\mathscr{G}} = K' \otimes_{K^p} A_{\mathscr{G}}/I$, where I is the ideal generated by $\{x_\lambda \otimes 1 - 1 \otimes x_\lambda\}_{\lambda \in \Lambda}$. Let d' denote the universal derivation of $K'/\mathbb{Z}$ and d the universal derivation of $A_{\mathscr{G}}/\mathbb{Z}$. Then

$$\Omega^1_{K' \otimes_K A_{\mathscr{G}}/\mathbb{Z}} = A_{\mathscr{G}} \otimes_K \Omega^1_{K'/\mathbb{Z}} \oplus K' \otimes_K \Omega^1_{A_{\mathscr{G}}/\mathbb{Z}}/U$$

where U is generated by $\{1 \otimes d'x_\lambda - 1 \otimes dx_\lambda\}_{\lambda \in \Lambda}$. Since the elements $d'x_\lambda$ ($\lambda \in \Lambda$) form part of a basis of $\Omega^1_{K'/\mathbb{Z}}$, the elements $1 \otimes d'x_\lambda - 1 \otimes dx_\lambda$ form also part of a basis of the free $K' \otimes_K A_{\mathscr{G}}$-module $A_{\mathscr{G}} \otimes_K \Omega^1_{K'/\mathbb{Z}} \oplus K' \otimes_K \Omega^1_{A_{\mathscr{G}}/\mathbb{Z}}$. Therefore $\Omega^1_{K' \otimes_K A_{\mathscr{G}}/\mathbb{Z}}$ is also a free $K' \otimes_K A_{\mathscr{G}}$-module, and $K' \otimes_K A_{\mathscr{G}}$ is regular by 7.5.

If K'/K is an inseparable field extension, then 7.12 is not true in general. For an example, take an imperfect field K, choose $\alpha \in K$ with $\alpha \notin K^p$, and put $A := K[X]/(X^p-\alpha)$. A is a field and therefore regular, but $A \otimes_K A$ contains the nonzero nilpotent element $x \otimes 1 - 1 \otimes x$, where x is the residue class of X in A.

Remember that a noetherian local ring R containing a field K is called <u>geometrically regular over K</u>, if $K' \otimes_K R$ is regular for each finite field extension K'/K. If char.$K = 0$ or, more generally, K is a perfect field, then a regular local ring $A_{\wp}$ is always geometrically regular over K by 7.12. For arbitrary ground fields K any localization of a polynomial algebra $K[X_1, \ldots, X_n]$ is geometrically regular over K. If A is an affine algebra over a field K, and $A_{\wp}$ is geometrically regular over K for some $\wp \in \mathrm{Spec}(A)$, then A_{η} is geometrically regular for each $\eta \in \mathrm{Spec}\, A$ with $\eta \subset \wp$. This follows easily from the regularity criteria.

<u>7.13. Lemma.</u>   Let A be an affine algebra over a field K and $R := A_{\wp}$ for some $\wp \in \mathrm{Spec}(A)$. Let $\overline{K}/K$ be an algebraic field extension.

a) For $\mathfrak{M} \in \mathrm{Max}(\overline{K} \otimes_K R)$ let $S := (\overline{K} \otimes_K R)_{\mathfrak{M}}$. Then $\dim S = \dim R$ and $\mathrm{Trdeg}(k(\mathfrak{M})/\overline{K}) = \mathrm{Trdeg}(k(\mathfrak{m})/K)$, where $\mathfrak{m} := \wp A_{\wp}$.

b) If R is geometrically regular over K, then $\overline{K} \otimes_K R$ is regular.

Proof.   a) $\overline{K} \otimes_K R$ is integral and faithfully flat over R. Therefore $\mathfrak{M} \cap R = \mathfrak{m}$ and $\dim S = \dim R$. Moreover, $S/\mathfrak{M}S$ is algebraic over $R/\mathfrak{m}$, hence the statement about transcendence degrees follows.

b) For $\mathfrak{M} \in \mathrm{Max}(\overline{K} \otimes_K R)$ there is a finite field extension K'/K with $K' \subset \overline{K}$ such that $\mathfrak{M}$ is already generated by the elements of $\mathfrak{M}' := \mathfrak{M} \cap (K' \otimes_K R)$. The ring $S' := (K' \otimes_K R)_{\mathfrak{M}'}$ is regular by assuption and we have $\dim R = \dim S' = \dim S$

by a). Since $\mathcal{m}S = \mathcal{m}'S$, the ring S is regular too.

7.14. Theorem (Criterion for geometric regularity in the affine case). For an affine algebra A/K and $\wp \in$ Spec(A) the following conditions are equivalent:

a) $A_\wp$ is geometrically regular over K.

b) $\Omega^1_{A_\wp/K}$ is a free $A_\wp$-module of rank $\dim_\wp A$.

c) $\mu_\wp(\Omega^1_{A/K}) \leq \dim_\wp A$.

Proof. Remember that $\dim_\wp A := \dim A_\wp + \dim A/\wp = \dim A_\wp + \mathrm{Trdeg}(k(\wp)/K)$.

a) $\to$ b). Let $\overline{K}$ be the algebraic closure of K, $\mathcal{m} \in \mathrm{Max}(\overline{K} \otimes_K A_\wp)$ and $S := (\overline{K} \otimes_K A_\wp)_{\mathcal{m}}$. By 7.13b) S is regular and therefore by 7.2 and 7.5

$$\Omega^1_{S/\overline{K}} = S \otimes_{A_\wp} \Omega^1_{A_\wp/K}$$

is a free S-module of rank $\dim S + \mathrm{Trdeg}(L/\overline{K})$, where L is the residue field of S. By 7.13a) $\dim S = \dim A_\wp$ and $\mathrm{Trdeg}(L/\overline{K}) = \mathrm{Trdeg}(k(\wp)/K)$. Since $S/A_\wp$ is faithfully flat, $\Omega^1_{S/\overline{K}}$ and $\Omega^1_{A_\wp/K}$ are free modules of the same rank, namely $\dim_\wp A$.

c) $\to$ a). Let K'/K be a finite field extension, $\mathcal{m} \in \mathrm{Max}(K' \otimes_K A_\wp)$ and $S := (K' \otimes_K A_\wp)_{\mathcal{m}}$. Using 7.13 we obtain $\mu(\Omega^1_{S/K'}) \leq \mu_\wp(\Omega^1_{A/K}) \leq \dim S + \mathrm{Trdeg}(k(\mathcal{m})/K')$. Since S is a localization of $K' \otimes_K A$, which is an algebra of finite type over K', we can apply 7.9 and obtain that S is regular. Consequently $A_\wp$ is geometrically regular over K.

7.15. Corollary. The set of all $\wp \in$ Spec A for which $A_\wp$ is geometrically regular over K, is open in Spec(A).

Similarly as in the proof of 7.7 we can assume that A is a domain, hence equidimensional. By 7.14c) we thus have to consider the set of all $\mathscr{g} \in \mathrm{Spec}\ A$ with $\mu_{\mathscr{g}}(\Omega^1_{A/k}) \leq \dim A$, which is open in Spec A. (It is also the set of all $\mathscr{g}$ for which $\Omega^1_{A_{\mathscr{g}}/K}$ is free of rank dim A).

7.16. Corollary. If $A_{\mathscr{g}}$ is regular and $k(\mathscr{g})/K$ separable, then $A_{\mathscr{g}}$ is geometrically regular over K.

By 6.5b) condition 7.14c) is satisfied.

7.17. Corollary. For an algebraic function field L/K the following conditions are equivalent:

a) L is geometrically regular over K.

b) For any algebraic field extension K'/K the ring $K' \otimes_K L$ is reduced.

c) L/K is separable.

Write $L = A_{\mathscr{g}}$ with an affine domain A over K and $\mathscr{g} = (0)$. Then 7.14b) is equivalent with the statement that $\dim_L \Omega^1_{L/K} = \mathrm{Trdeg}(L/K)$, which is also equivalent with L/K being separable, hence a) $\leftrightarrow$ c).

For a finite field extension K'/K the ring $K' \otimes_K L$ is a direct product of zerodimensional local rings. It is regular if and only if these local rings are fields, which is also equivalent with $K' \otimes_K L$ being reduced. a) $\leftrightarrow$ b) follows now easily.

7.18. Corollary (Jacobian criterion for geometric regularity). Given a presentation $A = K[X_1,\ldots,X_n]/(F_1,\ldots,F_m)$ and $\mathscr{g} \in \mathrm{Spec}(A)$, let $\dfrac{\partial F_k}{\partial \xi_i}$ denote the image of $\dfrac{\partial F_k}{\partial X_i}$ in $k(\mathscr{g})$ and $J(\mathscr{g}) = \left(\dfrac{\partial F_k}{\partial \xi_i}\right)_{\substack{k=1,\ldots,m \\ i=1,\ldots,n}}$. Then the following conditions are equivalent:

a) $A_{\mathfrak{p}}$ is geometrically regular over K.

b) $\mathrm{rank}(J(\mathfrak{p})) = n - \dim_{\mathfrak{p}} A$.

Starting with 7.14, one has simply to copy the arguments in the proof of 7.8 to obtain this criterion.

It shows that the notion of geometric regularity is more natural than that of regularity. The following theorem will show that a ring $A_{\mathfrak{p}}$ which is geometrically regular over K is "close" to being a localization of a polynomial algebra over K.

<u>7.19. Theorem</u> (Structure theorem for geometrically regular local rings). Let A be an equidimensional affine algebra over a field K with dim A = n. For $\mathfrak{p} \in \mathrm{Spec}(A)$ the following conditions are equivalent:

a) $A_{\mathfrak{p}}$ is geometrically regular over K.

b) There is a noetherian normalization $K[X_1,\ldots,X_n] \subset A$ such that $A_{\mathfrak{p}}/K[X_1,\ldots,X_n]_{\mathfrak{q}}$ ($\mathfrak{q} := \mathfrak{p} \cap K[X_1,\ldots,X_n]$) is étale.

If $\dim A_{\mathfrak{p}} > 0$, these conditions are also equivalent with

b') There is a noetherian normalization $K[X_1,\ldots,X_n] \subset A$ such that $A_{\mathfrak{p}}/K[X_1,\ldots,X_n]_{\mathfrak{q}}$ is étale and $k(\mathfrak{p}) = k(\mathfrak{q})$.

b') implies, in particular, that $A_{\mathfrak{p}}$ and $K[X_1,\ldots,X_n]_{\mathfrak{q}}$ have the same completion. Moreover,

$$A_{\mathfrak{p}} \cong K[X_1,\ldots,X_n]_{\mathfrak{q}}[Y]_{(\mathfrak{q},Y)}/(F)$$

where $F \in K[X_1,\ldots,X_n]_{\mathfrak{q}}[Y]$ is monic in Y, $F' \notin (\mathfrak{q},Y)$ and $F(O) \in \mathfrak{q}K[X_1,\ldots,X_n]_{\mathfrak{q}}$.

Proof. b) $\rightarrow$ a). Since $A_{\mathfrak{p}}$ is unramified over $B := K[X_1,\ldots,X_n]_{\mathfrak{q}}$, we have $\Omega^1_{A_{\mathfrak{p}}/B} = O$ by 6.8. From the exact sequence

$$A_{\mathfrak{p}} \otimes_B \Omega^1_{B/K} \to \Omega^1_{A_{\mathfrak{p}}/K} \to \Omega^1_{A_{\mathfrak{p}}/B} \to 0$$

we obtain $\mu_{\mathfrak{p}}(\Omega^1_{A/K}) \le n = \dim A$, and 7.14 implies that $A_{\mathfrak{p}}$ is geometrically regular over K.

We now assume that a) is true. Put $d := \dim A_{\mathfrak{p}}$ and $t := \text{Trdeg}(k(\mathfrak{p})/K)$; then $n = d+t$. Let $\bar{A}$ denote the image of A in $k(\mathfrak{p})$ and $\bar{a}$ the image of a, for each $a \in A$.

Choose a noetherian normalization $K[Y_1,\ldots,Y_n] \subset A$ with $\mathfrak{p} \cap K[Y_1,\ldots,Y_n] = (Y_{t'+1},\ldots,Y_n)$ for some $t' \le n$. Then $k(\mathfrak{p})$ is finite over $K(\bar{Y}_1,\ldots,\bar{Y}_{t'})$ and $\{\bar{Y}_1,\ldots,\bar{Y}_{t'}\}$ a transcendence basis of $k(\mathfrak{p})/K$, hence $t' = t$. We first show that after a suitable change of the normalization we can assume $\{d\bar{Y}_1,\ldots,d\bar{Y}_t\}$ to be part of a basis of $\Omega^1_{k(\mathfrak{p})/K}$.

If char.K = 0, this is already the case. So we assume that char.K =: p > 0. Choose $z_1,\ldots,z_m \in A$ such that $\{d\bar{z}_1,\ldots,d\bar{z}_m\}$ is a basis of $\Omega^1_{k(\mathfrak{p})/K}$. Observe that $m \ge t$. Let
$$z_1^r + f_1 z_1^{r-1} + \ldots + f_r = 0 \qquad (f_i \in K[Y_1,\ldots,Y_n])$$
be an equation of integral dependence of z_1 over $K[Y_1,\ldots,Y_n]$. Put $X_1 := z_1 - Y_1^{p^e}$ with some e > 0. If e is chosen large enough, then Y_1 is integral over $K[X_1,Y_2,\ldots,Y_n]$, hence A is integral over that ring. Moreover, $d\bar{X}_1 = d\bar{z}_1$. Similarly, we can replace $Y_2,\ldots,Y_t$ in the noetherian normalization of A by elements $X_2,\ldots,X_t$ with $d\bar{X}_i = d\bar{z}_i$ (i=2,...,t). Then $K[X_1,\ldots,X_t,Y_{t+1},\ldots,Y_n] \subset A$ is a noetherian normalization with $\mathfrak{p} \cap K[X_1,\ldots,X_t,Y_{t+1},\ldots,Y_n] = (Y_{t+1},\ldots,Y_n)$, and $\{d\bar{X}_1,\ldots,d\bar{X}_t\}$ is part of a basis of $\Omega^1_{k(\mathfrak{p})/K}$. $\{dX_1,\ldots,dX_t\}$ is then part of a basis of $\Omega^1_{A_{\mathfrak{p}}/K}$, and $\Omega^1_{k(\mathfrak{p})/K(\bar{X}_1,\ldots,\bar{X}_t)}$ is generated by n-t = d elements.

If $d = 0$, then $A_\wp = k(\wp)$ is a separable extension field of K. In this case $\Omega^1_{k(\wp)/K}(\overline{X}_1,\ldots,\overline{X}_t) = 0$, hence $k(\wp)$ is separably algebraic over $K(\overline{X}_1,\ldots,\overline{X}_t)$. We have shown that condition b) is satisfied in this case.

We now assume $d > 0$ and try to prove the stronger assertion b').

Since $\Omega^1_{k(\wp)/K}(\overline{X}_1,\ldots,\overline{X}_t)$ is generated by d elements, we know by 5.11 that the field extension $k(\wp)/K(\overline{X}_1,\ldots,\overline{X}_t)$ can be generated by d elements.

As $k(\wp)$ is generated over $K(\overline{X}_1,\ldots,\overline{X}_t)$ by $\overline{A}$, it is possible to find elements $t_1,\ldots,t_d \in A$ such that $k(\wp)/K(\overline{X}_1,\ldots,\overline{X}_t)$ is generated by $\overline{t}_1,\ldots,\overline{t}_d$. In the next step we show that such elements exist so that $\{dX_1,\ldots,dX_t,dt_1,\ldots,dt_d\}$ is a basis of $\Omega^1_{A_\wp/K}$. Suppose that $\{d\overline{X}_1,\ldots,d\overline{X}_t,d\overline{t}_1,\ldots,d\overline{t}_\delta\}$ is a basis of $\Omega^1_{k(\wp)/K}$ $(\delta \leq d)$.

Looking at the exact sequence

$$\wp A_\wp/\wp^2 A_\wp \to \Omega^1_{A_\wp/K}/\wp\,\Omega^1_{A_\wp/K} \to \Omega^1_{k(\wp)/K} \to 0$$

we see that elements $u_{\delta+1},\ldots,u_d \in \wp$ can be found such that $\{dX_1,\ldots,dX_t,dt_1,\ldots,dt_\delta,du_{\delta+1},\ldots,du_d\}$ is a basis of $\Omega^1_{A_\wp/K}$. If $\delta < d$, write

$$dt_{\delta+1} = \sum_{i=1}^{t} a_i dX_i + \sum_{i=1}^{\delta} b_i dt_i + \sum_{i=\delta+1}^{d} c_i du_i \quad (a_i,b_i,c_i \in A_\wp).$$

We can assume that $c_{\delta+1}$ is a unit of $A_\wp$, because, if necessary, we can replace $t_{\delta+1}$ by $t_{\delta+1} + u_{\delta+1}$ without disturbing the properties of $t_{\delta+1}$. But then $\{dX_1,\ldots,dX_t,dt_1,\ldots,dt_{\delta+1},du_{\delta+2},\ldots,du_d\}$ is also a basis of $\Omega^1_{A_\wp/K}$ and, repeating the process, we end up with a basis $\{dX_1,\ldots,dX_t,dt_1,\ldots,dt_d\}$ of $\Omega^1_{A_\wp/K}$.

We now change the normalization $K[X_1,\ldots,X_t,Y_{t+1},\ldots,Y_n]$ again so that $\{dX_1,\ldots,dX_d,dY_{t+1},\ldots,dY_n\}$ becomes a basis of $\Omega^1_{A_\wp/K}$ and $\{\overline{X}_1,\ldots,\overline{X}_t,\overline{Y}_{t+1},\ldots,\overline{Y}_n\}$ a system of genera-tors of $k(\wp)/K$.

Let
$$t_1^r + f_1 t_1^{r-1} + \ldots + f_r = 0 \qquad (f_i \in K[X_1,\ldots,X_t,Y_{t+1},\ldots,Y_n])$$
be an equation of integral dependence of t_1 over $K[X_1,\ldots,X_t,Y_{t+1},\ldots,Y_n]$. Put $X_{t+1} := t_1 - Y_{t+1}^q$ with a sufficiently large $q > 1$ such that Y_{t+1} (and hence A) becomes integral over $K[X_1,\ldots,X_{t+1},Y_{t+2},\ldots,Y_n]$. Since $Y_{t+1} \in \wp$, we have $\overline{X}_{t+1} = \overline{t}_1$ and, since $dX_{t+1} = dt_1 - qY_{t+1}^{q-1}dY_{t+1}$, we see that in the above basis of $\Omega^1_{A_\wp/K}$ we can replace dt_1 by dX_{t+1}. Repeating this process with t_2 and Y_{t+2} etc. we end up with a noetherian normalization $K[X_1,\ldots,X_n] \subset A$ such that $\{dX_1,\ldots,dX_n\}$ is a basis of $\Omega^1_{A_\wp/K}$ and $\overline{X}_1,\ldots,\overline{X}_n$ a system of generators of $k(\wp)/K$.

With $\eta := \wp \cap K[X_1,\ldots,X_n]$ we have $\Omega^1_{A_\wp/K[X_1,\ldots,X_n]_\eta} = 0$, hence $A_\wp$ is unramified over $K[X_1,\ldots,X_n]_\eta$. Clearly $k(\wp) = k(\eta)$ by construction. Since $A_\wp$ is regular, there is only one $\wp_0 \in \mathrm{Min}(A)$ with $\wp_0 \subset \wp$. As $A/\wp_0$ is integral over the image of $K[X_1,\ldots,X_n]$ in $A/\wp_0$ and $\dim A/\wp_0 = n$, we have $\wp_0 \cap K[X_1,\ldots,X_n] = (0)$. The composed map $K[X_1,\ldots,X_n] \to A/\wp_0 \to (A/\wp_0)_\wp = A_\wp$ is injective and hence also $K[X_1,\ldots,X_n]_\eta \to A_\wp$ is injective. Moreover, both rings have the same dimension. From $\wp A_\wp = \eta A_\wp$ and $k(\wp) = k(\eta)$ it follows that both rings also have the same completion and, in particular, that $A_\wp/K[X_1,\ldots,X_n]_\eta$ is flat. The last state-ment of the theorem follows from 6.16. This completes the proof of the theorem.

<u>7.20. Corollary.</u>   Let A/K be an affine algebra. Assume that A is a domain and $L := Q(A)$ is separable over K. Then there exists a noetherian normalization $K[X_1,\ldots,X_n] \subset A$ of A such that $L/K(X_1,\ldots,X_n)$ is separably algebraic.

Apply the theorem to the zero ideal of A.

The equivalence of a) and b') in theorem 7.19   (if $\dim A_{\gamma} > 0$) is a major ingredient of Lindel's solution (in the affine case) of a problem of Bass, Quillen and Suslin about projective modules over polynomial rings (Lindel [Lin]). For a discussion of geometric regularity of arbitrary noetherian rings we refer to Matsumura $[M_1]$. We close this paragraph by mentioning two unsolved problems connected with regularity of affine rings.

A <u>problem of Berger</u> [Be] asks the following: Let $R = A_{\gamma}$ be a localization of an affine domain A over a perfect field K. Assume that $\dim R = 1$ and $\Omega^1_{R/K}$ is torsionfree. Does it follow that R is regular?

Positive answers to this problem were given e.g.in the following cases:

a) R is a complete intersection or an almost complete intersection (Berger [Be]), see also 9.9,

b) R is "quasihomogeneous" (Scheja [Sch]),

c) R has deviation at most 3 (Ulrich [U]),

d) edim $R \leq 3$,or R is a Gorenstein ring with edim $R \leq 4$ (Herzog $[He_1]$),

e) R can be deformed into a complete intersection (Bassein [Ba], Buchweitz-Greuel [BG]),

f) R can be deformed into an almost complete intersection (Koch [Ko]).

See also Herzog-Waldi [HW] for results connected with this problem. The problem is related to (and, if R is Gorenstein, equivalent with) the question, whether a singular R of dimension 1 can be "rigid" (Pinkham [Pi]).

Assume now that K is a field of characteristic zero and $R = A_\wp$ a localization of an affine domain A over K. The <u>Zariski-Lipman problem</u> asks: If $Der_K(R)$ is a free R-module, does it follow that R is regular?

To this problem positive answers are known
a) if dim R = 1 (Lipman [Lip_1]),
b) for "hypersurface rings" (Scheja-Storch [SS_2]),
c) in the "graded case" (Hochster [Ho], Platte [Pl_1]),
d) for certain "rings of invariants" (Platte [Pl_2]),
e) for certain Cohen-Macaulay rings R with dim R $\geq$ 3 (Herzog [He_2]).

If char.K > O, there are easy counter-examples, see exercise 3). For the relation to "Zariski-differentials", see exercise 5).

<u>Exercises</u>

1) For an algebra R/R_o with a presentation of the form
$$R = R_o[X_1,\ldots,X_n]/(F_1,\ldots,F_m) = R_o[x_1,\ldots,x_n]$$ the canonical mapping $\Omega^1_{R_o[X_1,\ldots,X_n]/R_o} \to \Omega^1_{R/R_o}$ induces an injection
$$j : Der_{R_o}(R) \hookrightarrow Hom_R(\overset{n}{\underset{i=1}{\oplus}} RdX_i, R).$$
Let $(\frac{\partial}{\partial X_1},\ldots,\frac{\partial}{\partial X_n})$ be the dual basis to $(dX_1,\ldots,dX_n)$. Then the image of j is the set of all elements
$$\delta := \sum_{i=1}^{n} r_i \frac{\partial}{\partial X_i} \text{ with } r_1,\ldots,r_n \in R, \sum_{i=1}^{n} \frac{\partial F_k}{\partial x_i} r_i = O$$
(k=1,..,m). How can one describe δr for a given $r = F(x_1,\ldots,x_n) \in R$?

2) The affine R-algebra $A := \mathbb{R}[X,Y,Z]/(X^2+Y^2+Z^2-1)$ is regular. Use a result from analysis to show that $\Omega^1_{A/\mathbb{R}}$ and $\mathrm{Der}_R(A)$ are <u>not</u> free A-modules.

3) Let K be a perfect field of characteristic $p > 0$ and $A := K[X,Y,Z]/(Z^p+XY)$.

   a) Determine $\mathrm{Sing}(A)$.

   b) Show that $\mathrm{Der}_K(A)$ is a free A-module.

4) Let A be an affine algebra over a field K and $\wp \in \mathrm{Spec}(A)$.

   a) $A_\wp$ is geometrically regular over K if and only if

   $$\dim_{k(\wp)} T_\wp(A/K) = \dim_\wp A.$$

   (For the definition of $T_\wp(A/K)$, see §6, exercise 8).

   b) If $A_\wp$ is geometrically regular over K and K'/K an arbitrary field extension, then $K' \otimes_K A_\wp$ is regular.

5) For an algebra R/R_0 the dual $\mathrm{Der}_{R_0}(R)^*$ of the derivation module is called the <u>module of Zariski differentials</u>. Suppose R is local and Ω^1_{R/R_0} is finitely generated. Show that the following assertions are equivalent:

   a) $\mathrm{Der}_{R_0}(R)^*$ is a free R-module.

   b) $\mathrm{Der}_{R_0}(R)$ is a free R-module.

§ 8. Smooth Algebras

We shall consider now algebras S/R that are essentially
of finite type where R is an arbitrary noetherian ring.
S/R is called <u>smooth at the point $\mathfrak{p} \in \mathrm{Spec}(S)$</u>, if the
following conditions are satisfied:

a) $S_{\mathfrak{p}}/R_{\mathfrak{q}}$ is flat $(\mathfrak{q} := \mathfrak{p} \cap R)$.

b) $S_{\mathfrak{p}}/\mathfrak{q}S_{\mathfrak{p}}$ is geometrically regular over $k(\mathfrak{q})$.

S/R is called a <u>smooth algebra</u> or <u>S smooth over R</u>, if S/R
is smooth at all $\mathfrak{p} \in \mathrm{Spec}(S)$.

For example, each polynomial algebra $S = R[X_1, \ldots, X_n]$
is smooth over R. If in the general case S/R is étale at
$\mathfrak{p} \in \mathrm{Spec}(S)$, then S/R is also smooth at $\mathfrak{p}$, the ring $S_{\mathfrak{p}}/\mathfrak{q}S_{\mathfrak{p}}$
being a finite separable extension field of $k(\mathfrak{q})$. If $R = K$
is a field, smoothness of S at $\mathfrak{p}$ is equivalent with $S_{\mathfrak{p}}$
being geometrically regular over K.

In the general case $S_{\mathfrak{p}}/\mathfrak{q}S_{\mathfrak{p}}$ is a localization of an
affine $k(\mathfrak{q})$-algebra. Hence the differential criteria for
geometric regularity of § 7 can be applied. The purpose of
this section is to generalize these criteria to the present
situation.

Consider a presentation $S_{\mathfrak{p}} = P_{\mathfrak{n}}/I_{\mathfrak{n}}$, where $P := R[X_1, \ldots, X_n]$
is a polynomial algebra, $I \subset P$ is an ideal and $\mathfrak{n} \in \mathrm{Spec}(P)$.
Then $S_{\mathfrak{p}}/\mathfrak{q}S_{\mathfrak{p}} = P_{\mathfrak{n}}/\mathfrak{q}P_{\mathfrak{n}}/J$, where J is the image of $I_{\mathfrak{n}}$ in
$P_{\mathfrak{n}}/\mathfrak{q}P_{\mathfrak{n}}$. Using the generalized principal ideal theorem of
Krull, we obtain

$$(1) \qquad \mu(I_{\mathfrak{n}}) \geq \mu(J) \geq h(J) = \dim P_{\mathfrak{n}}/\mathfrak{q}P_{\mathfrak{n}} - \dim S_{\mathfrak{p}}/\mathfrak{q}S_{\mathfrak{p}}.$$

In case $R \subset S$ are integral domains with quotient
fields K and L respectively, we have

139

(2) $\dim S_{\mathfrak{p}} + \operatorname{Trdeg}(k(\mathfrak{p})/k(\mathfrak{q})) \leq \dim R_{\mathfrak{q}} + \operatorname{Trdeg}(L/K)$

where the equality sign holds, if $R_{\mathfrak{q}}$ is universally catenary or S is a polynomial algebra over R (B.1).

The formulas (1) and (2) will be frequently used in what follows. We first obtain a <u>smoothness criterion</u>.

<u>8.1. Theorem.</u>   The following conditions are equivalent:

a) S/R is smooth at $\mathfrak{p}$.

b) $S_{\mathfrak{p}}/R_{\mathfrak{q}}$ is flat and $\mu_{\mathfrak{p}}(\Omega^1_{S/R}) \leq \dim_{\mathfrak{p}} S_{\mathfrak{q}}/\mathfrak{q} S_{\mathfrak{q}}$.

c) $S_{\mathfrak{p}}/R_{\mathfrak{q}}$ is flat and $\Omega^1_{S_{\mathfrak{p}}/R}$ is a free $S_{\mathfrak{p}}$-module of rank $\dim_{\mathfrak{p}} S_{\mathfrak{q}}/\mathfrak{q} S_{\mathfrak{q}}$.

d) The canonical sequence of $S_{\mathfrak{p}}$-modules

$$0 \to I_{\mathfrak{n}}/I^2_{\mathfrak{n}} \to \Omega^1_{P_{\mathfrak{n}}/R}/I\Omega^1_{P_{\mathfrak{n}}/R} \to \Omega^1_{S_{\mathfrak{p}}/R} \to 0$$

is exact and splits.

If these conditions are satisfied, then $I_{\mathfrak{n}}$ is generated by a $P_{\mathfrak{n}}$-regular sequence of length $\dim P_{\mathfrak{n}}/\mathfrak{q} P_{\mathfrak{n}} - \dim S_{\mathfrak{p}}/\mathfrak{q} S_{\mathfrak{p}}$. If $\operatorname{char}.k(\mathfrak{q}) = 0$, then the conditions a)-d) are also equivalent with

e) $S_{\mathfrak{p}}/R_{\mathfrak{q}}$ is flat and $\Omega^1_{S_{\mathfrak{p}}/R}$ is a free $S_{\mathfrak{p}}$-module.

Proof. We have $\Omega^1_{S_{\mathfrak{p}}/\mathfrak{q} S_{\mathfrak{p}}/k(\mathfrak{q})} = \Omega^1_{S_{\mathfrak{p}}/R}/\mathfrak{q}\Omega^1_{S_{\mathfrak{p}}/R}$

and therefore by Nakayama's lemma

(3) $\mu_{\mathfrak{p}}(\Omega^1_{S/R}) = \mu(\Omega^1_{S_{\mathfrak{p}}/\mathfrak{q} S_{\mathfrak{p}}/k(\mathfrak{q})})$.

That this number is $\leq \dim_{\mathfrak{p}} S_{\mathfrak{q}}/\mathfrak{q} S_{\mathfrak{q}}$ is by 7.14 equivalent with $S_{\mathfrak{p}}/\mathfrak{q} S_{\mathfrak{p}}$ being geometrically regular over $k(\mathfrak{q})$. Thus a)↔b) is proved.

If $S_{\mathfrak{p}}/R_{\mathfrak{q}}$ is flat, then the exact sequence

$$0 \to I_{\mathfrak{n}} \to P_{\mathfrak{n}} \to S_{\mathfrak{p}} \to 0$$

induces an exact sequence

$$0 \to I_{\mathfrak{n}}/\mathfrak{g}I_{\mathfrak{n}} \to P_{\mathfrak{n}}/\mathfrak{g}P_{\mathfrak{n}} \to S_{\mathfrak{p}}/\mathfrak{g}S_{\mathfrak{p}} \to 0$$

hence $J = I_{\mathfrak{n}}/\mathfrak{g}I_{\mathfrak{n}}$. If we assume b), then $S_{\mathfrak{p}}/\mathfrak{g}S_{\mathfrak{p}}$ is regular
(as was already shown) and therefore $I_{\mathfrak{n}}/\mathfrak{g}I_{\mathfrak{n}}$ is generated by
elements that form part of a regular system of parameters
of the regular local ring $P_{\mathfrak{n}}/\mathfrak{g}P_{\mathfrak{n}}$. Moreover, we have

$$\mu(I_{\mathfrak{n}}/\mathfrak{g}I_{\mathfrak{n}}) = \dim P_{\mathfrak{n}}/\mathfrak{g}P_{\mathfrak{n}} - \dim S_{\mathfrak{p}}/\mathfrak{g}S_{\mathfrak{p}}.$$

Call this number α. By Nakayama and B.23, $I_{\mathfrak{n}}$ is then
generated by a $P_{\mathfrak{n}}$-regular sequence of length α too.

In the exact sequence

$$I_{\mathfrak{n}}/I_{\mathfrak{n}}^2 \to \Omega^1_{P_{\mathfrak{n}}/R}/I\Omega^1_{P_{\mathfrak{n}}/R} \to \Omega^1_{S_{\mathfrak{p}}/R} \to 0$$

the $S_{\mathfrak{p}}$-module $I_{\mathfrak{n}}/I_{\mathfrak{n}}^2$ is free of rank α, and $\Omega^1_{P_{\mathfrak{n}}/R}/I\Omega^1_{P_{\mathfrak{n}}/R}$ is
free of rank n. By assumption b)

$$\mu_{\mathfrak{p}}(\Omega^1_{S/R}) \le \dim_{\mathfrak{p}} S_{\mathfrak{g}}/\mathfrak{g}S_{\mathfrak{g}} = \dim S_{\mathfrak{p}}/\mathfrak{g}S_{\mathfrak{p}} + \operatorname{Trdeg}(k(\mathfrak{p})/k(\mathfrak{g})) = n-\alpha.$$

Let $\{t_1,\ldots,t_\alpha\}$ be a basis of $I_{\mathfrak{n}}/I_{\mathfrak{n}}^2$ and $\{\omega_1,\ldots,\omega_\beta\}$ a minimal set of generators of $\Omega^1_{S_{\mathfrak{p}}/R}$ ($\beta := \mu_{\mathfrak{p}}(\Omega^1_{S/R})$). The images
of $t_1,\ldots,t_\alpha$ together with a set of representatives of
$\omega_1,\ldots,\omega_\beta$ generate $\Omega^1_{P_{\mathfrak{n}}/R}/I\Omega^1_{P_{\mathfrak{n}}/R}$. Since $\alpha+\beta \le n$, these elements must be a basis of $\Omega^1_{P_{\mathfrak{n}}/R}/I\Omega^1_{P_{\mathfrak{n}}/R}$. It follows that
$\beta = n-\alpha$ and that $\{\omega_1,\ldots,\omega_\beta\}$ is a basis of $\Omega^1_{S_{\mathfrak{p}}/R}$. Moreover,
the sequence

$$0 \to I_{\mathfrak{n}}/I_{\mathfrak{n}}^2 \to \Omega^1_{P_{\mathfrak{n}}/R}/I\Omega^1_{P_{\mathfrak{n}}/R} \to \Omega^1_{S_{\mathfrak{p}}/R} \to 0$$

is exact and splits.

Thus we have seen that b) implies both c) and d). Since
c) $\to$ b) is trivial, it remains to be shown that d) $\to$ b).

If d) holds, then both $I_{\mathfrak{n}}/I_{\mathfrak{n}}^2$ and $\Omega^1_{S_{\mathfrak{p}}/R}$ are free $S_{\mathfrak{p}}$-modules.
Using (1) we see

(4) $\quad \mu_{\mathfrak{p}}(\Omega^1_{S/R}) \le n - \dim P_{\mathfrak{n}}/\mathfrak{g}P_{\mathfrak{n}} + \dim S_{\mathfrak{p}}/\mathfrak{g}S_{\mathfrak{p}} =$

$\quad\quad \dim S_{\mathfrak{p}}/\mathfrak{g}S_{\mathfrak{p}} + \mathrm{Trdeg}(k(\mathfrak{p})/k(\mathfrak{g})) = \dim_{\mathfrak{p}} S_{\mathfrak{g}}/\mathfrak{g}S_{\mathfrak{g}}$

and from (3) and 7.14 we conclude that $S_{\mathfrak{p}}/\mathfrak{g}S_{\mathfrak{p}}$ is geometrically regular over $k(\mathfrak{g})$ and that the equality sign holds in (4). Then $I_{\mathfrak{n}}$ is generated by $\dim P_{\mathfrak{n}}/\mathfrak{g}P_{\mathfrak{n}} - \dim S_{\mathfrak{p}}/\mathfrak{g}S_{\mathfrak{p}}$ elements. Their images in $P_{\mathfrak{n}}/\mathfrak{g}P_{\mathfrak{n}}$ form part of a regular system of parameters and hence a regular sequence. B.23 implies now that $S_{\mathfrak{p}}/R_{\mathfrak{g}}$ is flat.

If $\mathrm{char}.k(\mathfrak{g}) = 0$ and e) holds, then $\Omega^1_{S_{\mathfrak{p}}/\mathfrak{g}S_{\mathfrak{p}}/k(\mathfrak{g})}$ is a free $S_{\mathfrak{p}}/\mathfrak{g}S_{\mathfrak{p}}$-module, hence $S_{\mathfrak{p}}/\mathfrak{g}S_{\mathfrak{p}}$ is regular by 7.2 and geometrically regular over $k(\mathfrak{g})$, since $\mathrm{char}.k(\mathfrak{g}) = 0$. Thus we have shown that e) implies a).

<u>8.2. Corollary.</u>   The set of all $\mathfrak{p} \in \mathrm{Spec}(S)$ at which S/R is smooth, is open in $\mathrm{Spec}(S)$.

Consider a presentation $S=P/I$ where P is a ring of fractions of a polynomial algebra over R and $I \subset P$ an ideal. By 8.1d), S/R is smooth at $\mathfrak{p}$ if and only if the canonical sequence

$$0 \to I/I^2 \to \Omega^1_{P/R}/I\Omega^1_{P/R} \to \Omega^1_{S/R} \to 0$$

becomes split-exact after localization at $\mathfrak{p}$. It is known that these primes form an open set of $\mathrm{Spec}(S)$.

<u>8.3. Transitivity.</u> Suppose the algebras S/R and T/S are essentially of finite type, where R is noetherian, and let $\mathfrak{q} \in \mathrm{Spec}(T)$, $\mathfrak{p} := \mathfrak{q} \cap S$ be given. If T/S is smooth at $\mathfrak{q}$ and S/R is smooth at $\mathfrak{p}$, then T/R is smooth at $\mathfrak{q}$. If T/S and S/R are smooth, so is T/R.

Proof. Let $\mathfrak{g} := \mathfrak{q} \cap R = \mathfrak{p} \cap R$. Since $T_{\mathfrak{q}}/S_{\mathfrak{p}}$ and $S_{\mathfrak{p}}/R_{\mathfrak{g}}$ are flat, so is $T_{\mathfrak{q}}/R_{\mathfrak{g}}$. Moreover,

$$\dim T_{\mathfrak{q}} = \dim S_{\mathfrak{p}} + \dim T_{\mathfrak{q}}/\mathfrak{p}T_{\mathfrak{q}} = \dim R_{\mathfrak{r}} + \dim T_{\mathfrak{q}}/\mathfrak{r}T_{\mathfrak{q}}$$

and
$$\dim S_{\mathfrak{p}} = \dim R_{\mathfrak{r}} + \dim S_{\mathfrak{p}}/\mathfrak{r}S_{\mathfrak{p}}$$

consequently

(5)
$$\dim T_{\mathfrak{q}}/\mathfrak{r}T_{\mathfrak{q}} = \dim S_{\mathfrak{p}}/\mathfrak{r}S_{\mathfrak{p}} + \dim T_{\mathfrak{q}}/\mathfrak{p}T_{\mathfrak{q}}.$$

In the canonical exact sequence

$$T_{\mathfrak{q}} \otimes_{S_{\mathfrak{p}}} \Omega^1_{S_{\mathfrak{p}}/R} \to \Omega^1_{T_{\mathfrak{q}}/R} \to \Omega^1_{T_{\mathfrak{q}}/S} \to 0$$

the first module is free of rank

$$\dim S_{\mathfrak{p}}/\mathfrak{r}S_{\mathfrak{p}} + \operatorname{Trdeg}(k(\mathfrak{p})/k(\mathfrak{r}))$$

and the third is free of rank

$$\dim T_{\mathfrak{q}}/\mathfrak{p}T_{\mathfrak{q}} + \operatorname{Trdeg}(k(\mathfrak{q})/k(\mathfrak{p})).$$

Using (5) we obtain

$$\mu_{\mathfrak{q}}(\Omega^1_{T/R}) \le \dim T_{\mathfrak{q}}/\mathfrak{r}T_{\mathfrak{q}} + \operatorname{Trdeg}(k(\mathfrak{q})/k(\mathfrak{r})) = \dim_{\mathfrak{q}} T_{\mathfrak{r}}/\mathfrak{r}T_{\mathfrak{r}}.$$

8.1 shows now that T/R is smooth at $\mathfrak{q}$.

8.4. Example. Suppose $X_1,\ldots,X_n \in S$ are algebraically independent over $R \subset S$ and $S/R[X_1,\ldots,X_n]$ is étale. Then S/R is smooth.

8.5. Base change. Let R'/R be an algebra, where R' is a noetherian ring and put $S' := R' \otimes_R S$. If S/R is smooth at some $\mathfrak{p} \in \operatorname{Spec}(S)$, then S'/R' is smooth at any $\mathfrak{p}' \in \operatorname{Spec}(S')$ lying over $\mathfrak{p}$. In particular, if S/R is smooth, so is S'/R'.

Proof. Let $\mathfrak{r}' := \mathfrak{p}' \cap R'$, $\mathfrak{r} := \mathfrak{p}' \cap R = \mathfrak{p} \cap R$. Then $S'_{\mathfrak{p}'}$ is a localization of $R'_{\mathfrak{r}'} \otimes_{R_{\mathfrak{r}}} S_{\mathfrak{p}}$. Since flatness is compatible with base change and localization is flat, we conclude that $S'_{\mathfrak{p}'}/R'_{\mathfrak{r}'}$ is flat. Moreover, $S'_{\mathfrak{r}'}/\mathfrak{r}'S'_{\mathfrak{r}'} = k(\mathfrak{r}') \otimes_{k(\mathfrak{r})} S_{\mathfrak{r}}/\mathfrak{r}S_{\mathfrak{r}}$

and hence

$$\dim_{\mathfrak{p}'} S'_{\mathfrak{q}'}/\mathfrak{q}' S'_{\mathfrak{q}'} = \dim_{\mathfrak{p}} S_{\mathfrak{q}}/\mathfrak{q} S_{\mathfrak{q}}$$

(see (4) in appendix B). Since $\Omega^1_{S'_{\mathfrak{p}'}/R'} = S'_{\mathfrak{p}'} \otimes_{S_{\mathfrak{p}}} \Omega^1_{S_{\mathfrak{p}}/R}$ we have

$$\mu_{\mathfrak{p}'}(\Omega^1_{S'/R'}) \leq \mu_{\mathfrak{p}}(\Omega^1_{S/R}) = \dim_{\mathfrak{p}} S_{\mathfrak{q}}/\mathfrak{q} S_{\mathfrak{q}} = \dim_{\mathfrak{p}'} S'_{\mathfrak{q}'}/\mathfrak{q}' S'_{\mathfrak{q}'}$$

and 8.1 implies that S'/R' is smooth at $\mathfrak{p}'$.

<u>8.6. Tensor product.</u> Suppose the algebras S/R and T/R are essentially of finite type. If S/R and T/R are smooth, so is $S \otimes_R T/R$.

By 8.5, $S \otimes_R T$ is smooth over S and by 8.3, $S \otimes_R T$ is then also smooth over R.

<u>8.7. Proposition.</u> If under the assumptions of 8.1 the algebra S/R is equidimensional of dimension d (B.17), then the following conditions are equivalent:

a) S/R is smooth.

b) S/R is flat, and $\Omega^1_{S/R}$ is a projective S-module of rank d.

If we drop the assumption of equidimensionality, but assume R contains a field of characteristic O, then a) is equivalent with

b') S/R is flat, and $\Omega^1_{S/R}$ is a projective S-module.

That S/R is equidimensional of dimension d implies, in particular, that

$$\dim_{\mathfrak{p}} S_{\mathfrak{q}}/\mathfrak{q} S_{\mathfrak{q}} = d \text{ for each } \mathfrak{p} \in \mathrm{Spec}(S), \ \mathfrak{q} := \mathfrak{p} \cap R.$$

The proposition follows therefore directly from 8.1.

In certain situations it is possible to characterize smoothness of an algebra S/R in terms of the module of differentials alone.

<u>8.8. Proposition.</u>   Let S/R be an algebra of finite type where R and S are regular noetherian rings of finite Krull dimension. Assume that $\dim_{\mathfrak{p}} S = \dim S$ for all $\mathfrak{p} \in \operatorname{Spec}(S)$ and let $d := \dim S - \dim R$. Then the following conditions are equivalent:

a) S/R is equidimensional of dimension d and smooth.

b) $\Omega^1_{S/R}$ is a projective S-module of rank d.

Proof.   a) $\to$ b) follows from 8.7.

b) $\to$ a).. Choose a presentation $S = R[X_1,\ldots,X_n]/I$. For $\mathfrak{p} \in \operatorname{Spec}(S)$ let $\mathfrak{N}$ be the preimage of $\mathfrak{p}$ in $P := R[X_1,\ldots,X_n]$. Then $S_{\mathfrak{p}} = P_{\mathfrak{N}}/I_{\mathfrak{N}}$. Since $P_{\mathfrak{N}}$ and $S_{\mathfrak{p}}$ are regular local rings, $I_{\mathfrak{N}}$ is generated by a $P_{\mathfrak{N}}$-regular sequence of length $\dim P_{\mathfrak{N}} - \dim S_{\mathfrak{p}}$, and $I_{\mathfrak{N}}/I_{\mathfrak{N}}^2$ is a free $S_{\mathfrak{p}}$-module of that rank. Using the assumption about $\dim_{\mathfrak{p}} S$, we obtain

$$\dim P_{\mathfrak{N}} - \dim S_{\mathfrak{p}} \le \dim P - \dim P/\mathfrak{N} - \dim S + \dim S/\mathfrak{p} =$$
$$\dim R + n - \dim S = n - d.$$

Comparing the ranks of the free modules in the canonical sequence

$$0 \to I_{\mathfrak{N}}/I_{\mathfrak{N}}^2 \to \Omega^1_{P_{\mathfrak{N}}/R}/I\Omega^1_{P_{\mathfrak{N}}/R} \to \Omega^1_{S_{\mathfrak{p}}/R} \to 0$$

we see that this sequence is exact, hence split-exact, and from 8.1 we conclude that S/R is smooth.

In particular, S/R is flat, and each minimal prime of S contracts to a minimal prime of R. Moreover, by 8.1 we have with $\mathfrak{q} := \mathfrak{p} \cap R$

$$\dim_{\mathfrak{p}} S_{\mathfrak{q}}/\mathfrak{q} S_{\mathfrak{q}} = \operatorname{rank} \Omega^1_{S_{\mathfrak{p}}/R} = d$$

which shows that S/R is equidimensional of dimension d.

Observe that the assumptions about the dimensions are satisfied, if R and S are affine algebras over a field and

S is equidimensional.

The structure theorem 7.19 can be partially generalized as follows.

8.9. Theorem. Let S/R be an algebra of finite type where R is noetherian. For $\mathfrak{p} \in \mathrm{Spec}(S)$, $\mathfrak{q} := \mathfrak{p} \cap R$, the following conditions are equivalent:

a) The algebra S/R is smooth and equidimensional of dimension d at $\mathfrak{p}$.

b) There exists a quasinormalization $R[X_1,\ldots,X_d] \to S_f$ in the neighbourhood of $\mathfrak{p}$ (B.19) such that $S_f/R[X_1,\ldots,X_d]$ is étale.

Moreover, if a) holds and $\dim S_{\mathfrak{p}}/\mathfrak{q}S_{\mathfrak{p}} > 0$, we can arrange in b) that $k(\mathfrak{p}) = k(\mathfrak{a})$ where $\mathfrak{a} := \mathfrak{p}S_f \cap R[X_1,\ldots,X_d]$.

Proof. b) $\to$ a). By 8.4 the algebra S_f/R is smooth, and by B.20, S/R is equidimensional of dimension d at $\mathfrak{p}$.

a) $\to$ b). By assumption, there exists an $f \in S\setminus\mathfrak{p}$ such that each minimal prime of S_f contracts to a minimal prime of R and $\dim_{\mathfrak{a}}(S_f)_{\mathfrak{a}}/\mathfrak{a}(S_f)_{\mathfrak{a}} = d$ for each $\mathfrak{a} \in \mathrm{Spec}(S_f)$, $\mathfrak{a} := \mathfrak{a} \cap R$.

For simplicity, let us write S for S_f and $\mathfrak{p}$ for $\mathfrak{p}S_f$. Then $S_{\mathfrak{q}}/\mathfrak{q}S_{\mathfrak{q}}$ is an equidimensional affine $k(\mathfrak{q})$-algebra of dimension d and $S_{\mathfrak{p}}/\mathfrak{q}S_{\mathfrak{p}}$ is geometrically regular over $k(\mathfrak{q})$.

By 7.19 there exists a noetherian normalization $k(\mathfrak{q})[\xi_1,\ldots,\xi_d] \subset S_{\mathfrak{q}}/\mathfrak{q}S_{\mathfrak{q}}$ such that $S_{\mathfrak{q}}/\mathfrak{q}S_{\mathfrak{q}}$ is étale over $k(\mathfrak{q})[\xi_1,\ldots,\xi_d]$ at $\mathfrak{p}S_{\mathfrak{q}}/\mathfrak{q}S_{\mathfrak{q}}$, where we may assume that $\xi_1,\ldots,\xi_d$ are in the image of S in $S_{\mathfrak{q}}/\mathfrak{q}S_{\mathfrak{q}}$. Moreover, if $\dim S_{\mathfrak{p}}/\mathfrak{q}S_{\mathfrak{p}} > 0$, we can arrange that $k(\mathfrak{p}) = k(\overline{\mathfrak{a}})$ with $\overline{\mathfrak{a}} := \mathfrak{p}S_{\mathfrak{q}}/\mathfrak{q}S_{\mathfrak{q}} \cap k(\mathfrak{q})[\xi_1,\ldots,\xi_d]$.

Let $x_i \in S$ be a preimage of ξ_i $(i=1,\ldots,d)$. As was shown in the proof of B.20, a) → b), there is an $f_o \in S \setminus \mathfrak{p}$ such that $R[X_1, \ldots, X_d] \to S_{f_o}$ $(X_i \mapsto x_i)$ defines a quasinormalization in the neighbourhood of $\mathfrak{p}$. Since $S_{\mathfrak{q}}/\mathfrak{q} S_{\mathfrak{q}}$ is flat over $k(\mathfrak{q})[\xi_1, \ldots, \xi_d]$ at $\mathfrak{p} S_{\mathfrak{q}}/\mathfrak{q} S_{\mathfrak{q}}$, S/R is flat at $\mathfrak{p}$, and $R[X_1, \ldots, X_d]/R$ is flat, it follows that $S_{f_o}/R[X_1, \ldots, X_d]$ is flat at $\mathfrak{p} S_{f_o}$ (B.27). Since $S_{\mathfrak{q}}/\mathfrak{q} S_{\mathfrak{q}}$ is unramified over $k(\mathfrak{q})[\xi_1, \ldots, \xi_d]$ at $\mathfrak{p} S_{\mathfrak{q}}/\mathfrak{q} S_{\mathfrak{q}}$, so is $S_{f_o}/R[X_1, \ldots, X_d]$ at $\mathfrak{p} S_{f_o}$. Since the étale locus of an algebra of finite type is open, we may assume that $f_o \in S \setminus \mathfrak{p}$ was chosen so that $S_{f_o}/R[X_1, \ldots, X_d]$ is étale.

If $\dim S_{\mathfrak{p}}/\mathfrak{q} S_{\mathfrak{p}} > 0$, we obtain $k(\mathfrak{p}) = k(\mathfrak{q})$ with $\mathfrak{q} := \mathfrak{p} S_{f_o} \cap R[X_1, \ldots, X_d]$, since $k(\mathfrak{q}) = k(\overline{\mathfrak{q}})$ and since $k(\overline{\mathfrak{q}}) = k(\mathfrak{p})$ by construction.

The following is an algebraic analogue to Sard's lemma of analysis.

<u>8.10. Theorem.</u>   Let $R \subseteq S$ be affine domains over a field k of characteristic O, and assume S is regular. Then there is a dense open set $U \subset \mathrm{Spec}(R)$ such that for each $\mathfrak{q} \in U$ and any $\mathfrak{p} \in \mathrm{Spec}(S)$ lying over $\mathfrak{q}$ the algebra S/R is smooth at $\mathfrak{p}$.

Proof.   Reg(R) is open in Spec(R) by 7.7 and not empty, since $(O) \in \mathrm{Reg}(R)$. We may therefore assume that R is regular. Let $K := Q(R)$, $L := Q(S)$ and $d := \dim S - \dim R = \mathrm{Trdeg}(L/K)$. Since L/K is separable, $\Omega^1_{L/K}$ has dimension d, and $\mu_{\mathfrak{p}}(\Omega^1_{S/R}) \geq d$ for each $\mathfrak{p} \in \mathrm{Spec}(S)$.

The set $A := \{\mathfrak{p} \in \mathrm{Spec}(S) \mid \mu_{\mathfrak{p}}(\Omega^1_{S/R}) \geq d+1\}$ is closed in Spec(S), that is, $A = \mathcal{V}(J)$ with some ideal J of S. Let $\alpha : \mathrm{Spec}(S) \to \mathrm{Spec}(R)$ be the map corresponding to the inclusion $R \subset S$. We shall show that $\alpha(A)$ is not dense in

Spec(R). We then choose $f \in R \setminus \{0\}$ such that $D(f) \cap \alpha(A) = \phi$.
For each $\mathfrak{p} \in \alpha^{-1}(D(f))$ we then have $\mu_{\mathfrak{p}}(\Omega^1_{S/R}) = d$. Since
dim S_f = dim S, dim R_f = dim R, we conclude from 7.3 and
8.8 that S_f/R_f is smooth. Thus $U := D(f)$ meets the requirements of the theorem.

Suppose $\alpha(A)$ were dense in Spec(R). Then the map
Spec(S/J) $\to$ Spec(R) has dense image, which implies that
there is a prime ideal $\mathfrak{p}_0 \in \mathcal{V}(J)$ with $\mathfrak{p}_0 \cap R = (0)$. Consider the following commutative diagram with exact rows

$$
\begin{array}{ccccccc}
k(\mathfrak{p}_0) \otimes_K \Omega^1_{K/k} & \to & k(\mathfrak{p}_0) \otimes_{S_{\mathfrak{p}_0}} \Omega^1_{S_{\mathfrak{p}_0}/k} & \to & k(\mathfrak{p}_0) \otimes_{S_{\mathfrak{p}_0}} \Omega^1_{S_{\mathfrak{p}_0}/K} & \to & 0 \\
\| & & \downarrow & & \downarrow & & \\
0 \to k(\mathfrak{p}_0) \otimes_K \Omega^1_{K/k} & \longrightarrow & \Omega^1_{k(\mathfrak{p}_0)/k} & \longrightarrow & \Omega^1_{k(\mathfrak{p}_0)/K} & \to & 0
\end{array}
$$

where the row on the bottom is exact, because the field
extensions $k(\mathfrak{p}_0)/K$ and K/k are separable, k being a field
of characteristic 0 (5.12).

We see that $k(\mathfrak{p}_0) \otimes_K \Omega^1_{K/k} \to k(\mathfrak{p}_0) \otimes_{S_{\mathfrak{p}_0}} \Omega^1_{S_{\mathfrak{p}_0}/k}$ is injective.
But the first vector space has dimension dim R, the second
has dimension dim S, and $k(\mathfrak{p}_0) \otimes_{S_{\mathfrak{p}_0}} \Omega^1_{S_{\mathfrak{p}_0}/K}$ has dimension
$>d$ = dim S - dim R, because $\mathfrak{p}_0 \in A$. This contradiction
shows that $\alpha(A)$ cannot be dense in Spec(R), so the proof
is complete.

The theorem tells us, in particular, that for each $\mathfrak{q}$
in the dense open set U the fiber $S_{\mathfrak{q}}/\mathfrak{q}S_{\mathfrak{q}}$ of the map
Spec(S) $\to$ Spec(R) is smooth over $k(\mathfrak{q})$.

Exercises

1) Under the assumptions of 8.1 the following conditions
   are equivalent:

   a) S/R is smooth at $\mathfrak{p} \in \mathrm{Spec}(S)$.

   b) For each $S_{\mathfrak{p}}$-module M the canonical sequence
   $$0 \to \mathrm{Der}_R(S_{\mathfrak{p}},M) \to \mathrm{Der}_R(P_{\mathfrak{n}},M) \to \mathrm{Hom}_{S_{\mathfrak{p}}}(I_{\mathfrak{n}}/I_{\mathfrak{n}}^2,M) \to 0$$
   is exact.

   c) S/R is flat at $\mathfrak{p}$ and $\dim_{k(\mathfrak{p})} T_{\mathfrak{p}}(S/R) \leq \dim_{\mathfrak{p}} S_{\mathfrak{q}}/\mathfrak{q}S_{\mathfrak{q}}$.

2) Under the assumptions of 8.1 let S/R be flat and
   $\mathrm{Spec}(S) \to \mathrm{Spec}(R)$ a closed surjective map. Then the set
   of all $\mathfrak{q} \in \mathrm{Spec}(R)$ for which $S_{\mathfrak{q}}/\mathfrak{q}S_{\mathfrak{q}}$ is smooth over $k(\mathfrak{q})$
   is open in $\mathrm{Spec}(R)$.

3) Let S/R be an algebra of finite type where R is a
   noetherian ring. Let $I \subset S$ be an ideal and $\overline{S} := S/I$.
   For $\mathfrak{p} \in \mathrm{Spec}(S)$ with $\mathfrak{p} \supset I$, let $\overline{\mathfrak{p}}$ be the image of $\mathfrak{p}$ in $\overline{S}$.
   If S/R is smooth at $\mathfrak{p}$ and $\overline{S}/R$ is smooth at $\overline{\mathfrak{p}}$, then the
   canonical sequence of $k(\mathfrak{p})$-vector spaces
   $$0 \to T_{\overline{\mathfrak{p}}}(\overline{S}/R) \to T_{\mathfrak{p}}(S/R) \to \mathrm{Hom}_{S_{\mathfrak{p}}}(I_{\mathfrak{p}}/I_{\mathfrak{p}}^2,k(\mathfrak{p})) \to 0$$
   is exact.

4) Let S/R be a flat algebra of finite type, where $(R,\mathfrak{m})$ is
   a complete noetherian local ring. Suppose there are only
   finitely many prime ideals of $S/\mathfrak{m}S$ at which $S/\mathfrak{m}S$ is not
   smooth over $R/\mathfrak{m}$ ("the special fiber $S/\mathfrak{m}S$ has only isolated singularities"). Then the set of all $\mathfrak{q} \in \mathrm{Spec}(R)$
   for which $S_{\mathfrak{q}}/\mathfrak{q}S_{\mathfrak{q}}$ is smooth over $k(\mathfrak{q})$ is open in $\mathrm{Spec}(R)$.
   (Hint: With some ideal I of S let $V(I)$ be the set of all
   $\mathfrak{p} \in \mathrm{Spec}(S)$ at which S/R is not smooth. Then $S/\mathfrak{m}S + I$ is
   a finite dimensional vector space over $R/\mathfrak{m}$. Conclude that

S/I is finite over R. Consider the image of $\mathcal{V}(I)$ in Spec(R)).

5) a) Let S/R be a regular algebra, that is, S/R is flat and $S_{\mathfrak{p}}/\mathfrak{q}S_{\mathfrak{p}}$ is a regular local ring for all $\mathfrak{p} \in$ Spec(S), $\mathfrak{q} := \mathfrak{p} \cap R$. For an ideal I of S assume that S/I too is a regular algebra over R. Show that for each $\mathfrak{p} \in$ Spec(S) $I_{\mathfrak{p}}$ is generated by a regular sequence (apply B.23).

b) Suppose S/R is of finite type and smooth, and R is noetherian. Show that the kernel of $S \otimes_R S \to S$ ($a \otimes b \to ab$) is locally generated by a regular sequence.

§ 9. Differential Modules of Complete Intersections

The relevant facts about complete intersections are contained in appendix C. Here we shall study their differential modules. If R is a ring and M is an R-module, $pd_R(M)$ will always denote the projective dimension of M. Suppose S/R is an algebra that is essentially of finite type. Choose a presentation

$$S = R[X_1, \ldots, X_n]_N / I$$

with a multiplicatively closed set N of $R[X_1, \ldots, X_n]$. Write $P := R[X_1, \ldots, X_n]_N$.

9.1. Lemma.   If $pd_R(S) < \infty$, then $pd_P(S) < \infty$.

Proof.   Let x_i be the image of X_i in S (i=1,...,n). We identify $S \otimes_R P$ with $S[X_1, \ldots, X_n]_N$. Then

$$S = S[X_1, \ldots, X_n]_N / (X_1 - x_1, \ldots, X_n - x_n)$$

and S, as an $S \otimes_R P$-module, has a free resolution of length n, since $\{X_1 - x_1, \ldots, X_n - x_n\}$ is a regular sequence of $S[X_1, \ldots, X_n]_N$. From $pd_R(S) < \infty$ and the flatness of P/R we conclude that $pd_P(S \otimes_R P) < \infty$. But then also $pd_P(S) < \infty$, since S has a finite $S \otimes_R P$-free resolution.

Algebras that are locally complete intersections can be characterized in terms of the projective dimension of certain differential modules. Let S/R be an algebra of finite type with a presentation

$$(1) \qquad S = P/I, \quad P := R[X_1, \ldots, X_n].$$

9.2. Theorem.   Suppose R is noetherian and S is reduced.
For $\mathfrak{p} \in \mathrm{Spec}(S)$, $\mathfrak{q} := \mathfrak{p} \cap R$, assume that $pd_{R_\mathfrak{q}}(S_\mathfrak{p}) < \infty$. Let $\delta : R \to R\delta R$ be a derivation that is admissible for each

$\mathfrak{p}_o \in \mathrm{Min}(S)$ with $\mathfrak{p}_o \subseteq \mathfrak{p}$ (see 6.19), and assume $R_\mathfrak{q}\delta R_\mathfrak{q}$ is a free $R_\mathfrak{q}$-module. Then the following conditions are equivalent:

a) S/R is locally at $\mathfrak{p}$ a complete intersection.

b) $S_\mathfrak{p}/R_\mathfrak{q}$ is flat and $\mathrm{pd}_{S_\mathfrak{p}}(\Omega^1_{S_\mathfrak{p}/\delta}) \leq 1$.

Proof. For each $\mathfrak{p}_o \in \mathrm{Min}(S)$ with $\mathfrak{p}_o \subseteq \mathfrak{p}$ the algebra $S_{\mathfrak{p}_o}/R_{\mathfrak{q}_o}$ ($\mathfrak{q}_o := \mathfrak{p}_o \cap R$) is flat in each of the cases a) and b), hence $\mathfrak{q}_o \in \mathrm{Min}(R)$. Since S is reduced, it follows that $R_{\mathfrak{q}_o}$ and $S_{\mathfrak{p}_o}$ are fields. $\Omega^1_{S_{\mathfrak{p}_o}/\delta}$ is an $S_{\mathfrak{p}_o}$-vector space of dimension $r + \mathrm{Trdeg}(S_{\mathfrak{p}_o}/R_{\mathfrak{q}_o})$ with $r := \dim_{R_{\mathfrak{q}_o}} R_{\mathfrak{q}_o}\delta R_{\mathfrak{q}_o}$, because δ is admissible for $\mathfrak{p}_o$.

Let $\mathfrak{q}_o$ be the preimage of $\mathfrak{p}_o$ in P. Then $S_{\mathfrak{p}_o} = P_{\mathfrak{q}_o}/I_{\mathfrak{q}_o}$, $S_{\mathfrak{p}_o} \otimes_P \Omega^1_{P/\delta}$ is a vector space of dimension $r+n$, and $I_{\mathfrak{q}_o}/I^2_{\mathfrak{q}_o}$ is a vector space of dimension

$$h(I_{\mathfrak{q}_o}) = \dim P_{\mathfrak{q}_o} = n - \mathrm{Trdeg}(S_{\mathfrak{p}_o}/R_{\mathfrak{q}_o}).$$

We conclude that the canonical sequence

$$0 \to I_{\mathfrak{q}_o}/I^2_{\mathfrak{q}_o} \to S_{\mathfrak{p}_o} \otimes_P \Omega^1_{P/\delta} \to \Omega^1_{S_{\mathfrak{p}_o}/\delta} \to 0$$

is exact. Since $L := Q(S_\mathfrak{p})$ is the direct product of the $S_{\mathfrak{p}_o}$, the canonical sequence

$$0 \to L \otimes_P I/I^2 \to L \otimes_P \Omega^1_{P/\delta} \to \Omega^1_{L/\delta} \to 0$$

is exact as well.

a) $\to$ b). Let $\mathfrak{q}$ be the preimage of $\mathfrak{p}$ in P. Since S/R is locally at $\mathfrak{p}$ a complete intersection, the $S_\mathfrak{p}$-module $I_\mathfrak{q}/I^2_\mathfrak{q}$ is free, and hence $I_\mathfrak{q}/I^2_\mathfrak{q} \to L \otimes_{S_\mathfrak{p}} I_\mathfrak{q}/I^2_\mathfrak{q}$ is an injection. But then the canonical sequence

$$0 \to I_\mathfrak{q}/I^2_\mathfrak{q} \to S_\mathfrak{p} \otimes_P \Omega^1_{P/\delta} \to \Omega^1_{S_\mathfrak{p}/\delta} \to 0$$

is exact. Since $S_{\not{p}} \otimes_P \Omega^1_{P/\delta}$ is a free $S_{\not{p}}$-module, we have shown that $pd_{S_{\not{p}}}(\Omega^1_{S_{\not{p}}/\delta}) \leq 1$.

b) $\to$ a). Under the hypothesis that $pd_{S_{\not{p}}}(\Omega^1_{S_{\not{p}}/\delta}) \leq 1$ the image of the canonical map

$$\alpha \; : \; I_{\mathcal{O}\!\!\!/}/I^2_{\mathcal{O}\!\!\!/} \to S_{\not{p}} \otimes_P \Omega^1_{P/\delta}$$

is a free $S_{\not{p}}$-module. Then

$$I_{\mathcal{O}\!\!\!/}/I^2_{\mathcal{O}\!\!\!/} \cong S^h_{\not{p}} \oplus K$$

with some $h \in \mathbb{N}$ and $K := \ker \alpha$. If $\not{p}_o \subset \not{p}$ is a minimal prime of S and $\mathcal{O}\!\!\!/_o$ its preimage in P, then

$$h = \dim_{S_{\not{p}_o}} (S_{\not{p}_o} \otimes_P I/I^2) = h(I_{\mathcal{O}\!\!\!/_o}) = \dim P_{\mathcal{O}\!\!\!/_o},$$

hence $I_{\mathcal{O}\!\!\!/}$ does not contain a regular sequence of length h+1. From 9.1 we obtain that $pd_{P_{\mathcal{O}\!\!\!/}}(S_{\not{p}}) < \infty$, and hence $pd_{P_{\mathcal{O}\!\!\!/}}(I_{\mathcal{O}\!\!\!/}) < \infty$. By the theorem of Ferrand and Vasconcelos (C.2) $I_{\mathcal{O}\!\!\!/}$ is generated by a $P_{\mathcal{O}\!\!\!/}$-regular sequence, and by C.6 S/R is locally at $\not{p}$ a complete intersection.

Suppose that under the assumptions of 9.2 $S_{\not{p}}/R_{\not{8}}$ is flat and for each $\not{p}_o \in \text{Min}(S)$ with $\not{p}_o \subset \not{p}$ the field extension $S_{\not{p}_o}/R_{\not{8}_o}$ ($\not{8}_o := \not{p}_o \cap R$) is separable. Then the trivial derivation $\delta = 0$ satisfies the admissibility assumption of theorem 9.2. In this case S/R is a complete intersection at $\not{p}$ if and only if $pd_{S_{\not{p}}}(\Omega^1_{S_{\not{p}}/R_{\not{8}}}) \leq 1$. For other situations in which a derivation δ meeting the requirements of the theorem exists, see 6.21, 6.22, and 6.25.

<u>9.3. Corollary.</u>   Under the assumptions of 9.2 suppose that $pd_R(S) < \infty$, δ is admissible for all $\not{p} \in \text{Min}(S)$, and $R\delta R$ is a projective R-module.

a) S/R is locally a complete intersection if and only if

S/R is flat and $pd_S(\Omega^1_{S/\delta}) \leq 1$.

b) The set of all $\not{p} \in Spec(S)$ at which S/R is locally a
complete intersection is open in Spec(S).

Proof. Since $pd_R(S) < \infty$, we also have $pd_P(S) < \infty$ by 9.1,
and hence $pd_{P_{O\!\!\!/}}(S_{\not{p}}) < \infty$ for each $\not{p} \in Spec(S)$, where $O\!\!\!/$ is
the preimage of $\not{p}$ in P. Only this condition was used in
the above proof. Therefore a) follows by applying the
local-global principle.

The set in b) is the intersection of the set of all $\not{p}$ at
which S/R is flat and the set of all $\not{p}$ with $pd_{S_{\not{p}}}(\Omega^1_{S_{\not{p}}/\delta}) \leq 1$.
Since both of these sets are open, claim b) follows.

<u>9.4. Corollary.</u>   Let A be a reduced affine algebra over a
field K, and let $K_O \subset K$ be a subfield such that the universal derivation of K/K_O is admissible for all $\not{p}_O \in Min(A)$.
Then A is locally a complete intersection if and only if
$pd_A(\Omega^1_{A/K_O}) \leq 1$.

It is an open problem whether it is enough to require
that $pd_A(\Omega^1_{A/K_O}) < \infty$. For results in this direction, see
Vasconcelos $[Va_2]$, Herzog $[He_3]$ and Platte $[Pl_3]$.

Global complete intersections (see C.9) are characterized in the next theorem.

<u>9.5. Theorem</u> (Mohan Kumar [MK]). Let S/R be an algebra of
finite type where R is noetherian and S is reduced. Suppose
$pd_R(S) < \infty$, and let $\delta : R \to R\delta R$ be a derivation that is admissible for all $\not{p}_O \in Min(S)$. Assume $R\delta R$ is a free R-module.
Then the following statements are equivalent:

a) S/R is globally a complete intersection.

b) S/R is flat, and there is an exact sequence

$$0 \to F_1 \to F_0 \to \Omega^1_{S/\delta} \to 0$$

with free S-modules F_i (i=0,1) of finite rank.

Proof. In both of the cases a) or b) S/R is locally a complete intersection. If we assume this, it remains to be shown that S/R is globally a complete intersection if and only if $\Omega^1_{S/\delta}$ has a free resolution $0 \to F_1 \to F_0 \to \Omega^1_{S/\delta} \to 0$ as in b).

Choose a presentation (1) for S/R. In the proof of 9.2 it was shown that the canonical sequence

$$(2) \qquad 0 \to I/I^2 \to S \otimes_P \Omega^1_{P/\delta} \to \Omega^1_{S/\delta} \to 0$$

is locally exact and hence exact (S/R being locally a complete intersection). Here $S \otimes_P \Omega^1_{P/\delta}$ is a free S-module of rank r+n, if RδR is free of rank r.

a) $\to$ b). If S/R is a global complete intersection, there exists a presentation (1) where I is generated by a quasi-regular sequence of P, hence I/I^2 is a free S-module. Putting $F_1 := I/I^2$ and $F_0 := S \otimes_P \Omega^1_{P/\delta}$ we are done.

b) $\to$ a). Conversely, assume there is a free resolution

$$0 \to F_1 \to F_0 \to \Omega^1_{S/\delta} \to 0$$

as in b). By Schanuel's lemma

$$I/I^2 \oplus F_0 \cong F_1 \oplus S \otimes_P \Omega^1_{P/\delta}.$$

Hence there exist m,m' $\in \mathbb{N}$ such that

$$I/I^2 \oplus S^m \cong S^{m'}.$$

Consider now the presentation

(3) $$S = R[X_1,\dots,X_n,Y_1,\dots,Y_m]/J$$

with $J = (I,Y_1,\dots,Y_m)$. Then $J/J^2 \cong I/I^2 \oplus S^m \cong S^{m'}$ and we conclude from C.12 that S/R is globally a complete inter-section, q.e.d.

Under the assumptions of 9.4 it follows from 9.5 that the following statements are equivalent:

a) A/K is a global complete intersection.

b) There exists a resolution

$$0 \to F_1 \to F_o \to \Omega^1_{A/K_o} \to 0$$

with free A-modules F_i (i=0,1) of finite rank.

In case A is regular and of dimension 1 there is a more precise result (recall §7, exercise 2)):

9.6. <u>Theorem.</u>   Let A be a regular affine domain over a field K. If char.K =: p > 0, let K_o be a subfield of K that is admissible for A, if char.K = 0, let $K_o := K$. If dim A = 1, the following statements are equivalent:

a) Ω^1_{A/K_o} is a free A-module.

b) A/K is a global complete intersection.

c) For any presentation $A = K[X_1,\dots,X_n]/I$ of A/K the ideal I is generated by a quasiregular sequence.

Proof.   It remains to be shown that b) implies both a) and c). Since A is assumed to be regular, Ω^1_{A/K_o} is a projective A-module by 7.2 and 7.5. Given a presentation $A = K[X_1,\dots,X_n]/I$ of A/K as a complete intersection, the canonical sequence (with $P := K[X_1,\ldots,X_n]$)

$$(4) \qquad 0 \to I/I^2 \to A \otimes_P \Omega^1_{P/K_0} \to \Omega^1_{A/K_0} \to 0$$

is exact, as was shown in the proof of 9.2. Here it even splits, and I/I^2 is a free A-module of rank n-1, since dim A = 1.

By a theorem of Serre ([Ku], p.118) $\Omega^1_{A/K_0} \cong A^m \oplus Q$ for some $m \in \mathbb{N}$ and a projective A-module Q of rank 1. Using (4) we see that $A \otimes_P \Omega^1_{P/K_0} \cong A^{n-1+m} \oplus Q$. But this module is also free of rank n+m. Then $Q \cong \Lambda^{n+m}(A \otimes_P \Omega^1_{P/K_0}) \cong A$, and we have shown that Ω^1_{A/K_0} is a free A-module.

Consider now an arbitrary presentation $A = K[Y_1,\ldots,Y_t]/J$ of A/K and the exact sequence (4) corresponding to it. Schanuel's lemma then tells us that for some $r \in \mathbb{N}$

$$I/I^2 \oplus A^{t+r} \cong J/J^2 \oplus A^{n+r}$$

and the same argument as above shows that J/J^2 is a free A-module of rank t-1. By results of Mohan Kumar ([Ku], p.209, Cor.2.9 and p.215,Cor.3.8) J is generated by a regular sequence, q.e.d.

For a reduced affine algebra A over a field K which is locally a complete intersection there is a close connection between the Serre conditions for A and those for Ω^1_{A/K_0}. Remember that for $k \in \mathbb{N}$ a noetherian ring A is said to satisfy Serre's condition (R_k) if $A_\wp$ is a regular local ring for all $\wp \in \mathrm{Spec}(A)$ with dim $A_\wp \leq k$ (A is regular in codimension k).

A finitely generated module M over A is said to satisfy condition (S_k), if

$$\mathrm{depth}(M_\wp) \geq \mathrm{Min}\{k, \dim A_\wp\} \quad \text{for all } \wp \in \mathrm{Spec}(A).$$

M is Cohen-Macaulay if and only if M satisfies (S_k) for

all $k \in \mathbb{N}$.

9.7. <u>Proposition.</u>   Let A be a reduced affine algebra over a field K. If char.K = 0, put $K_0 := K$, if char.K > 0, let K_0 be a subfield of K that is admissible for A. If A is locally a complete intersection, then for $k \in \mathbb{N}$ the following statements are equivalent:

a) A satisfies (R_k).

b) Ω^1_{A/K_0} satisfies (S_k).

Proof.   By 9.4 we have $\mathrm{pd}_A \Omega^1_{A/K_0} \leq 1$ and by the formula of Auslander-Buchsbaum (c.f.[Ku], p.202, prop. 1.12)

$$\mathrm{pd}_{A_{\wp}} \Omega^1_{A_{\wp}/K_0} + \mathrm{depth}\ \Omega^1_{A_{\wp}/K_0} = \mathrm{depth}\ A_{\wp}$$

for each $\wp \in \mathrm{Spec}(A)$. Here depth $A_{\wp}$ = dim $A_{\wp}$, since complete intersections are Cohen-Macaulay, and $\mathrm{pd}_{A_{\wp}} \Omega^1_{A_{\wp}/K_0} = 0$ if and only if $A_{\wp}$ is regular (7.2 and 7.5). For any $\wp \in \mathrm{Spec}(A)$ we therefore have

$$\dim A_{\wp} = \mathrm{depth}\ \Omega^1_{A_{\wp}/K_0} + 1, \text{ or } A_{\wp} \text{ is regular.}$$

From this the equivalence of a) and b) immediately follows.

9.8. <u>Corollary.</u>   The following statements are equivalent:

a) A is normal.

b) A is regular in codimension 1.

c) Ω^1_{A/K_0} is torsion free.

Proof.   A finitely generated module over a reduced noetherian ring is clearly torsion free if and only if it satisfies (S_1). Therefore b) and c) are equivalent by 9.7. Since A is Cohen-Macaulay, a) and b) are equivalent by Serre's criterion for normality (Matsumura [M_1], Thm. 39).

9.9. Corollary (Berger [Be]). If dim A = 1, the following statements are equivalent:

a) Ω^1_{A/K_0} is torsion free.

b) A is regular.

9.10. Corollary. Under the assumptions of 9.7 the following statements are equivalent:

a) Ω^1_{A/K_0} is reflexive.

b) A is regular in codimension 2.

Proof. By 9.7, b) is equivalent with Ω^1_{A/K_0} satisfying (S_2). Since A is normal, it is known that this condition is equivalent with Ω^1_{A/K_0} being reflexive.

There exist generalizations of 9.8 and 9.10 for the modules Ω^p_{A/K_0} with $p \geq 1$, see Vetter [Ve]. For further work related to the content of this section, see Bruns [Bru] and Bruns-Vetter [BV].

Exercises

1) Let K be a field of characteristic 0 and

$$A := K[X,Y,Z]/(X^3-Y(Z+1),Y^2-XZ,X^2Y-Z(Z+1)).$$

Show that A is a regular domain of dimension 1, but A/K is not a global complete intersection.

2) Let K be a field of characteristic $\neq 2$ and let $A := K[X,Y,Z]/(Z^2-XY)$. Show that A is not regular, but $\Omega^1_{A/K}$ is torsion free.

§ 10. The Kähler Differents (Jacobian Ideals) of an Algebra

These ideals are the Fitting ideals (appendix D) of the differential module of an algebra. We wish to study which informations about the algebra they contain.

Let S/R be an algebra, and let $\delta : R \to R\delta R$ be a derivation such that $\Omega^1_{S/\delta}$ is a finite S-module.

10.1. Definition. The i-th Fitting ideal of $\Omega^1_{S/\delta}$

$$\vartheta^{(i)}(S/\delta) := F_i(\Omega^1_{S/\delta}) \qquad (i \in \mathbb{N})$$

is called i-th <u>Kähler different</u> (<u>Jacobian ideal</u>) of S/δ. In case δ is the trivial derivation of R we write $\vartheta^{(i)}(S/R)$ for $\vartheta^{(i)}(S/\delta)$.

Clearly

$$\vartheta^{(o)}(S/\delta) \subset \vartheta^{(1)}(S/\delta) \subset \ldots \subset \vartheta^{(i)}(S/\delta) \subset \ldots$$

and

$$\vartheta^{(i)}(S/\delta) = S \text{ for } i \geq \mu(\Omega^1_{S/\delta}).$$

10.2. Example. Suppose S/R is essentially of finite type, and let

$$S = R[X_1, \ldots, X_n]_N / I$$

be a presentation of S/R. Choose a system $\{F_\lambda\}_{\lambda \in \Lambda}$ of generators of I. We then have by 4.19 a presentation of the differential module $\Omega^1_{S/R}$

$$0 \to K \to SdX_1 \oplus \ldots \oplus SdX_n \to \Omega^1_{S/R} \to 0$$

where K is generated by the elements

$$\sum_{i=1}^{n} \frac{\partial F_\lambda}{\partial x_i} dX_i \qquad (\lambda \in \Lambda)$$

x_i denoting the image of X_i in S. The Jacobian matrix

$$\left(\frac{\partial F_\lambda}{\partial x_i}\right)_{\substack{i=1,\ldots,n \\ \lambda \in \Lambda}}$$

is a relation matrix of $\Omega^1_{S/R}$ with respect to $\{dx_1,\ldots,dx_n\}$. Thus $\vartheta^{(i)}(S/R)$ is the ideal of S generated by all $(n-i)$-minors of the Jacobian matrix. In particular, $\vartheta^{(o)}(S/R)$ is generated by the maximal minors of that matrix.

For $S = R[x] = R[X]/(f)$ we have
$\vartheta^{(o)}(S/R) = (f'(x))$ and $\vartheta^{(i)}(S/R) = S$ for $i > 0$.

The general rules about Fitting ideals immediately allow us to make the following statements.

10.3. <u>Rules.</u>   a) (Base change). Let R'/R be an algebra and $S' := R' \otimes_R S$. Then we have in S' the relation

$$\vartheta^{(i)}(S'/R') = S' \cdot \vartheta^{(i)}(S/R) \qquad (i \in \mathbb{N}).$$

b) For an ideal $I \subset R$ and $\overline{S} := S/IS$, $\overline{R} := R/I$ we have

$$\vartheta^{(i)}(\overline{S}/\overline{R}) = \overline{\vartheta^{(i)}(S/R)} \qquad (i \in \mathbb{N})$$

where $\overline{\vartheta^{(i)}(S/R)}$ denotes the image of $\vartheta^{(i)}(S/R)$ in $\overline{S}$.

c) For each multiplicatively closed set $N \subset S$

$$\vartheta^{(i)}(S_N/R) = \vartheta^{(i)}(S/R)_N \qquad (i \in \mathbb{N}).$$

d) If S_1/R and S_2/R are algebras such that $\Omega^1_{S_k/R}$ is finitely generated (k=1,2), then

$$\vartheta^{(i)}(S_1 \otimes_R S_2/R) = \sum_{\rho+\sigma=i} \vartheta^{(\rho)}(S_1/R) \otimes_R \vartheta^{(\sigma)}(S_2/R)$$

and, in particular,

$$\vartheta^{(o)}(S_1 \otimes_R S_2/R) = \vartheta^{(o)}(S_1/R) \otimes_R \vartheta^{(o)}(S_2/R).$$

a) follows from $\Omega^1_{S'/R'} = R' \otimes_R \Omega^1_{S/R} = S' \otimes_S \Omega^1_{S/R}$ and D.4b), b) is a special case of a). Assertion c) comes from

D.4c) and d) comes from

$$\Omega^1_{S_1 \otimes_R S_2/R} = S_2 \otimes_R \Omega^1_{S_1/R} \oplus S_1 \otimes_R \Omega^1_{S_2/R} \text{ and D.15.}$$

D.7 allows the following statements:

10.4. Proposition. If $\Omega^1_{S/\delta}$ has rank r, then

$$\vartheta^{(0)}(S/\delta) = \ldots = \vartheta^{(r-1)}(S/\delta) = 0 \text{ and } \vartheta^{(i)}(S/\delta) \neq (0) \text{ for } i \geq r.$$

10.5. Corollary. Let S be a domain with quotient field L, and let K be the quotient field of the image of R in S. Then the following statements are equivalent:

a) $\vartheta^{(0)}(S/R) \neq (0)$.

b) $\Omega^1_{S/R}$ is a torsion module.

c) $\Omega^1_{L/K} = 0$.

If L/K is a finitely generated field extension, then a) - c) are equivalent with

d) L/K is finite and separable.

More generally, if in the situation of 10.5 the field extension L/K is finitely generated, then by the results of §5

$$\vartheta^{(0)}(S/R) = \ldots = \vartheta^{(r-1)}(S/R) = (0) \text{ and } \vartheta^{(r)}(S/R) \neq (0)$$

with

$$r := \begin{cases} \mathrm{Trdeg}(L/K) & \text{if char.} K = 0 \\ \mathrm{p\text{-}deg}(L/K) & \text{if char.} K =: p > 0. \end{cases}$$

L/K is separable if and only if

$$\vartheta^{(t-1)}(S/R) = (0) \text{ and } \vartheta^{(t)}(S/R) \neq (0)$$

with $t := \mathrm{Trdeg}(L/K)$.

A special case of D.9 is

<u>10.6. Proposition.</u>   For $\wp \in \mathrm{Spec}(S)$ and $n \in \mathbb{N}$ the following statements are equivalent:

a) $\mu_\wp(\Omega^1_{S/\delta}) = n$.

b) $\vartheta^{(n-1)}(S/\delta) \subset \wp$ and $\vartheta^{(n)}(S/\delta) \not\subset \wp$.

Classically differents were introduced in algebraic number theory (by Dedekind) in order to characterize ramification. Here we have

<u>10.7. Theorem</u> (Ramification criterion). Let S/R be an algebra which is essentially of finite type. For its ramification locus we have

$$V_{S/R} = \mathcal{V}(\vartheta^{(0)}(S/R)).$$

In other words: S/R is ramified at $\wp \in \mathrm{Spec}(S)$ if and only if $\vartheta^{(0)}(S/R) \subset \wp$.

By 6.10 we have $V_{S/R} = \mathrm{Supp}(\Omega^1_{S/R})$ and by D.11 $\mathrm{Supp}(\Omega^1_{S/R}) = \mathcal{V}(\vartheta^{(0)}(S/R))$.

A special case of D.13 is

<u>10.8. Proposition.</u> $\Omega^1_{S/\delta}$ is locally free of rank r if and only if $\vartheta^{(i)}(S/\delta) = (0)$ for $i=0,\ldots,r-1$ and $\vartheta^{(r)}(S/\delta) = S$.

Since regularity of affine algebras is connected with projectivity of their differential modules, the regularity criteria of § 7 can also be expressed in terms of Kähler differents.

<u>10.9. Theorem</u> (Regularity criterion). Let A be a reduced and equidimensional affine algebra over a field K. If $\mathrm{char}.K = 0$, let $K_0 := K$. If $\mathrm{char}.K =: p > 0$, let K_0 be a subfield of K which is admissible for A. Let $\vartheta^{(r)}(A/K_0)$ be the

smallest non-vanishing Kähler different of A/K_0 (i.e.
$\vartheta^{(i)}(A/K_0) = 0$ for $i=0,\ldots,r-1$ and $\vartheta^{(r)}(A/K_0) \neq 0$). Then
the following statements are equivalent:

a) $\mathfrak{p} \in \mathrm{Reg}(A)$.

b) $\vartheta^{(r)}(A/K_0) \not\subset \mathfrak{p}$.

Proof. Let $L := Q(A)$. The assumptions guarantee that Ω^1_{L/K_0}
is free of rank r. b) is then equivalent with
$\vartheta^{(r)}(A_\mathfrak{p}/K_0) = A_\mathfrak{p}$ which by D.8 is equivalent with $\Omega^1_{A_\mathfrak{p}/K_0}$
being free. By 7.2 and 7.5 this is equivalent with $A_\mathfrak{p}$
being regular.

As we see
$$\mathrm{Sing}(A) = \vartheta(\vartheta^{(r)}(A/K_0))$$
which shows again the closedness of the singular locus in
our case.

Let A be an affine algebra over a field K. We call
$\mathfrak{p} \in \mathrm{Spec}(A)$ an <u>isolated singularity</u> of A if there is an
$f \in A \setminus \mathfrak{p}$ such that $\mathrm{Sing}(A_f) = \{\mathfrak{p}_f\}$. Clearly $\mathfrak{p}$ is then a maximal
ideal of A.

<u>10.10. Corollary.</u>   Under the assumptions of 10.9 the
following conditions are equivalent:

a) $\mathrm{Sing}(A)$ consists only of isolated singularities.

b) $A/\vartheta^{(r)}(A/K_0)$ is a finite dimensional vector space over K.

Proof. b) is equivalent with the fact that
$\vartheta(\vartheta^{(r)}(A/K_0)) = \mathrm{Sing}(A)$ consists only of (finitely many)
maximal ideals and this is equivalent with a).

<u>10.11. Example.</u>   Let $A = K[X_1,\ldots,X_n]/(f)$ be a "hypersur-

face ring", and assume K is perfect and f $\neq$ 0 a polynomial
without a multiple irreducible factor. Then $\vartheta^{(n-1)}(A/K)$ is
the smallest non-vanishing Kähler different of A/K and

$$A/\vartheta^{(n-1)}(A/K) = K[X_1,\ldots,X_n]/(f,\frac{\partial f}{\partial X_1},\ldots,\frac{\partial f}{\partial X_n}).$$

A has only isolated singularities if and only if
$A/\vartheta^{(n-1)}(A/K)$ is finite dimensional over K.

The _smoothness criterion_ 8.1 can be translated by means
of D.9 as follows.

10.12. Theorem. Let S/R be an algebra that is essentially
of finite type where R is noetherian. For $\mathfrak{p} \in \operatorname{Spec}(S)$,
$\mathfrak{q} := \mathfrak{p} \cap R$, let $d := \dim_{\mathfrak{p}} S_{\mathfrak{q}}/\mathfrak{q} S_{\mathfrak{q}}$. Then the following statements are equivalent:
a) S/R is smooth at $\mathfrak{p}$.
b) $S_{\mathfrak{p}}/R_{\mathfrak{q}}$ is flat, and $\vartheta^{(d)}(S/R) \not\subset \mathfrak{p}$.
c) $S_{\mathfrak{p}}/R_{\mathfrak{q}}$ is flat, $\vartheta^{(i)}(S/R) \subset \mathfrak{p}$ for i=0,...,d-1, and
$\vartheta^{(d)}(S/R) \not\subset \mathfrak{p}$.

10.13. Corollary. If S/R is flat and equidimensional of
dimension d, then $V(\vartheta^{(d)}(S/R))$ is the set of all $\mathfrak{p} \in \operatorname{Spec}(S)$
at which S/R is not smooth.

This follows immediately from 10.12.

Combining 9.3 with D.19 we obtain a criterion for an
algebra to be locally a complete intersection.

10.14. Theorem. Let S/R be an algebra of finite type
where R is noetherian, S is reduced, and $\operatorname{pd}_R(S) < \infty$. With

$L := Q(S)$ assume that $\Omega^1_{L/R} = 0$. Then the following statements are equivalent:

a) S/R is locally a complete intersection.

b) S/R is flat and $\vartheta^{(0)}(S/R)$ is an invertible ideal.

Proof. We may assume S/R is flat. The condition $\Omega^1_{L/R} = 0$ is then equivalent with $S_{\mathfrak{p}_0}$ being separably algebraic over $R_{\mathfrak{p}_0}$ ($\mathfrak{p}_0 := \mathfrak{p}_0 \cap R$) for each $\mathfrak{p}_0 \in \mathrm{Min}(S)$. Therefore the trivial derivation of R is admissible for all these $\mathfrak{p}_0$. Moreover, $\Omega^1_{S/R}$ is a torsion module. By 9.3, S/R is locally a complete intersection if and only if $\mathrm{pd}_S(\Omega^1_{S/R}) \leq 1$. By D.19 this is equivalent with $\vartheta^{(0)}(S/R)$ being invertible.

As an application we obtain in a special case the "purity of the branch locus".

<u>10.15. Corollary.</u> Let S/R be locally a complete intersection. Then the following conditions are equivalent:

a) S/R is unramified.

b) S/R is unramified at each $\mathfrak{p} \in \mathrm{Spec}(S)$ with $h(\mathfrak{p}) = 1$.

By 10.7 condition b) is equivalent with the fact that $\vartheta^{(0)}(S/R)$ is not contained in any prime ideal of height 1 of S. Since $\vartheta^{(0)}(S/R)$ is locally generated by a non-zero-divisor (10.14), this can only happen if $\vartheta^{(0)}(S/R) = S$ (principal ideal theorem), that is, S/R is unramified at all $\mathfrak{p} \in \mathrm{Spec}(S)$.

There is obviously also a local version of theorem 10.14 corresponding to 9.2. It is used in the proof of the following proposition.

10.16. Proposition. Let S/R be an algebra of finite type where R is noetherian and S is reduced. Assume that in the neighbourhood of $\mathfrak{p} \in \mathrm{Spec}(S)$ there exists a quasinormalization $R[X_1,\ldots,X_d] \to S_f$ (B.19). Let $\mathfrak{q} := \mathfrak{p} \cap R[X_1,\ldots,X_d]$, $\mathfrak{r} := \mathfrak{p} \cap R$, and suppose that $\dim S_{\mathfrak{p}}/\mathfrak{r}S_{\mathfrak{p}} = \dim R[X]_{\mathfrak{q}}/\mathfrak{r}R[X]_{\mathfrak{q}}$. Put $L := Q(S_{\mathfrak{p}})$, $L_o := Q(R[X_1,\ldots,X_d]_{\mathfrak{q}})$, and assume that $\Omega^1_{L/L_o} = 0$ and $\mathrm{pd}_{R[X_1,\ldots,X_d]}(S_{\mathfrak{p}}) < \infty$. Then the following conditions are equivalent:

a) S/R is locally at $\mathfrak{p}$ a complete intersection.

b) $S_{\mathfrak{p}}/R_{\mathfrak{r}}$ is flat and $\vartheta^{(o)}(S_{\mathfrak{p}}/R[X_1,\ldots,X_d])$ is a principal ideal generated by some non-zero-divisor of $S_{\mathfrak{p}}$.

Due to the above remark one has only to point out that S/R is locally at $\mathfrak{p}$ a complete intersection if and only if $S_f/R[X_1,\ldots,X_d]$ is at $\mathfrak{p}_f$ (see C.23).

Which connections exist between the Kähler different $\vartheta^{(o)}$, the Dedekind different ϑ_D (defined in G.9a), and the Noether different ϑ_N (defined in G.1)? Let us now write ϑ_K for $\vartheta^{(o)}$.

10.17. Proposition. Let R be a noetherian ring, and let the algebra S/R be finite and locally a complete intersection. Assume $L := Q(S)$ is étale over $K := Q(R)$. Then

$$\vartheta_K(S/R) = \vartheta_D(S/R) = \vartheta_N(S/R).$$

Proof. Since S/R is projective, we have $\vartheta_D(S/R) = \vartheta_N(S/R)$ by G.11b). After localization in R we may assume R is local. Then S/R has a presentation

$$S = R[X_1,\ldots,X_n]/(t_1,\ldots,t_n) = R[x_1,\ldots,x_n]$$

as a complete intersection, and by G.3 and 10.2 the different $\vartheta_N(S/R)$ agrees with $\vartheta_K(S/R)$.

Under somewhat weaker assumptions one has the following relation between the Kähler and Noether differents.

10.18. <u>Proposition.</u> Let S/R be an algebra of finite type and let $m := \mu(\Omega^1_{S/R})$. Then

$$\vartheta_N(S/R)^m \subset (\text{Ann}\,\Omega^1_{S/R})^m \subset \vartheta_K(S/R) \subset \vartheta_N(S/R) \subset \text{Ann}\,\Omega^1_{S/R}.$$

Proof. Write $S = R[X_1,\ldots,X_n]/\mathcal{U} = R[x_1,\ldots,x_n]$. By G.2

$$\vartheta_N(S/R) = \{f(x_1,\ldots,x_n) \mid f(X_1,\ldots,X_n) \in (\mathcal{U}S[X] : (X_1-x_1,\ldots,X_n-x_n))\}.$$

For $f_1,\ldots,f_n \in \mathcal{U}$ let us show that

$$\frac{\partial(f_1,\ldots,f_n)}{\partial(x_1,\ldots,x_n)} \in \vartheta_N(S/R)$$

from which we obtain that $\vartheta_K(S/R) \subset \vartheta_N(S/R)$. Indeed, in $S[X]$ there are relations

$$f_i = \sum_{j=1}^{n} h_{ij} \cdot (X_j-x_j) \qquad (h_{ij} \in S[X]).$$

Putting $H := \det(h_{ij})$ we conclude from Cramer's rule that

$$H \cdot (X_i-x_i) \in (f_1,\ldots,f_n)S[X] \subset \mathcal{U}S[X].$$

Since $H(x_1,\ldots,x_n) = \dfrac{\partial(f_1,\ldots,f_n)}{\partial(x_1,\ldots,x_n)}$ the claim follows.

Let I be the kernel of $\mu : S \otimes_R S \to S$ ($a \otimes b \mapsto ab$). Then $\vartheta_N(S/R) = \mu(\text{Ann}_{S\otimes_R S}I)$ by the definition of ϑ_N and $\Omega^1_{S/R} \cong I/I^2$ by 1.21a). Therefore $\vartheta_N(S/R) \subset \text{Ann}\,\Omega^1_{S/R}$. By D.14 we have $(\text{Ann}\,\Omega^1_{S/R})^m \subset \vartheta_K(S/R)$, q.e.d.

10.19. <u>Corollary.</u> $\text{Ann}\,\Omega^1_{S/R}$, $\vartheta_K(S/R)$ and $\vartheta_N(S/R)$ have the same prime divisors, namely the ramified primes of S/R.

10.20. <u>Corollary</u>.   Let S/R be a finite projective algebra such that L := Q(S) is étale over K := Q(R). With $m := \mu(\Omega^1_{S/R})$ we then have

$$\vartheta_D(S/R)^m \subset \vartheta_K(S/R) \subset \vartheta_D(S/R).$$

Proof.   Under the present assumptions $\vartheta_D(S/R) = \vartheta_N(S/R)$ by G.11b), and the formula follows therefore from 10.18.

10.21. <u>Corollary</u> (Dedekind).   Under the assumptions of 10.20

$$\mathfrak{V}(\vartheta_D(S/R)) = V_{S/R}.$$

10.22. <u>Proposition</u>.   Let S/R be an algebra of finite type such that L := Q(S) is étale over K := Q(R). Then

$$\vartheta_K(S/R) \subset \vartheta_D(S/R)$$

hence the prime divisor of $\vartheta_D(S/R)$ are ramified.

Proof.   Follows from $\vartheta_N(S/R) \subset \vartheta_D(S/R)$ (G.11a) and 10.18.

The differents ϑ_K and ϑ_D are in general distinct (see exercise 4)). There is an example of Waldi [Wa] showing that this can happen already for "almost complete intersections".

Exercises

1) Let S/R be a ring extension of finite type where R and S are integral domains. Suppose $\{X_1,\ldots,X_n\}$ and $\{Y_1,\ldots,Y_n\}$ are systems of elements of S which are algebraically independent over R. Let $L := Q(S)$, $K := Q(R)$, $K_X := Q(R[X_1,\ldots,X_n])$ and $K_Y := Q(R[Y_1,\ldots,Y_n])$. Assume that

$$\Omega^1_{L/K_X} = \Omega^1_{L/K_Y} = 0.$$

Then:

a) $\Omega^1_{L/K} = L\,dX_1 \oplus \ldots \oplus L\,dX_n = L\,dY_1 \oplus \ldots \oplus L\,dY_n$.

b) $\vartheta^{(0)}(S/R[X]) = \Delta^X_Y \cdot \vartheta^{(0)}(S/R[Y])$

where $\Delta^X_Y \in L$ is defined by the equation

$$dX_1\ldots dX_n = \Delta^X_Y \cdot dY_1\ldots dY_n$$

in $\Omega^n_{L/K}$.

2) Let S/R and R/R_0 be algebras, and let $\delta : R_0 \to R_0\delta R_0$ be a derivation such that $\Omega^1_{S/\delta}$ and $\Omega^1_{R/\delta}$ are finitely generated. The product formula

$$\vartheta^{(0)}(S/R) \cdot \vartheta^{(0)}(R/\delta) = \vartheta^{(0)}(S/\delta)$$

holds, if one of the following conditions is satisfied:

a) S/R is finite and locally a complete intersection.

b) For each $\mathfrak{M} \in \mathrm{Max}(S)$ the canonical sequence

$$0 \to S_{\mathfrak{M}} \otimes_R \Omega^1_{R/L} \to \Omega^1_{S_{\mathfrak{M}}/L} \to \Omega^1_{S_{\mathfrak{M}}/R} \to 0$$

is exact and there is an isomorphism $\Omega^1_{S_{\mathfrak{M}}/R} \cong S^t_{\mathfrak{M}}/U$ where U is generated by t elements.

3) Let k be a field, $R := k[X_1,\ldots,X_n]/(F) = k[x_1,\ldots,x_n]$ a hypersurface ring, and $K := Q(R)$. Assume $\Omega^1_{K/k}$ is a free K-module of rank $n-1$, and let $T(\Omega^1_{R/k})$ denote the torsion of the

differential module $\Omega^1_{R/k}$. Show that

$$T(\Omega^1_{R/k}) \cong R : \vartheta^{(n-1)}(R/k)/R = R : \left(\frac{\partial F}{\partial x_1}, \ldots, \frac{\partial F}{\partial x_n}\right)/R$$

(Hint: Appendix D, exercise).

4) Let S/R be a finite Gorenstein algebra where R is noetherian and $L := Q(S)$ is étale over $K := Q(R)$. Show that $\vartheta_K(S/R) = \vartheta_D(S/R)$ if and only if S/R is locally a complete intersection (Use G.10 and 10.14).

§ 11. Universally Finite Differential Algebras

In previous sections we have seen how differential algebras and differential modules can be used to define invariants of algebras and prove structure theorems about them. Frequently it was necessary to impose finiteness conditions on the differential modules under consideration. But already for a simple ring like the power series ring R over a field K of characteristic O the differential module $\Omega^1_{R/K}$ is <u>not</u> finitely generated (5.5a)). Fortunately,a modification of $\Omega^1_{R/K}$, the "universally finite" differential module, which we will discuss here and in the following sections, allso to handle such rings like affine algebras.

Let R/R_O be an algebra and (Ω,δ) a differential algebra of R/R_O. We say that Ω is <u>finite</u>, if Ω is finitely generated as an R-module. For this it is sufficient that Ω^1 be finitely generated. Then $\Omega^P = O$ for $p > \mu(\Omega^1)$. A derivation $\delta : R \to M$ of R/R_O into an R-module M is called <u>finite</u>, if $R\delta R$ is finitely generated as an R-module. The differential algebra $\Omega = R \ltimes R\delta R$ associated with δ (2.5c) is then finite as well.

<u>11.1. Definition.</u> Ω is called <u>universally finite</u>, if Ω is finite and each finite differential algebra of R/R_O is a homomorphic image of Ω (with respect to an R-homomorphism).

If a universally finite differential algebra of R/R_O exists, it is unique up to a canonical R-isomorphism. It will then be denoted $\widetilde{\Omega}_{R/R_O}$.

That it need not exist in general can easily be seen:

<u>11.2. Example.</u> If R is a field, then $\tilde{\Omega}_{R/R_O}$ exists if and only if Ω_{R/R_O} is finite. In this case $\tilde{\Omega}_{R/R_O} = \Omega_{R/R_O}$.

We have only to realize that $\tilde{\Omega}_{R/R_O}$ does not exist, if Ω_{R/R_O} is not finite. In this case, for each $n \in \mathbb{N}$ there is a subspace U_n of the R-vector space Ω^1_{R/R_O} with $\dim \Omega^1_{R/R_O}/U_n = n$. Let I_n be the ideal of Ω_{R/R_O} generated by U_n and dU_n, and put $\Omega_n := \Omega_{R/R_O}/I_n$. This is a finite differential algebra with $\dim(\Omega_n)^1 = n$. Therefore $\tilde{\Omega}_{R/R_O}$ cannot exist.

Before discussing the existence of $\tilde{\Omega}_{R/R_O}$, we first modify the notion of universal extension of a differential algebra or derivation.

<u>11.3. Definition.</u> Let $\rho : R \to S$ be a homomorphism of R_O-algebras. A <u>universally finite ρ-extension</u> of Ω is a pair $(\tilde{\Omega}_\rho, \tilde{\varphi}_\rho)$ where $\tilde{\Omega}_\rho$ is a finite differential algebra of S/R_O and $\tilde{\varphi}_\rho : \Omega \to \tilde{\Omega}_\rho$ a ρ-homomorphism such that the following universal property holds: If $\psi : \Omega \to \Omega^*$ is an arbitrary ρ-homomorphism of Ω into a finite differential algebra Ω^* of S/R_O, then there is an S-homomorphism $h : \tilde{\Omega}_\rho \to \Omega^*$ (automatically satisfying the relation $\psi = h \circ \tilde{\varphi}_\rho$).

If Ω_ρ is finite, then clearly $\tilde{\Omega}_\rho$ exists and $\tilde{\Omega}_\rho = \Omega_\rho$. Moreover, if $\tilde{\Omega}_{R/R_O}$ exists, then $\tilde{\Omega}_{R/R_O}$ is the universally finite extension of the trivial differential algebra of R_O.

<u>11.4. Definition.</u> Let $\rho : R \to S$ be a homomorphism of R_O-algebras and $\delta : R \to R\delta R$ a derivation of R/R_O. An R_O-derivation $d : S \to N$ of S into an S-module N is called a <u>universally finite ρ-extension of δ</u>, if the following holds:

a) d is a ρ-extension of δ (1.24) and finite (i.e. SdS finitely generated).

b) If $\Delta : S \to N'$ is an arbitrary finite ρ-extension of δ, then there is exactly one S-linear map h : N $\to$ N' with $\Delta := h \circ d$.

If the universally finite ρ-extension d : S $\to$ N of δ exists, we write N =: $\tilde{\Omega}^1_{S/\delta}$ and call this the <u>universally finite module of differentials of S/δ</u>. In case δ is the trivial derivation of R we write $\tilde{\Omega}^1_{S/R}$ instead of $\tilde{\Omega}^1_{S/\delta}$.

About the existence of universally finite extensions we have the following general fact:

<u>11.5. Proposition.</u>   Let (Ω,δ) be a differential algebra of R/R_o and $\rho : R \to S$ a homomorphism of R_o-algebras. The following conditions are equivalent:

a) $\tilde{\Omega}_\rho$ exists.

b) Let $\{U_\lambda\}_{\lambda\in\Lambda}$ be the family of submodules of Ω^1_ρ such that Ω^1_ρ/U_λ is finitely generated, and put $U := \underset{\lambda\in\Lambda}{\cap} U_\lambda$. Then Ω^1_ρ/U is finitely generated as well.

c) The universally finite ρ-extension of the derivation $\delta : R \to \Omega^1$ exists.

If one of the conditions a)-c) is satisfied, then

$$\tilde{\Omega}_\rho = \Omega_\rho/I \quad \text{with } I := (U,dU)$$

and d : S $\to$ $\tilde{\Omega}^1_\rho$ is the universally finite ρ-extension of δ:

$$\tilde{\Omega}^1_{S/\delta} = \tilde{\Omega}^1_\rho .$$

Proof.   a) $\to$ c). Let $\Delta : S \to N$ be a finite ρ-extension of δ and $\Omega^* := S \ltimes S\Delta S$ the differential algebra associated with it. There is a ρ-homomorphism $\psi : \Omega \to \Omega^*$, since Δ is an extension of δ. By the universal property of $\tilde{\Omega}_\rho$ there is also an S-homomorphism $h : \tilde{\Omega}_\rho \to \Omega^*$ with $\psi = h \circ \varphi_\rho$. Let $\ell := h^1$ be the restriction of h to $\tilde{\Omega}_\rho^1$. Then $\Delta = \ell \circ d$, where d is the differentiation of $\tilde{\Omega}_\rho$. Since $\tilde{\Omega}_\rho^1 = SdS$, only one linear map ℓ with $\Delta = \ell \circ d$ can exists. Therefore $d : S \to \tilde{\Omega}_\rho^1$ is a universally finite ρ-extension of δ.

c) $\to$ b). We have $\Omega_{S/\delta}^1 = \Omega_\rho^1$ and $\tilde{\Omega}_{S/\delta}^1 = \Omega_{S/\delta}^1/V$ for some submodule V of $\Omega_{S/\delta}^1$. By the universal property of $\tilde{\Omega}_{S/\delta}^1$ it follows that $V \subset U_\lambda$ for all λ and hence $V \subset U$. But then Ω_ρ^1/U is finitely generated too.

b) $\to$ a). We wish to show that under the assumption b) Ω_ρ/I with $I := (U, dU)$ satisfies the universal property of 11.3. It is clear that Ω_ρ/I is a finite differential algebra and the canonical map $\varphi : \Omega \to \Omega_\rho/I$ a ρ-homomorphism.

Given a finite differential algebra Ω^* of S/R_O and a ρ-homomorphism $\psi : \Omega \to \Omega^*$, there is an S-homomorphism $h : \Omega_\rho \to \Omega^*$. Since $(\Omega^*)^1$ is finite, we have $U \subset \ker h$, and consequently there is an induced S-homomorphism $\Omega_\rho/I \to \Omega^*$.

The proposition implies that $\tilde{\Omega}_{R/R_O}$ exists if and only if $\tilde{\Omega}_{R/R_O}^1$ exists (in which case $\tilde{\Omega}_{R/R_O}^1 = (\tilde{\Omega}_{R/R_O})^1$). Moreover, we can deduce that $\tilde{\Omega}_\rho$ has the following stronger universal property:

<u>11.6. Corollary.</u> Assume that $\tilde{\Omega}_\rho$ exists. Let $\sigma : S \to T$ be a homomorphism of R_O-algebras and $\psi : \Omega \to \Omega^*$ a $\sigma \circ \rho$ -homomorphism of Ω into a differential algebra (Ω^*, d^*) of T/R_O. Suppose that Sd^*S is a finite S-module. Then there is a σ-homomorphism $\tilde{\Omega}_\rho \to \Omega^*$.

This follows from the fact that the kernel of the canonical map $\Omega_\rho \to \Omega^*$ contains the ideal I defined in 11.5.

11.7. Corollary   (Functorial property of $\tilde{\Omega}_\rho$).
Suppose that under the assumptions of 11.6 the differential
algebra $\tilde{\Omega}_{\sigma \circ \rho}$ exists, and let d_T be its differentiation.
There is a σ-homomorphism

$$\tilde{\Omega}_\rho \to \tilde{\Omega}_{\sigma \circ \rho}$$

if and only if $Sd_T S$ is a finite S-module.

11.8. Corollary.   Let $\rho : R \to S$ be a homomorphism of R_0-
algebras. Assume that $\tilde{\Omega}_{R/R_0}$ and $\tilde{\Omega}_{S/R_0}$ exist, and let d_S be
the differentiation of $\tilde{\Omega}_{S/R_0}$. There is a ρ-homomorphism

$$\tilde{\Omega}_{R/R_0} \to \tilde{\Omega}_{S/R_0}$$

if and only if $Rd_S R$ is a finite R-module.

The assumption about $Rd_S R$ is satisfied, for example, if
R is noetherian and S a finite R-module, or if S is a homomorphic image of R.

Sometimes the existence of a universally finite extension follows from the following transitive law:

11.9. Proposition.   Let $\rho : R \to S$ and $\sigma : S \to T$ be homomorphisms of R_0-algebras, Ω a differential algebra of R/R_0
and assume $\tilde{\Omega}_\rho$ exists.
a) If S is noetherian and T is a finite S-module, then $\tilde{\Omega}_{\sigma \circ \rho}$
exists and

$$\tilde{\Omega}_{\sigma \circ \rho} = (\tilde{\Omega}_\rho)_\sigma$$

the universal σ-extension of $\tilde{\Omega}_\rho$.
b) If $T = S/I$ for some ideal I of S and σ is the canonical
epimorphism, then $\tilde{\Omega}_{\sigma \circ \rho}$ exists and $\tilde{\Omega}_{\sigma \circ \rho} = \tilde{\Omega}_\rho/(I, dI)$.

Proof. Let $\psi : \Omega \to \Omega^*$ be a $\sigma \circ \rho$ -homomorphism of Ω into a finite differential algebra (Ω^*, d^*) of T/R_O. Under the assumptions of a) the S-module Sd^*S is finite. By 11.6 there is a σ-homomorphism $\tilde{\Omega}_\rho \to \Omega^*$ and hence a T-homomorphism $(\tilde{\Omega}_\rho)_\sigma \to \Omega^*$. Similarly, under the assumption of b), we obtain $\tilde{\Omega}_{\sigma \circ \rho} = (\tilde{\Omega}_\rho)_\sigma$, and the formula of b) now follows from 4.12.

<u>11.10. Corollary.</u>   Under the assumptions of 11.9b) let $\delta : R \to R\delta R$ be an R_O-derivation for which $\tilde{\Omega}^1_{S/\delta}$ exists. Then $\tilde{\Omega}^1_{S/I/\delta}$ exists and the sequence

$$I/I^2 \overset{\alpha}{\to} \tilde{\Omega}^1_{S/\delta}/I\tilde{\Omega}^1_{S/\delta} \overset{\beta}{\to} \tilde{\Omega}^1_{S/I/\delta} \to 0$$

is exact. Here β is induced by the functorial map $\tilde{\Omega}^1_{S/\delta} \to \tilde{\Omega}^1_{S/I/\delta}$ and α is defined as in 4.17.

The proof is analogous to that of 4.17.

The existence of $\tilde{\Omega}_{R/R_O}$ implies the existence of $\tilde{\Omega}_{R/\mathfrak{m}/R_O}$ for each $\mathfrak{m} \in \mathrm{Max}(R)$. Using 11.2 we obtain

<u>11.11. Corollary.</u>   If $\tilde{\Omega}_{R/R_O}$ exists, then $\Omega_{R/\mathfrak{m}/R_O}$ is a finite-dimensional vector space over $R/\mathfrak{m}$ for all $\mathfrak{m} \in \mathrm{Max}(R)$.

There is a local-global-principle for the existence of $\tilde{\Omega}_{R/R_O}$ (due to Bingener):

<u>11.12. Theorem.</u>   Suppose R is noetherian and (Ω, d) is a differential algebra of R/R_O. Then $\Omega = \tilde{\Omega}_{R/R_O}$ if and only if $\Omega_\mathfrak{m} = \tilde{\Omega}_{R_\mathfrak{m}/R_O}$ for all $\mathfrak{m} \in \mathrm{Max}(R)$.

Proof. Write $\Omega = \Omega_{R/R_O}/I$ with a homogeneous differentially closed ideal $I \subset \Omega_{R/R_O}$. Let Ω^* be a finite differential algebra of R/R_O and $\alpha : \Omega_{R/R_O} \to \Omega*$ the canonical map. If $\Omega_\mathfrak{m} = \Omega_{R_\mathfrak{m}/R_O}/I_\mathfrak{m}$ is the universally finite differential algebra of $R_\mathfrak{m}/R_O$, then $I_\mathfrak{m}$ is contained in the kernel of $\alpha_\mathfrak{m} : \Omega_{R_\mathfrak{m}/R_O} \to \Omega_\mathfrak{m}*$. If this is so for all $\mathfrak{m} \in \mathrm{Max}(R)$, then $I \subset \ker \alpha$ and α induces an R-homomorphism $\Omega \to \Omega*$, hence $\Omega = \tilde{\Omega}_{R/R_O}$.

Conversely, assume $\Omega = \tilde{\Omega}_{R/R_O}$, and let $\mathfrak{m}$ be a maximal ideal of R. To prove $\Omega_\mathfrak{m} = \tilde{\Omega}_{R_\mathfrak{m}/R_O}$ it suffices by 11.5 to show that $d_\mathfrak{m} : R_\mathfrak{m} \to \Omega_\mathfrak{m}^1$ is a universally finite derivation of $R_\mathfrak{m}/R_O$. Let $\delta : R_\mathfrak{m} \to N$ be an arbitrary finite derivation of $R_\mathfrak{m}/R_O$ with $N = R_\mathfrak{m}\delta R_\mathfrak{m}$. For all $i \in \mathbb{N}$ we have $R_\mathfrak{m}/\mathfrak{m}^i R_\mathfrak{m} = R/\mathfrak{m}^i$ and the $R/\mathfrak{m}^i$-module $N/\mathfrak{m}^i N$ is finite, hence also a finite R-module. Let δ_i be the composition $R \to R_\mathfrak{m} \xrightarrow{d_\mathfrak{m}} N \to N/\mathfrak{m}^i N$. By the universal property of d there is an R-linear map $h_i : \Omega^1 \to N/\mathfrak{m}^i N$ with $\delta_i r = h_i(dr)$ for all $r \in R$.

h_i induces an $R_\mathfrak{m}$-linear map $h_i' : \Omega_\mathfrak{m}^1 \to N/\mathfrak{m}^i N$. Let $h' : \Omega_\mathfrak{m}^1 \to \prod_{i\in\mathbb{N}} N/\mathfrak{m}^i N$ be the map with $h'(d_\mathfrak{m} r) = (h_i'(d_\mathfrak{m} r))_{i\in\mathbb{N}}$ for all $r \in R$ and $j : N \to \prod_{i\in\mathbb{N}} N/\mathfrak{m}^i N$ the map induced by the canonical epimorphisms $N \to N/\mathfrak{m}^i N$. Since $\bigcap_{i\in\mathbb{N}} \mathfrak{m}^i N = O$ by Krull's intersection theorem, j is injective. Moreover, the diagram

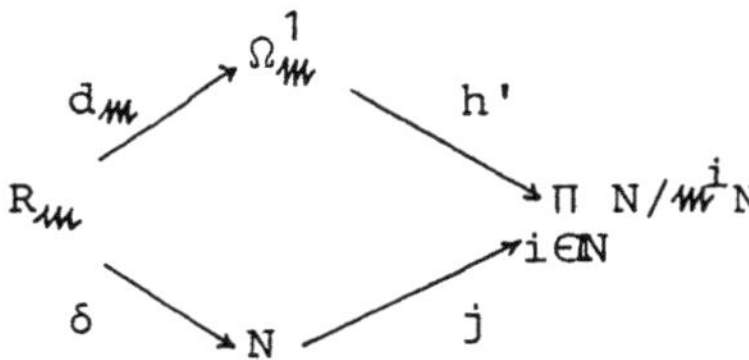

is commutative, hence $h'(\Omega_\mathfrak{m}^1) \subset j(N)$ and h' induces an $R_\mathfrak{m}$-linear map $h : \Omega_\mathfrak{m}^1 \to N$ with $\delta = h \circ d_\mathfrak{m}$, q.e.d.

Concerning the existence of $\tilde{\Omega}_{R/R_0}$ for semilocal rings R we have the following criterion.

<u>11.13. Proposition.</u> Let R/R_0 be an algebra, where R is a noetherian semilocal ring with Jacobson radical $\mathfrak{m}$. Then $\tilde{\Omega}_{R/R_0}$ exists if and only if $\Omega^1_{R/R_0} / \underset{i \in \mathbb{N}}{\cap} \mathfrak{m}^i \Omega^1_{R/R_0}$ is a finite R-module. In this case

$$\tilde{\Omega}_{R/R_0} = \Omega_{R/R_0} / \underset{i \in \mathbb{N}}{\cap} \mathfrak{m}^i \Omega_{R/R_0} .$$

Proof. Let U_λ be a submodule of Ω^1_{R/R_0} such that $M := \Omega^1_{R/R_0} / U_\lambda$ is a finite R-module. Then $\underset{i \in \mathbb{N}}{\cap} \mathfrak{m}^i M = 0$ by Krull's intersection theorem, hence $\underset{i \in \mathbb{N}}{\cap} \mathfrak{m}^i \Omega^1_{R/R_0} \subset U_\lambda$ and $\underset{i \in \mathbb{N}}{\cap} \mathfrak{m}^i \Omega^1_{R/R_0}$ is contained in the intersection U of all the U_λ. Suppose now that $\Omega^1_{R/R_0} / \underset{i \in \mathbb{N}}{\cap} \mathfrak{m}^i \Omega^1_{R/R_0}$ is a finite R-module. Then $U = \underset{i \in \mathbb{N}}{\cap} \mathfrak{m}^i \Omega^1_{R/R_0}$ and $\tilde{\Omega}_{R/R_0}$ exists by 11.5. Since $dU \subset \underset{i \in \mathbb{N}}{\cap} \mathfrak{m}^i \Omega_{R/R_0}$ we have (again by 11.5) $\tilde{\Omega}_{R/R_0} = \Omega_{R/R_0} / (U,dU)$ $= \Omega_{R/R_0} / \underset{i \in \mathbb{N}}{\cap} \mathfrak{m}^i \Omega_{R/R_0} .$

Conversely, assume now that $\tilde{\Omega}_{R/R_0}$ exists. Let $\mathfrak{m}_1, \ldots, \mathfrak{m}_t$ be the maximal ideals of R. Then $R/\mathfrak{m} = R/\mathfrak{m}_1 \times \ldots \times R/\mathfrak{m}_t$ by the Chinese remainder theorem. Put $\mathfrak{k}_i := R/\mathfrak{m}_i$ (i=1,...,t). By 11.11, $\Omega^1_{\mathfrak{k}_i/R_0}$ is a finite $\mathfrak{k}_i$-vector space, hence $\Omega^1_{R/\mathfrak{m}/R_0} = \Omega^1_{\mathfrak{k}_1/R_0} \times \ldots \times \Omega^1_{\mathfrak{k}_t/R_0}$ is a finite R-module. The exact sequence $\mathfrak{m}/\mathfrak{m}^2 \to \Omega^1_{R/R_0} / \mathfrak{m}\Omega^1_{R/R_0} \to \Omega^1_{R/\mathfrak{m}/R_0} \to 0$ shows that $\Omega^1_{R/R_0} / \mathfrak{m}\Omega^1_{R/R_0}$ is a finite R-module, hence so is $\Omega^1_{R/R_0} / \mathfrak{m}^i \Omega^1_{R/R_0}$ for all $i \in \mathbb{N}$. This implies that $\Omega^1_{R/\mathfrak{m}^i/R_0}$ is a finite R-module: $\Omega^1_{R/\mathfrak{m}^i/R_0} = \tilde{\Omega}^1_{R/\mathfrak{m}^i/R_0}$ ($i \in \mathbb{N}$).

The canonical map $\Omega^1_{R/R_O} \to \tilde{\Omega}^1_{R/R_O}$ induces a surjective R-linear map $\alpha : \Omega^1_{R/R_O} / \bigcap_{i \in \mathbb{N}} \mathfrak{m}^i \Omega^1_{R/R_O} \to \tilde{\Omega}^1_{R/R_O}$. We wish to show that α is an isomorphism. We write $\Omega := \Omega^1_{R/R_O} / \bigcap_{i \in \mathbb{N}} \mathfrak{m}^i \Omega^1_{R/R_O}$ and take $\omega \in \Omega$, $\omega \neq 0$. Then $\omega \notin \mathfrak{m}^n \Omega$ for some $n \in \mathbb{N}$. Consider the commutative diagram of canonical maps

$$
\begin{array}{ccccc}
\Omega^1_{R/R_O} \longrightarrow \Omega & & \xrightarrow{\;\;\alpha\;\;} & & \tilde{\Omega}^1_{R/R_O} \\
& \searrow \quad \downarrow & & \downarrow & \\
& \Omega^1_{R/\mathfrak{m}^{n+1}/R_O} & = & \tilde{\Omega}^1_{R/\mathfrak{m}^{n+1}/R_O} & .
\end{array}
$$

Since $\Omega^1_{R/\mathfrak{m}^{n+1}/R_O} = \Omega^1_{R/R_O} / \mathfrak{m}^{n+1} \Omega^1_{R/R_O} + Rd\mathfrak{m}^{n+1}$ and $Rd\mathfrak{m}^{n+1} \subset \mathfrak{m}^n \Omega^1_{R/R_O}$, the image of ω in $\Omega^1_{R/\mathfrak{m}^{n+1}/R_O}$ is different from zero, hence $\alpha(\omega) \neq 0$.

11.14. Corollary. Suppose R is a local ring with maximal ideal $\mathfrak{m}$, and let $K := R/\mathfrak{m}$, $\mathfrak{g} := \mathfrak{m} \cap R_O$, and $K_O := k(\mathfrak{g})$. Assume $\tilde{\Omega}_{R/R_O}$ exists.

a) If K/K_O is separable, then the canonical sequence

$$
O \to \mathfrak{m}/\mathfrak{m}^2 + \mathfrak{g}R \to \tilde{\Omega}^1_{R/R_O} / \mathfrak{m}\tilde{\Omega}^1_{R/R_O} \to \Omega^1_{K/R_O} \to O
$$

is exact.

b) If char.$K =: p > 0$ and $\mathfrak{q} := \mathfrak{m} \cap R_O[R^p]$, then there is a canonical exact sequence

$$
O \to \mathfrak{m}/\mathfrak{m}^2 + \mathfrak{q}R \to \tilde{\Omega}^1_{R/R_O} / \mathfrak{m}\tilde{\Omega}^1_{R/R_O} \to \Omega^1_{K/R_O} \to O.
$$

Proof. By 11.13 we have $\tilde{\Omega}^1_{R/R_O} / \mathfrak{m}\tilde{\Omega}^1_{R/R_O} = \Omega^1_{R/R_O} / \mathfrak{m}\Omega^1_{R/R_O}$, and thus the sequences are those of 6.5 and 6.7.

__11.15. Corollary.__ Let R and S be noetherian semilocal R_o-algebras and $\varphi : R \to S$ an R_o-homomorphism that maps the radical $\mathcal{M}$ of R into the radical $\mathcal{N}$ of S. If $\tilde{\Omega}_{R/R_o}$ and $\tilde{\Omega}_{S/R_o}$ exist, there is a φ-homomorphism

$$\varphi^* : \tilde{\Omega}_{R/R_o} \to \tilde{\Omega}_{S/R_o}.$$

Proof. The φ-homomorphism $\Omega_{R/R_o} \to \Omega_{S/R_o}$ maps $\bigcap_{i \in \mathbb{N}} \mathcal{M}^i \Omega_{R/R_o}$ into $\bigcap_{i \in \mathbb{N}} \mathcal{N}^i \Omega_{S/R_o}$ and therefore induces φ^*.

__11.16. Corollary.__ If R is an artinian semilocal ring, then $\tilde{\Omega}_{R/R_o}$ exists if and only if Ω_{R/R_o} is finite.

We have $\mathcal{M}^i = 0$ for some $i \in \mathbb{N}$ and hence $\bigcap_{i \in \mathbb{N}} \mathcal{M}^i \Omega^1_{R/R_o} = 0$.

Finally let us look what happens with $\tilde{\Omega}_{R/R_o}$, if we enlarge the ground ring R_o.

__11.17. Proposition.__ Let $\rho : R'_o \to R$ be a homomorphism of R_o-algebras. If $\tilde{\Omega}_{R/R_o}$ exists, so does $\tilde{\Omega}_{R/R'_o}$, and

$$\tilde{\Omega}_{R/R'_o} \cong \tilde{\Omega}_{R/R_o} / (\{d\rho(a)\}_{a \in R'_o}).$$

In particular, there is a canonical exact sequence

$$R \otimes_{R'_o} \Omega^1_{R'_o/R_o} \to \tilde{\Omega}^1_{R/R_o} \to \tilde{\Omega}^1_{R/R'_o} \to 0.$$

Proof. A finite differential algebra Ω^* of R/R'_o is also a finite differential algebra of R/R_o. The canonical R-homomorphism $\tilde{\Omega}_{R/R_o} \to \Omega^*$ maps $I := (\{d\rho(a)\}_{a \in R'_o})$ onto 0 and induces therefore an R-homomorphism $\tilde{\Omega}_{R/R_o}/I \to \Omega^*$. This proves that $\tilde{\Omega}_{R/R'_o} = \tilde{\Omega}_{R/R_o}/I$. Looking at 1-forms we immediately obtain the exact sequence of 11.17.

For an algebra R/R_0 for which $\tilde{\Omega}^1_{R/R_0}$ exists, the Kähler differents $\vartheta^{(i)}(R/R_0)$ are defined to be the Fitting ideals $F_i(\tilde{\Omega}^1_{R/R_0})$ ($i \in \mathbb{N}$). This extends the definition given in § 10 to a larger class of algebras. Clearly many properties of $\tilde{\Omega}^1_{R/R_0}$ can be expressed in terms of the $\vartheta^{(i)}(R/R_0)$, as was shown in appendix D and §10.

Exercises

1) Let R/R_0 be an algebra for which $\tilde{\Omega}_{R/R_0}$ exists. Then for each <u>finitely generated</u> R-module N there is a canonical isomorphism

$$\mathrm{Der}_{R_0}(R,N) \cong \mathrm{Hom}_R(\tilde{\Omega}_{R/R_0},N).$$

In particular, for the tangent space at $\mathfrak{x} \in \mathrm{Spec}(R)$ we have

$$T_{\mathfrak{x}}(R/R_0) \cong \mathrm{Hom}_R(\tilde{\Omega}_{R/R_0},k(\mathfrak{x}))$$

and $\dim_{k(\mathfrak{x})} T_{\mathfrak{x}}(R/R_0) = \mu_{\mathfrak{x}}(\tilde{\Omega}_{R/R_0})$. (see §6, exercise 8). $\mathrm{Spec}(R) \to \mathbb{N}$ ($\mathfrak{x} \mapsto \dim_{k(\mathfrak{x})} T_{\mathfrak{x}}(R/R_0)$) is an upper semi-contineous function.

2) A module M over a ring R is called <u>prefinite</u>, if for $m_1, m_2 \in M$ with $m_1 \neq m_2$ there is a linear map $\ell : M \to N$ into a finite R-module N with $\ell(m_1) \neq \ell(m_2)$. A derivation $d : R \to M$ of an algebra R/R_0 into an R-module M is called prefinite, if $R\delta R$ is prefinite. Show that a "universally prefinite" derivation of R/R_0 always exists.

§ 12. Differential Algebras and Completion

The universally finite differential algebra exists under rather mild assumptions for certain complete algebras. We shall show this after having studied the behaviour of differential algebras under completion in general.

Let R/R_O be an algebra, (Ω, d) a differential algebra of R/R_O and I an ideal of R. Endow R and Ω with the I-topology.

12.1. Remark. $d : \Omega \to \Omega$ is a continuous map. If Ω is separated (i.e. $\bigcap_{n \in \mathbb{N}} I^n\Omega = 0$) and $\omega = \lim_{n \to \infty} \omega_n$ for a sequence $\{\omega_n\}_{n \in \mathbb{N}}$, $\omega_n \in \Omega$, then $d\omega = \lim_{n \to \infty} d\omega_n$.

This is clear, since $d(I^{n+1}\Omega) \subset I^n\Omega$ for all $n \in \mathbb{N}$.

Suppose now that R is a noetherian ring and Ω is a finite differential algebra of R/R_O. Let $\hat{R}$ and $\hat{\Omega} := \hat{R} \otimes_R \Omega$ denote the completions of R and Ω with respect to the I-topology. Extend d by continuity to a mapping $\hat{d} : \hat{\Omega} \to \hat{\Omega}$. Then $(\hat{\Omega}, \hat{d})$ becomes a differential algebra of $\hat{R}/R_O$. Clearly $\hat{\Omega} = \bigoplus_{n \in \mathbb{N}} \widehat{\Omega^n}$ with the completion $\widehat{\Omega^n} = \hat{R} \otimes_R \Omega^n$ of Ω^n, hence $\hat{\Omega}$ is a graded $\hat{R}$-algebra. By the continuity of addition, multiplication and scalar multiplication (with scalars from R_O) it follows immediately that $\hat{d}$ is an R_O-linear map of degree 1, satisfying the axioms a),c)-e) of 2.1. As for 2.1b), we observe that $\hat{\Omega}$, as an $\hat{R}$-algebra, is already generated by $\{\hat{d}r\}_{r \in R}$, because $\hat{\Omega} = \hat{R} \otimes_R \Omega$.

12.2. Definition. We call $(\hat{\Omega}, \hat{d})$ the <u>completion</u> of the (finite) differential algebra (Ω, d).

Obviously the canonical map $\Omega \to \hat{\Omega}$ is a γ-homomorphism of differential algebras, where $\gamma : R \to \hat{R}$ is the canonical homomorphism into the completion.

12.3. Remark. Let $\varepsilon_n : R \to R/I^n$ be the canonical epimorphism and $\Omega_{\varepsilon_n} = \Omega/(I^n, dI^n)$ the universal ε_n-extension of Ω. Then

$$\hat{\Omega} = \varprojlim_n \Omega_{\varepsilon_n}$$

the projective limit of the differential algebras Ω_{ε_n}.

We have $\hat{\Omega} = \hat{R} \otimes_R \Omega = \varprojlim \Omega/I^n\Omega$. Since $I^{n+1}\Omega \subset (I^{n+1}, dI^{n+1}) \subset I^n\Omega$, the last projective limit is the same as $\varprojlim \Omega_{\varepsilon_n}$. Clearly $\hat{d}$ can be identified with the differentiation of the projective limit of the Ω_{ε_n}.

We now show the existence of the universally finite extension for complete algebras in a special case.

12.4. Proposition. Let $\rho: R \to S$ be a homomorphism of noetherian R_0-algebras, I an ideal of S and $\hat{S}$ the completion of S with respect to the I-topology. Let $\gamma: S \to \hat{S}$ denote the canonical homomorphism and $\hat{\rho} := \gamma \circ \rho$. Assume Ω is a differential algebra of R/R_0, for which the universal ρ-extension Ω_ρ is a finite differential algebra. Then $\tilde{\Omega}_{\hat{\rho}}$ exists and

$$\tilde{\Omega}_{\hat{\rho}} = \widehat{\Omega_\rho}.$$

Proof. Clearly $\widehat{\Omega_\rho}$ is a finite differential algebra of $\hat{S}/R_0$. The composition $\Omega \to \Omega_\rho \to \widehat{\Omega_\rho}$ is a $\hat{\rho}$-homomorphism. Let Ω^* be a finite differential algebra of $\hat{S}/R_0$ and $\varphi : \Omega \to \Omega^*$ a $\hat{\rho}$-homomorphism. By the universal property of Ω_ρ there is a γ-homomorphism $\psi : \Omega_\rho \to \Omega^*$ such that

184

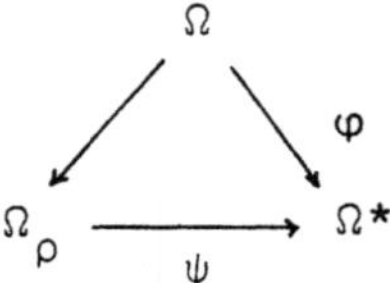

is commutative. Since Ω^* is a complete $\hat{S}$-module, ψ induces an $\hat{S}$-linear map $h : \hat{\Omega}_\rho \to \Omega^*$, for which

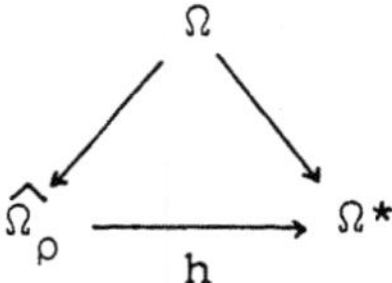

is commutative. By continuity, h is an $\hat{S}$-homomorphism of differential algebras. Therefore $\hat{\Omega}_\rho$ has the universal property of $\tilde{\Omega}_{\hat\rho}$.

12.5. Corollary. Let I be an ideal of the noetherain ring R and $\hat{R}$ the completion of R in the I-topology.

a) If Ω is a finite differential algebra of R/R_0, then $\tilde{\Omega}_{\hat{R}}$ exists and

$$\tilde{\Omega}_{\hat{R}} = \hat{\Omega}.$$

b) If Ω_{R/R_0} is finite, then $\tilde{\Omega}_{\hat{R}/R_0}$ exists and

$$\tilde{\Omega}_{\hat{R}/R_0} = \widehat{\Omega_{R/R_0}}.$$

a) follows from 12.4, if we put S = R, b) is a special case of a).

12.6. Corollary. Under the assumptions of 12.4 let Ω be a finite differential algebra of R/R_0. Let $\hat{R}$ be the completion of R with respect to the $(I \cap R)$-topology and $\sigma : \hat{R} \to \hat{S}$ the homomorphism induced by $\rho : R \to S$. Then $\tilde{\Omega}_{\hat\rho}$ is also the universally finite σ-extension of $\hat{\Omega}$.

Proof. $\Omega \to \Omega_\rho$ induces a σ-homomorphism $\hat{\Omega} \to \hat{\Omega}_\rho$, and one immediately checks that it has the universal property of a universally finite σ-extension of $\hat{\Omega}$.

12.7. Example. For $S := R[X_1,\ldots,X_n]$ and $I := (X_1,\ldots,X_n)$ we have $\hat{S} = R[\![X_1,\ldots,X_n]\!]$. If (Ω,δ) is a finite differential algebra of R/R_o, then Ω_S is a finite differential algebra of S/R_o, and hence $\tilde{\Omega}_{\hat{S}}$ exists:

$$\tilde{\Omega}_{\hat{S}} = \hat{\Omega}_S = \hat{S} \otimes_S \Omega_S = \bigoplus_{r=0}^{n} \bigoplus_{1 \le i_1 < \ldots < i_r \le n} (\hat{S} \otimes_R \Omega)\, dX_{i_1}\ldots dX_{i_r}.$$

The differentiation $\hat{d}$ of this algebra is the continuous extension of the differentiation d of Ω_S. For a power series

$$f = \Sigma r_{\nu_1\ldots\nu_n} X_1^{\nu_1}\ldots X_n^{\nu_n} \qquad (r_{\nu_1\ldots\nu_n} \in R)$$

the derivative $\hat{d}f$ is given by

$$\hat{d}f = \Sigma X_1^{\nu_1}\ldots X_n^{\nu_n} \otimes \delta r_{\nu_1\ldots\nu_n} + \sum_{i=1}^{n} \frac{\partial f}{\partial X_i}\, dX_i$$

where $\Sigma X_1^{\nu_1}\ldots X_n^{\nu_n} \otimes \delta r_{\nu_1\ldots\nu_n} \in \hat{S} \otimes_R \Omega$ denotes the limit of

the partial sums $\displaystyle\sum_{\nu_1+\ldots+\nu_n \le N} X_1^{\nu_1}\ldots X_n^{\nu_n} \otimes \delta r_{\nu_1\ldots\nu_n}$ and $\dfrac{\partial f}{\partial X_i}$

is the formal partial derivative of the series f.

In particular, if $\Omega = R$, then

$$\tilde{\Omega}_{\hat{S}/R} = \bigoplus_{r=0}^{n} \bigoplus_{1 \le i_1 < \ldots < i_r \le n} \hat{S}\, dX_{i_1}\ldots dX_{i_r}$$

and

$$\hat{d}f = \sum_{i=1}^{n} \frac{\partial f}{\partial X_i}\, dX_i \quad \text{for each } f \in \hat{S}.$$

This natural derivation, which was already introduced in 1.4, is the universally finite derivation of $\hat{S}/R$. As was mentioned in 5.5.a), the universal derivation of $\hat{S}/R$ need not be finite and is, moreover, difficult to describe.

The next two propositions make use of the following lemma.

<u>12.8. Lemma.</u>   Suppose R is separated and complete for the I-topology and $P \subset R$ a closed subring, $J := I \cap P$. Let M be an R-module, which is separated for the I-topology. Let $\omega_1, \ldots, \omega_n$ be elements of M, whose residue classes in M/JM generate this module as a P-module. Then $\omega_1, \ldots, \omega_n$ generate M as a P-module.

Proof.   By assumption $M = P\omega_1 + \ldots + P\omega_n + JM$. Each $m \in M$ can be written $m = \rho_{o1}\omega_1 + \ldots + \rho_{on}\omega_n + m_1$ with $\rho_{oj} \in P$ $(j=1,\ldots,n)$ and $m_1 \in JM$. By induction we have
$$m = (\sum_{i=o}^{k} \rho_{i1})\omega_1 + \ldots + (\sum_{i=o}^{k} \rho_{in})\omega_n + m_{k+1} \text{ with } \rho_{ij} \in J^i \text{ and}$$
$m_{k+1} \in J^{k+1}M$. Since $J^i \subset I^i$ the series $\sum_{i=o}^{\infty} \rho_{ij}$ converges to an element $\rho_j \in R$. Since P is closed in R, we have $\rho_j \in P$ $(j=1,\ldots,n)$. By construction $m-(\rho_1\omega_1+\ldots+\rho_n\omega_n) \in \bigcap_{i \in \mathbb{N}} I^iM = O$, whence $M = P\omega_1 + \ldots + P\omega_n$.

<u>12.9. Theorem.</u>   Let R/R_o be an algebra, where R is a noetherian semilocal ring with the maximal ideals $\mathfrak{m}_1, \ldots, \mathfrak{m}_t$. Let $\mathfrak{m} := \mathfrak{m}_1 \cap \ldots \cap \mathfrak{m}_t$ be the Jacobson radical of R, and assume R is complete for the $\mathfrak{m}$-topology. Let $\mathfrak{k}_i := R/\mathfrak{m}_i$ $(i=1,\ldots,t)$. Then $\tilde{\Omega}_{R/R_o}$ exists if and only if $\dim_{\mathfrak{k}_i} \Omega^1_{\mathfrak{k}_i/R_o} < \infty$ for $i=1,\ldots,t$.

Proof.   The existence of $\tilde{\Omega}_{R/R_o}$ implies by 11.11 that $\dim_{\mathfrak{k}_i} \Omega^1_{\mathfrak{k}_i/R_o} < \infty$ $(i=1,\ldots,t)$. Conversely, suppose now that these conditions are satisfied. Since R is complete, we have $R = R_1 \times \ldots \times R_t$, where R_i is the completion of R with respect to the $\mathcal{m}_i$-topology $(i=1,\ldots,t)$. The exact sequence

$$\mathcal{m}_i/\mathcal{m}_i^2 \to \Omega^1_{R_i/R_0}/\mathcal{m}_i\Omega^1_{R_i/R_0} \to \Omega^1_{k_i/R_0} \to 0 \quad \text{shows that}$$

$\Omega^1_{R_i/R_0}/\mathcal{m}_i\Omega^1_{R_i/R_0}$ is a finite k_i-vector space. This implies that

$$\Omega^1_{R/R_0}/\mathcal{m}\,\Omega^1_{R/R_0} = \Omega^1_{R_1/R_0}/\mathcal{m}_1\Omega^1_{R_1/R_0} \times \ldots \times \Omega^1_{R_t/R_0}/\mathcal{m}_t\Omega^1_{R_t/R_0}$$

is a finite R-module.

Let $\Omega := \Omega^1_{R/R_0}\big/\bigcap_{n\in\mathbb{N}}\mathcal{m}^n\Omega^1_{R/R_0}$ be the associated separated module of Ω^1_{R/R_0}. From 12.8 we conclude that Ω is finitely generated. Hence $\widetilde{\Omega}_{R/R_0}$ exists by 11.13.

<u>12.10. Corollary.</u> Let R be semilocal and noetherian with Jacobson radical $\mathcal{m}$, and let $\hat{R}$ be its completion with respect to the $\mathcal{m}$-topology. If $\widetilde{\Omega}_{R/R_0}$ exists, so does $\widetilde{\Omega}_{\hat{R}/R_0}$, and

$$\widetilde{\Omega}_{\hat{R}/R_0} = \widehat{\widetilde{\Omega}_{R/R_0}}.$$

Proof. The existence of $\widetilde{\Omega}_{\hat{R}/R_0}$ follows from 11.11 and the theorem. Moreover, $\widetilde{\Omega}_{\hat{R}/R_0} = \Omega_{\hat{R}/R_0}\big/\bigcap_{n\in\mathbb{N}}\mathcal{m}^n\Omega_{\hat{R}/R_0}$ by 11.13. This $\hat{R}$-module is finite and complete, hence $\widetilde{\Omega}_{\hat{R}/R_0} =$

$$\varprojlim \Omega_{\hat{R}/R_0}/\mathcal{m}^{n+1}\Omega_{\hat{R}/R_0} = \varprojlim \Omega_{\hat{R}/\mathcal{m}^{n+1}\hat{R}/R_0} = \varprojlim \Omega_{R/\mathcal{m}^{n+1}/R_0}$$

$$= \varprojlim \widetilde{\Omega}_{R/\mathcal{m}^{n+1}/R_0} = \widehat{\widetilde{\Omega}_{R/R_0}}.$$

For complete semilocal rings R of prime characteristic $\widetilde{\Omega}_{R/R_0}$ is a universal differential algebra of R over an enlarged ground ring:

<u>12.11. Proposition.</u> Let R/R_0 be an algebra, where R is a complete semilocal noetherian ring of prime characteristic p. Let $\overline{R_0[R^p]}$ denote the topological closure of the subring $R_0[R^p]$ in R. Assume $\widetilde{\Omega}_{R/R_0}$ exists. Then $\widetilde{\Omega}_{R/R_0} = \Omega_{R/\overline{R_0[R^p]}}$.

Proof. There is a topological isomorphism $R \cong R_1 \times \ldots \times R_t$, where the R_i are complete local rings. The image of $\overline{R_o[R^p]}$ under this isomorphism is the product of the $\overline{R_o[R_i^p]}$. We shall show that R_i is a finite $\overline{R_o[R_i^p]}$-module $(i=1,\ldots,t)$ and consequently R a finite $R_o[R^p]$-module. By 11.17 we know that $\tilde{\Omega}_{R/R_o} = \tilde{\Omega}_{R/R_o[R^p]}$ and by continuity $\tilde{\Omega}_{R/R_o[R^p]} = \tilde{\Omega}_{R/\overline{R_o[R^p]}} = \Omega_{R/\overline{R_o[R^p]}}$, the last equality following from the finiteness of $R/\overline{R_o[R^p]}$.

The existence of $\tilde{\Omega}_{R/R_o}$ implies that $\Omega^1_{\mathscr{k}_i/R_o}$ is a finite $\mathscr{k}_i$-vector space, where $\mathscr{k}_i$ is the residue class field of R_i $(i=1,\ldots,t)$. $\overline{R_o[R_i^p]}$ is a local ring, its image $\mathscr{k}_o^{(i)}$ in $\mathscr{k}_i$ contains $\mathscr{k}_i^p$, and $\Omega^1_{\mathscr{k}_i/R_o} = \Omega^1_{\mathscr{k}_i/\mathscr{k}_o^{(i)}}$. Since this vector space is finite dimensional, we have $[\mathscr{k}_i : \mathscr{k}_o^{(i)}] < \infty$. If $\mathscr{w}_i$ is the maximal ideal of $\overline{R_o[R_i^p]}$, then $R_i/\mathscr{w}_i R_i$ is a finite vector space over $\mathscr{k}_o^{(i)}$. From 12.8 we obtain that R_i is indeed a finite module over $\overline{R_o[R_i^p]}$.

12.12. Example. Let K be a field of characteristic $p > 0$ and $R = K[\![X_1,\ldots,X_n]\!]$ the ring of formal power series in $X_1,\ldots,X_n$ over K. For a subfield $K_o \subset K$ with $K^p \subset K_o$ we have $\overline{K_o[R^p]} = K_o[\![X_1^p,\ldots,X_n^p]\!]$, the subring of R consisting of all series in $X_1^p,\ldots,X_n^p$ with coefficients in K_o.

If $[K:K_o] < \infty$, then $\tilde{\Omega}_{R/K_o}$ exists by 12.9 and

$$\tilde{\Omega}_{R/K_o} = \Omega_{R/\overline{K_o[R^p]}}$$

by 12.11. With a p-basis $\{\alpha_1,\ldots,\alpha_m\}$ of K/K_o $\{\alpha_1,\ldots,\alpha_m\} \cup \{X_1,\ldots,X_n\}$ is a p-basis of $K[\![X_1,\ldots,X_n]\!]$ over $K_o[\![X_1^p,\ldots,X_n^p]\!]$. Therefore $\tilde{\Omega}^1_{R/K_o}$ is a free R-module with the basis $\{d\alpha_1,\ldots,d\alpha_m,dX_1,\ldots,dX_n\}$ and $\tilde{\Omega}_{R/K_o}$ is the exterior algebra of this module.

By Cohen's structure theorem any complete noetherian local ring $(R, \mathcal{M})$ is a homomorphic image of a power series ring $K[X_1, \ldots, X_n]$, where K is a "Cohen ring", that is, a field or a complete discrete valuation ring, whose maximal ideal is generated by a prime number p (the characteristic of the residue field k of K). For a subring $K_o \subset K$ the universally finite differential algebras $\tilde{\Omega}_{R/K_o}$ and $\tilde{\Omega}_{K[X_1, \ldots, X_n]/K_o}$ exist if and only if $\dim_k \Omega^1_{k/K_o} < \infty$ (12.9).

If we write $R = K[X_1, \ldots, X_n]/I$ with an ideal I of $K[X_1, \ldots, X_n]$, then

12.13.
$$\tilde{\Omega}_{R/K_o} = \tilde{\Omega}_{K[X_1, \ldots, X_n]/K_o}/(I, dI)$$

by 11.9. In case K is a field or $K_o = K$, the structure of $\tilde{\Omega}_{K[X_1, \ldots, X_n]/K_o}$ is very simple (12.7 and 12.12). For general K_o the differential algebra $\tilde{\Omega}_{K[X_1, \ldots, X_n]/K_o}$ is much more complicated, as can be seen from the following statements.

12.14. <u>Proposition.</u>   Let K be a Cohen ring, which is not a field, and $S := K[X_1, \ldots, X_n]$. Let $\mathcal{M} = (p)$ be the maximal ideal of K and $k := K/\mathcal{M}$. Assume that $K_o \subset K$ is a subring, which is local with maximal ideal $\mathcal{M}_o = K_o \cdot p$ and such that $\dim \Omega^1_{k/k_o} < \infty$, where $k_o := K_o/\mathcal{M}_o$. Then

a) $\mu(\tilde{\Omega}^1_{S/K_o}) = n + \dim_k \Omega^1_{k/k_o}$.

b) If k/k_o is finitely generated and $L := Q(S)$, then

$$\dim_L (L \otimes_S \tilde{\Omega}^1_{S/K_o}) = n + \mathrm{Trdeg}(k/k_o).$$

Proof. a) By 11.14b) there is an exact sequence

$$0 \to \mathfrak{m}/\mathfrak{m}^2 + \mathfrak{y}S \to \tilde{\Omega}^1_{S/K_o}/\mathfrak{m}\tilde{\Omega}^1_{S/K_o} \to \Omega^1_{k/k_o} \to 0$$

where $\mathfrak{m} := (p, X_1, \ldots, X_n)$ is the maximal ideal of S and $\mathfrak{y} := \mathfrak{m} \cap K_o[S^p]$. Write $\overline{S} := S/pS = k[\![X_1, \ldots, X_n]\!]$ and let $\overline{\mathfrak{m}} := (X_1, \ldots, X_n)$ denote the maximal ideal of $\overline{S}$. Under the canonical epimorphism $S \to \overline{S}$ the ideal $\mathfrak{y}$ is mapped into $\overline{\mathfrak{m}} \cap k_o[\overline{S}^p]$, hence $\mathfrak{y}S \subset (p, X_1^p, \ldots, X_n^p)$ and there is a vector space isomorphism $\mathfrak{m}/\mathfrak{m}^2 + \mathfrak{y}S \cong \overline{\mathfrak{m}}/\overline{\mathfrak{m}}^2$. The above exact sequence and Nakayama's lemma show that $\mu(\tilde{\Omega}_{S/K_o}) = n + \dim_k \Omega^1_{k/k_o}$.

b) Under the assumptions of b) let $\{\tau_1, \ldots, \tau_m\}$ be a transcendence basis of k/k_o and $\{t_1, \ldots, t_m\}$ a system of representatives of the τ_i in K. Since K_o is a discrete valuation ring, $\{t_1, \ldots, t_m\}$ is algebraically independent over K_o. Let $\mathfrak{z}$ be the prime ideal generated by p in $K_o[t_1, \ldots, t_m]$ and $K' := K_o[t_1, \ldots, t_m]_{\mathfrak{z}}$. This is a subring of K, over which K is a finite module, since the corresponding extension of residue fields is finite. Therefore S is a finite module over $S' := K'[\![X_1, \ldots, X_n]\!]$. S' is the completion of $R := K_o[t_1, \ldots, t_m]_{\mathfrak{z}}[X_1, \ldots, X_n]$ with respect to $I := (p, X_1, \ldots, X_n)$. By 12.5b) we have $\tilde{\Omega}_{S'/K_o} = S' \otimes_R \Omega_{R/K_o}$, therefore $\tilde{\Omega}_{S'/K_o}$ is a free S'-module of rank $n+m$.

L is a finite extension of $L' := Q(S')$, and L/L' is separable, since K is a ring of characteristic zero. By 11.9a) $\tilde{\Omega}_{S/K_o}$ is the universal S-extension of $\tilde{\Omega}_{S'/K_o}$, consequently $L \otimes_S \tilde{\Omega}_{S/K_o}$ is the universal L-extension of $L' \otimes_{S'} \tilde{\Omega}_{S'/K_o}$. Since L/L' is separable, $L \otimes_S \tilde{\Omega}_{S/K_o}$ has the same dimension as $L' \otimes_{S'} \tilde{\Omega}_{S'/K_o}$, namely $n+m = n + \mathrm{Trdeg}(k/k_o)$.

12.15. Corollary.   Under the assumptions of 12.14b) the following conditions are equivalent:

a) $\tilde{\Omega}^1_{S/K_o}$ is a free S-module.

b) k/k_o is separable.

Proof. Since S is local, $\tilde{\Omega}^1_{S/K_o}$ is free if and only if $\mu(\tilde{\Omega}^1_{S/K_o}) = \dim_L(L \otimes_S \tilde{\Omega}^1_{S/K_o})$. By 12.14 this equation holds if and only if $\dim_k \Omega^1_{k/k_o} = \mathrm{Trdeg}(k/k_o)$, that is, if and only if k/k_o is separable (5.10).

12.16. Corollary.   Let S be a complete regular local ring, whose residue field k is finitely generated over its prime field. Then the following conditions are equivalent:

a) $\tilde{\Omega}^1_{S/\mathbb{Z}}$ is a free S-module.

b) S is isomorphic to a formal power series algebra over a Cohen ring (i.e. absolutely unramified in the case of characteristics).

Proof. For equicharacteristic complete regular local rings both a) and b) are true. In the case of mixed characteristics b) implies a) by 12.15. Suppose now that char.S = 0, char.k =: $p > 0$, and S is not a power series algebra over a Cohen ring. Then $pS \subset \mathfrak{m}^2$, where $\mathfrak{m}$ is the maximal ideal of S. There is a presentation $S = K[\![X_1, \ldots, X_d]\!]/(p-F)$, where K is a Cohen ring and $F \in (X_1, \ldots, X_d)^2$. Then

$$\tilde{\Omega}^1_{S/\mathbb{Z}} = S \otimes \tilde{\Omega}^1_{K[\![X_1, \ldots, X_d]\!]/\mathbb{Z}} / \langle 1 \otimes dF \rangle$$

and since $dF \in (X_1, \ldots, X_d)\tilde{\Omega}^1_{K[\![X_1, \ldots, X_d]\!]/\mathbb{Z}}$, the S-module $\tilde{\Omega}^1_{S/\mathbb{Z}}$ is not free.

Exercises

1) Let R/R_0 be an algebra, where R is a noetherian semilocal ring with Jacobson-radical $\mathcal{W}$. Suppose R is complete with respect to the $\mathcal{W}$-topology and $\tilde{\Omega}^1_{R/R_0}$ exists. Let $R \hat{\otimes}_{R_0} R$ be the complete tensor product (i.e. $R \hat{\otimes}_{R_0} R$ is the completion of $R \otimes_{R_0} R$ with respect to $\mathcal{M} := \mathcal{W} \otimes_{R_0} R + R \otimes_{R_0} \mathcal{W}$). Let I be the kernel of the canonical map $R \hat{\otimes}_{R_0} R \to R$ (induced by $R \otimes_{R_0} R \to R$, $a \otimes b \mapsto ab$). Show that $\tilde{\Omega}^1_{R/R_0} \cong I/I^2$.

2) Let R/R_0 be an algebra, where R is a noetherian semilocal ring with Jacobson radical $\mathcal{W}$, and let $\hat{R}$ be the completion of R with respect to the $\mathcal{W}$-topology. Assume $\tilde{\Omega}^1_{R/R_0}$ exists. Then the different $\vartheta^{(i)}(\hat{R}/R_0)$ is defined (see end of §11), and

$$\vartheta^{(i)}(\hat{R}/R_0) = \hat{R} \otimes_R \vartheta^{(i)}(R/R_0) = \hat{R} \cdot \vartheta^{(i)}(R/R_0)$$

for all $i \in \mathbb{N}$.

3) Let K be a field and $R = K[\![X_1,\ldots,X_n]\!]/I$. Let x_i be the image of X_i in R, and let $\{F_1,\ldots,F_m\}$ be a system of generators of I. Then $\vartheta^{(i)}(R/K)$ is defined for all $i \in \mathbb{N}$ and is the ideal generated by all $(n-i)$-rowed minors of the Jacobian matrix

$$\left(\frac{\partial F_j}{\partial x_k}\right)_{\substack{j=1,\ldots,m \\ k=1,\ldots,n}}$$

§ 13. Differential Modules of Semianalytic Algebras

In this section k is a complete noetherian local ring whose maximal ideal is generated by $p \cdot 1$, where p is the characteristic of the residue field of k. In particular, k can be a Cohen ring (§ 12).

Let R/k be an algebra. R is called an <u>analytic k-algebra</u> (a <u>semianalytic k-algebra</u>), if there is a power series algebra $P = k⟦X_1, \ldots, X_n⟧$ and a k-homomorphism $\alpha : P \to R$ such that R/P is finite (essentially of finite type). R is called a <u>local analytic k-algebra</u>, if R is an analytic k-algebra and a local ring.

Any analytic k-algebra is a complete noetherian semi-local ring, and henceforth a direct product of finitely many complete noetherian local k-algebras (local analytic k-algebras) whose residue fields are finite over k.

Examples of analytic k-algebras are the homomorphic images of $k⟦X_1, \ldots, X_n⟧$. If R is any complete noetherian local k-algebra whose residue field is finite over k, and $\{x_1, \ldots, x_d\}$ is a system of parameters of R, then there is a k-homomorphism $\alpha : k⟦X_1, \ldots, X_d⟧ \to R$ ($\alpha(X_i) = x_i$) and R is finite over $k⟦X_1, \ldots, X_d⟧$ (12.8) , hence R is a local analytic k-algebra. If k is a field, then α is injective. In this case we call $k⟦X_1, \ldots, X_n⟧ \hookrightarrow R$ a <u>noetherian normalization</u> of the local analytic k-algebra R. The localizations of analytic k-algebras are examples of semianalytic k-algebras.

In this and the next section the statements about differential modules of affine algebras and their localizations (§ 7) will be generalized to the case of semiana-

lytic algebras. We first study some general properties of
such algebras (cf. [BKKN], §2).

13.1. <u>Proposition</u> (F.K.Schmidt).   Let $(R,\mathcal{M})$ be a complete
local (or more generally a henselian local) domain which
is not a field. Let L be a field with $R \subset L$ and $V \subset L$ a
discrete valuation ring with $Q(V) = L$. Then $R \subset V$.

Proof.   Let $p := \mathrm{char}(R/\mathcal{M})$. For each $y \in \mathcal{M}$ and each $n \in \mathbb{N}$
with $n \not\equiv 0 \bmod p$, the polynomial $X^n - (1+y) \in R[X]$ has a
zero $x \in R$ by Hensel's lemma. Let v be the normed discrete
valuation on L that belongs to V. Then $v(x) = \frac{1}{n} v(1+y)$.
Since n can be made arbitraryly large, this is only possible
if $v(1+y) = O$. Then $y \in V$, and it is shown that $\mathcal{M} \subset V$.

   For an arbitrary $z \in R$ consider yz^i with $y \in \mathcal{M} \setminus \{O\}$ and
$i \in \mathbb{N}$. If $v(z) < O$, then $v(yz^i) = v(y) + iv(z) < O$ for
large i. This contradicts what was shown above, since
$yz^i \in \mathcal{M}$. We conclude that $v(z) \geq O$, hence $R \subset V$.

13.2. <u>Proposition.</u>   Let R be a complete noetherian local
domain with quotient field K, and let L/K be a finitely
generated field extension. Then:
a) The integral closure $\overline{R}$ of R in L is finite over R.
b) If S is another complete local domain, not a field,
and if $S \subset L$, then $S \subset \overline{R}$.

Proof.   a) The algebraic closure $\overline{K}$ of K in L is finite
over K, and $\overline{R}$ is the integral closure of R in $\overline{K}$. Since R is
complete and noetherian, $\overline{R}/R$ is finite (Matsumura $[M_1]$, p.234).

b) By 13.1 we have $S \subset V$ for each discrete valuation ring V with $Q(V) = L$. We shall show that $\overline{R}$ is an intersection of such V, from which $S \subset \overline{R}$ follows.

If $\overline{R} \neq \overline{K}$, then $\overline{R}$, being a normal noetherian domain, is the intersection of all discrete valuation rings V_o with $\overline{R} \subset V_o$ and $Q(V_o) = \overline{K}$. Any such V_o is of the form $V_o = V \cap \overline{K}$ where V is a discrete valuation ring with $Q(V) = L$. If $\overline{R} = \overline{K}$, then, as is wellknown, $\overline{R}$ is the intersection of all discrete valuation rings V with $\overline{R} \subset V$ and $Q(V) = L$.

13.3. Theorem. Let S be a reduced semianalytic k-algebra, R an analytic k-algebra with $R \subset S$ such that S/R is essentially of finite type. Let $\overline{R}$ denote the integral closure of R in S. Then:

a) $\overline{R}/R$ is finite.

b) If P is an arbitrary analytic k-algebra and $\alpha : P \to S$ a k-homomorphism, then $\alpha(P) \subset \overline{R}$.

Proof. Let $\mathscr{p}_1, \ldots, \mathscr{p}_t$ be the minimal prime ideals of S, and write $S_i := S/\mathscr{p}_i$, $R_i := R/\mathscr{p}_i \cap R$ $(i=1,\ldots,t)$. Let $\overline{R}_i$ be the integral closure of R_i in S_i and $\overline{\overline{R}}_i$ the integral closure of R_i in $L_i := Q(S_i)$.

a) Since S is reduced, the canonical map $S \to \prod_{i=1}^{t} S_i$ is injective, and $R \subset \prod_{i=1}^{t} R_i$ where $\prod_{i=1}^{t} R_i$ is finite over R. By 13.2 $\overline{\overline{R}}_i/R_i$ is finite, therefore also $\overline{R}_i/R_i$ and $\prod_{i=1}^{t} \overline{R}_i/R$ are finite. $\prod \overline{R}_i$ is the integral closure of $\prod R_i$ in $\prod S_i$. We then have $\overline{R} \subset \prod \overline{R}_i$ and can conclude that $\overline{R}/R$ is finite.

b) For the image P_i of P in S_i we shall show that $P_i \subset \overline{R}_i$ $(i=1,\ldots,t)$. Then $\alpha(P) \subset \prod_{i=1}^{t} \overline{R}_i \cap S = \overline{R}$.

If P_i is a field, then P_i is necessarily isomorphic to one of the residue fields of P with respect to a maximal ideal. Then P_i is finite over the image k' of k in S_i. Since $k \subset R_i$, we conclude that $P_i \subset \overline{R_i}$.

If P_i is not a field, then P_i is finite over a complete noetherian local domain $P_i' \subset S_i$. By 13.2 we have $P_i' \subset \overline{R_i}$ and hence $P_i \subset \overline{R_i}$.

13.4. Corollary. Any reduced semianalytic k-algebra S contains a unique maximal analytic k-algebra A (i.e. all subalgebras of S/k that are analytic k-algebras are contained in A). If R is an arbitrary analytic k-algebra with $R \subset S$ such that S/R is essentially of finite type, then A is the integral closure of R in S.

Proof. If R is given as in the corollary, then its integral closure $\overline{R}$ in S is finite over R by 13.3a), and hence is an analytic k-algebra. If $R' \subset S$ is another analytic k-algebra, then $R' \subset \overline{R}$ by 13.3b), and $\overline{R}' \subset \overline{R}$. If in addition S/R' is essentially of finite type, then by symmetry we also have $\overline{R} \subset \overline{R}'$, and hence $\overline{R} = \overline{R}'$.

For a reduced semianalytic k-algebra S we denote by A(S) the maximal analytic subalgebra of S. If S is not reduced, such algebra need not exist (exercise 2). If S is a domain, then A(S) is a local domain, because A(S) is always a direct product of local rings.

13.5. Corollary. Let $\varphi : S \to S'$ be a homomorphism of reduced semianalytic k-algebras. Then $\varphi(A(S)) \subset A(S')$. Hence φ

induces a k-homomorphism

$$A(\varphi) : A(S) \to A(S').$$

This is an immediate consequence of 13.3b).

For each reduced semianalytic k-algebra S let $D_k(S)$ denote the universal S-extension of $\widetilde{\Omega}^1_{A(S)/k}$, the universally finite differential module of A(S)/k, whose existence is guaranteed by 12.9. The $D_k(S)$ will turn out to be the appropriate differential modules for studying semianalytic algebras.

If R is an arbitrary analytic k-algebra such that S is essentially of finite type over R, then by the transitive law for universal extension (3.17) $D_k(S)$ is also the universal S-extension of $\widetilde{\Omega}^1_{R/k}$. $D_k(S)$ is a functor of S: Under the assumptions of 13.5 there is an S-module homomorphism

$$D_k(\varphi) : D_k(S) \to D_k(S') \qquad (ds \mapsto d\varphi(s))$$

obtained from the functorial A(S)-linear map $\widetilde{\Omega}^1_{A(S)/k} \to \widetilde{\Omega}^1_{A(S')/k}$ by extension to S and S' (3.15).

In what follows we shall first study $D_k(S)$ in the case the semianalytic k-algebra S is a field. We write L for S, and we denote the quotient field of A(S) by K. L/K is then a finitely generated field extension. The image of k in L is a field or a complete discrete valuation ring (a Cohen ring). We may assume without loss of generality that $k \subset L$.

13.6. Proposition.   a) We always have

$$\dim_L D_k(L) \geq \dim_K D_k(K) + \mathrm{Trdeg}(L/K).$$

If L/K is separable, then the equality sign holds.

b) If k is not a field, then

$$\dim_L D_k(L) = \dim A(L) - 1 + \operatorname{Trdeg}(L/K).$$

c) If k is a field of characteristic O, then

$$\dim_L D_k(L) = \dim A(L) + \operatorname{Trdeg}(L/K).$$

d) If k is a field of characteristic p > O, then

$$\dim_L D_k(L) \geq \dim A(L) + \operatorname{Trdeg}(L/K).$$

Proof. Let $\delta : K \to D_k(K)$ be the derivation belonging to $D_k(K)$. By transitivity $D_k(L)$ is the universal L-extension of $D_k(K)$. Consider the exact sequence (defined in § 5)

$$(1) \qquad O \to T(L/\delta) \to L \otimes_K D_k(K) \to D_k(L) \to \Omega^1_{L/K} \to O.$$

If L/K is separable, then $T(L/\delta) = O$ by 5.12 and we conclude that $\dim_L D_k(L) = \dim_K D_k(K) + \operatorname{Trdeg}(L/K)$.

If $k \subset L$ is not a field, then k must be a discrete valuation ring whose maximal ideal is generated by a prime number p, the characteristic of the residue field of k. There is a system of parameters of A(L) which is of the form $\{p, x_1, \ldots, x_{d-1}\}$, where $d := \dim A(L)$. The k-homomorphism $\varphi : k[\![X_1, \ldots, X_{d-1}]\!] \to A(L)$ with $\varphi(X_i) = x_i$ $(i=1, \ldots, d-1)$ is injective, since A(L) is finite over $k[\![X_1, \ldots, X_{d-1}]\!]$ and both rings have the same dimension. K is a finite extension field of $Q(k[\![X_1, \ldots, X_{d-1}]\!])$ and

$$D_k(K) \cong K \otimes_{k[\![X_1, \ldots, X_{d-1}]\!]} \Omega^1_{k[\![X_1, \ldots, X_{d-1}]\!]/k}$$ is a K-vector space of dimension d-1. We obtain the formula in b).

Now let k be a field. Then A(L) is finite over $k[\![X_1, \ldots, X_d]\!]$, $d := \dim A(L)$, where $X_1, \ldots, X_d$ corresponds to a system of parameters of A(L). Write $K' := Q(k[\![X_1, \ldots, X_d]\!])$. If char.k = O, then K/K' is finite and separable, hence $D_k(K) \cong K \otimes_{K'} D_k(K')$ which is a vector space of dimension

$d = \dim A(L)$. We obtain the formula in c). If k is a field

of characteristic $p > 0$, then by 5.12b) we have

$$\dim_L T(L/\delta) \leq p\text{-deg}(L/K) - \mathrm{Trdeg}(L/K) = \dim_L \Omega^1_{L/K} - \mathrm{Trdeg}(L/K)$$

and the exact sequence (1) gives us the inequality in a).

If we apply (1) again, with L replaced by K, and K replaced

by K', we obtain

$$\dim_K D_k(K) \geq \dim A(L)$$

and henceforth the inequality in d).

We want to have some more precise information in case k

is a field of positive characteristic. Assume at first that

k is an arbitrary field. $k((X_1,\ldots,X_n))$ will always denote

the fraction field of the power series ring $k[\![X_1,\ldots,X_n]\!]$.

Let L/k be an extension of fields. L is called a _semianalytic_

extension field of k, if there is a k-homomorphism

$k((X_1,\ldots,X_n)) \to L$ such that $L/k((X_1,\ldots,X_n))$ is finitely

generated. If $L/k((X_1,\ldots,X_n))$ is finite, we call L an _ana-_

lytic extension field of k.

Semianalytic extension fields of k can be defined as

semianalytic k-algebras that are fields: If a semiana-

lytic k-algebra L is given, which is a field, and if

$\{x_1,\ldots,x_d\}$ is a system of parameters of A(L), then there

is an injection $k[\![X_1,\ldots,X_d]\!] \to A(L)$ $(X_i \mapsto x_i)$, and L is a

finitely generated extension field of $k((X_1,\ldots,X_d))$. Simi-

larly the analytic extension fields of k are the quotient

fields of analytic k-algebras which are domains. If L/k is

a semianalytic field extension, then Q(A(L)) is the unique

maximal analytic extension field of k contained in L.

13.7. <u>Definition</u>.   Let K/k be an analytic field extension. Suppose K is finite over $k((X_1,\ldots,X_n)) \subset K$. Then n is called the <u>analytic transcendence degree</u> of K/k:

$$n =: a\,\mathrm{Trdeg}(K/k)$$

and $\{X_1,\ldots,X_n\}$ is called an <u>analytic transcendence basis</u> of K/k. $\{X_1,\ldots,X_n\}$ is called <u>separating</u>, if $K/k((X_1,\ldots,X_n))$ is separable. K/k is called <u>analytically separable</u>, if K/k has a separating analytic transcendence basis.

Observe that the number n above is an invariant of K/k. For, if K is finite over $k((Y_1,\ldots,Y_m)) \subset K$, put $R := k[\![X_1,\ldots,X_n]\!]$ and $R' := k[\![Y_1,\ldots,Y_m]\!]$. Then $A(K)$ is (by 13.4) the integral closure of both R and R', hence

$$n = \dim R = \dim A(K) = \dim R' = m.$$

It is easy to see that $k((X_1,\ldots,X_n))/k$ is a separable field extension (§ 5, exercise 5)). By the transitivity of separability it follows that, if K/k is analytically separable, then K/k is separable as well. The converse is not true in general.

Nor is it true that a $\mathrm{Trdeg}(K'/k) \leq a\mathrm{Trdeg}(K/k)$ when K' is a subfield of K that is an analytic extension field of k. For any $n \in \mathbb{N}$ there is an injective k-homomorphism $k[\![X_1,\ldots,X_n]\!] \rightarrow k[\![X_1,X_2]\!]$ (Abhyankar [Ab]). Passing to the quotient fields gives a counter-example.

13.8. <u>Proposition</u>.   Let L/k be a semianalytic field extension and $K := Q(A(L))$.

a) We always have

$$\dim_L D_k(L) \geq a\mathrm{Trdeg}(K/k) + \mathrm{Trdeg}(L/K).$$

b) If K/k is analytically separable and L/K separable, the equality sign holds.

Proof. a) follows trivially from 13.6.

b) If $\{X_1,\ldots,X_d\}$ is a separating analytic transcendence basis of K/k, then $\{dX_1,\ldots,dX_d\}$ is a basis of $D_k(K)$ and hence b) follows from 13.6a).

Now let k be a field of characteristic $p > 0$ and $k_0 \subset k$ a subfield with $k^p \subset k_0$ and $[k:k_0] < \infty$. Each semianalytic k-algebra S is also a semianalytic k_0-algebra. If S is reduced, then $D_{k_0}(S)$ is defined. Suppose that S is essentially of finite type over the power series algebra $P=k[\![X_1,\ldots,X_d]\!]$. Since $D_{k_0}(S)$ is the universal extension of $\tilde{\Omega}^1_{P/k_0}$ and $\tilde{\Omega}^1_{P/k_0} = \Omega^1_{P/k_0}[\![X_1^p,\ldots,X_d^p]\!]$ by 12.12, we have

$$(2) \qquad D_{k_0}(S) = \Omega^1_{S/k_0}[\![X_1^p,\ldots,X_d^p]\!].$$

This fact will frequently be used in the sequel. We first show

<u>13.9. Proposition.</u>   Let L/k be a semianalytic field extension, $k_0 \subset k$ a subfield with $k^p \subset k_0$, $[k:k_0] < \infty$, and let $K := Q(A(L))$.

a) We always have

$$\dim_L D_{k_0}(L) \geq \mathrm{aTrdeg}(K/k) + \mathrm{Trdeg}(L/K) + p\text{-}\deg(k/k_0).$$

b) If $[k:k^p] < \infty$ and $k_0 := k^p$, the formula holds with the equality sign.

Proof.   Choose a power series algebra $k[\![X_1,\ldots,X_d]\!] \subset L$ over which L is essentially of finite type, put $K' := k((X_1,\ldots,X_d))$ and $K_0 := k_0((X_1^p,\ldots,X_d^p))$. Then $d = \mathrm{aTrdeg}(K/k)$ and $D_{k_0}(L) = \Omega^1_{L/K_0}$ by (2). For $k_0 = k^p$ we have $K_0 = K'^p$. Since $p\text{-}\deg(K'/K_0) = d+p\text{-}\deg(k/k_0) = \mathrm{aTrdeg}(K/k) + p\text{-}\deg(k/k_0)$, the proposition turns out to be a special case of 5.13.

In the case of <u>analytic</u> field extensions a more precise statement is possible.

<u>13.10. Theorem.</u>   For an analytic field extension K/k the following statements are equivalent:

a) K/k is analytically separable.

b) $\dim_K D_k(K) \leq \mathrm{aTrdeg}(K/k)$.

c) $\dim_K D_k(K) = \mathrm{aTrdeg}(K/k)$.

Over a perfect field k any analytic extension K/k is analytically separable.

Proof. By 13.9 it only remains to be shown that b) implies a).

Let $A := A(K)$, and let $\mathfrak{m}$ be the maximal ideal of A. There are $d := \dim A = \mathrm{aTrdeg}(K/k)$ elements $y_1,\ldots,y_d \in A$ with $D_k(K) = Kdy_1 +\ldots+ Kdy_d$. If $d = 0$, then K/k is algebraic, $\Omega^1_{K/k} = D_k(K) = 0$, and a) follows.

If $d > 0$, then $\mathfrak{m} \neq (0)$. For $a \in \mathfrak{m}\smallsetminus\{0\}$ we have $d(a^p y_i) = a^p dy_i$ $(i=1,\ldots,d)$, and therefore we may assume that $y_1,\ldots,y_d \in \mathfrak{m}$.

By changing the y_i successively we shall construct elements $y_1,\ldots,y_d \in \mathfrak{m}$ such that

α) $\{y_1,\ldots,y_d\}$ is a system of parameters of A,

β) $\{dy_1,\ldots,dy_d\}$ a system of generators of $D_k(K)$.

It then follows that $K/k((y_1,\ldots,y_d))$ is finite and $\Omega^1_{K/k((y_1,\ldots,y_d))} = 0$, hence K separably algebraic over $k((y_1,\ldots,y_d))$. This will show that K/k is analytically separable.

Suppose that $\dim A/(y_1,\ldots,y_\delta) = d-\delta$ for some $\delta < d$. (For $\delta = 0$ this is certainly the case). Write $\bar{A} := A/(y_1,\ldots,y_\delta)$ and denote the image of elements or ideals of A in $\overline{A}$ by an overbar. Let $\text{Min}(\overline{A}) = \{\overline{g}_1,\ldots,\overline{g}_s\}$ and suppose $\overline{y}_{\delta+1} \notin \bigcup_{i=1}^{\sigma} \overline{g}_i$, $\overline{y}_{\delta+1} \in \bigcap_{i=\sigma+1}^{s} \overline{g}_i$. Since $\dim \overline{A} > 0$, we have $\overline{m} \notin \text{Min}(\overline{A})$, and we can find $\overline{b} \in \bigcap_{i=1}^{\sigma} \overline{g}_i \cap \overline{m}$, $\overline{b} \notin \bigcup_{j=\sigma+1}^{s} \overline{g}_j$. Let $b \in m$ be a preimage of $\overline{b}$, and put $y'_{\delta+1} := y_{\delta+1} + b^p$. Then $dy'_{\delta+1} = dy_{\delta+1}$ and $\dim A/(y_1,\ldots,y_\delta,y'_{\delta+1}) = d-(\delta+1)$. By induction we obtain elements satisfying the conditions α) and β) above.

13.11. Definition. Suppose L/k is a semianalytic field extension and $K := Q(A(L))$. Let k_o be given as in 13.9. Then k_o is called <u>admissible</u> for L/k if
$$\dim_L D_{k_o}(L) = \text{aTrdeg}(K/k) + \text{Trdeg}(L/K) + p\text{-deg}(k/k_o).$$

Let $k[X_1,\ldots,X_d] \hookrightarrow A(K)$ be a noetherian normalization of $A(K)$, hence $d = \dim A(K) = \text{aTrdeg}(K/k)$. By (2) we have $D_{k_o}(L) = \Omega^1_{L/k_o((X_1^p,\ldots,X_d^p))}$. Moreover,
$$p\text{-deg}(k((X_1,\ldots,X_d))/k_o((X_1^p,\ldots,X_d^p))) = p\text{-deg}(k/k_o) + d.$$
Therefore k_o is admissible for L/k if and only if $k_o((X_1^p,\ldots,X_d^p))$ is admissible for the algebraic function field $L/k((X_1,\ldots,X_d))$ in the sense of 5.19b). By 5.22 this is the case if and only if $k_o((X_1^p,\ldots,X_d^p))$ and L^p are linearly disjoint over $k((X_1,\ldots,X_d))^p = k^p((X_1^p,\ldots,X_d^p))$. From this criterion it is seen that, if k_o is admissible for L/k, then so is any subfield $k'_o \subset k_o$ with $k^p \subset k'_o$ and $[k_o:k'_o] < \infty$.

13.12. Examples. a) If K/k is analytically separable and L/K separable, then k is admissible for L/k (13.8b)).

b) If $[k:k^p] < \infty$, then $k_o = k^p$ is admissible for any semianalytic field extension (13.9b)).

c) If K/k is an analytic field extension, then k is admissible for K/k if and only if K/k is analytically separable (13.10).

13.13. Theorem (Existence of admissible fields).

Let L/k be a semianalytic field extension and $k' \subset k$ a subfield with $k^p \subset k'$ and $[k:k'] < \infty$. Then there exists a subfield $k_o \subset k'$ which is admissible for L/k.

Proof. Let $\{X_1,\ldots,X_d\}$ be an analytic transcendence basis of Q(A(L)) over k and put $K := k((X_1,\ldots,X_d))$. Let $\mathfrak{Z}$ denote the set of all subfields k_α of k' with $k^p \subset k_\alpha$ and $[k':k_\alpha] < \infty$. Write $K_\alpha := k_\alpha((X_1^p,\ldots,X_d^p))$ for each $k_\alpha \in \mathfrak{Z}$. We shall show the existence of a $k_\alpha \in \mathfrak{Z}$ such that K_α and L^p are linearly disjoint over K^p. Then k_α is admissible for L/k.

Let $\{Y_1,\ldots,Y_t\}$ be a transcendence basis of L/K and $T := K(Y_1,\ldots,Y_t)$. It suffices to find $k_\alpha \in \mathfrak{Z}$ such that $K_\alpha[T^p]$ and L^p are linearly disjoint over T^p, because then

$$K_\alpha[L^p] \cong K_\alpha[T^p] \otimes_{T^p} L^p \cong (K_\alpha \otimes_{K^p} T^p) \otimes_{T^p} L^p \cong K_\alpha \otimes_{K^p} L^p$$

hence K_α and L^p are linearly disjoint over K^p as well.

For $k_\alpha \in \mathfrak{Z}$ let $v_\alpha := [K_\alpha[L^p]:K_\alpha[T^p]]$. Clearly $v_\alpha \le [L^p:T^p]$ and $v_{\alpha'} \le v_\alpha$ if $k_{\alpha'} \in \mathfrak{Z}$, $k_{\alpha'} \subset k_\alpha$. Choose $k_{\alpha_o} \in \mathfrak{Z}$ such that $v_{\alpha'} = v_{\alpha_o}$ for each $k_{\alpha'} \in \mathfrak{Z}$ with $k_{\alpha'} \subset k_{\alpha_o}$.

We shall prove that

$$(3) \qquad \bigcap_{k_{\alpha'} \subset k_{\alpha_o}} K_{\alpha'}[T^p] = T^p.$$

It is enough to show for any $a \in K_{\alpha_o}[T^p]$, $a \notin T^p$ the existence of a $k_{\alpha'} \in \mathcal{J}$ with $k_{\alpha'} \subset k_{\alpha_0}$ and a $\notin K_{\alpha'}[T^p]$.

Write $a = \frac{f}{g}$ with $f,g \in k_{\alpha_0}[\![X_1^p,\ldots,X_d^p]\!][Y_1^p,\ldots,Y_t^p]$. It

suffices to show that $ag^p \notin K_{\alpha'}[T^p]$ for a suitable $k_{\alpha'} \subset k_{\alpha_0}$,

$k_{\alpha'} \in \mathcal{J}$, hence we can assume $a \in k_{\alpha_0}[\![X_1^p,\ldots,X_d^p]\!][Y_1^p,\ldots,Y_t^p]$.

Write a as a power series in $\{X_1^p,\ldots,X_d^p,Y_1^p,\ldots,Y_t^p\}$. There

is at least one coefficient b of this series, which is in k_{α_0}

and not in k^p, since a $\notin T^p$. Let $\{\beta_\lambda\}_{\lambda \in \Lambda}$ be a p-basis of

k_{α_0} over $k^p[b]$ and $k_{\alpha'} := k^p(\{\beta_\lambda\}_{\lambda \in \Lambda})$. Then

$a \notin k_{\alpha'}[\![X_1^p,\ldots,X_d^p]\!][Y_1^p,\ldots,Y_t^p]$. We conclude that a $\notin K_{\alpha'}[T^p]$

since

$k_{\alpha_0}[\![X_1^p,\ldots,X_d^p]\!][Y_1^p,\ldots,Y_t^p] \cap K_{\alpha'}[T^p] = k_{\alpha'}[\![X_1^p,\ldots,X_d^p]\!][Y_1^p,\ldots,Y_t^p]$

as is seen by comparing coefficients. We thus have estab-

lished (3).

Next we prove that $\nu_{\alpha_0} = [L^p:T^p]$. Then elements of L^p

that are linearly independent over T^p are so over $K_{\alpha_0}[T^p]$

as well, hence $K_{\alpha_0}[T^p]$ and L^p are linearely disjoint over

T^p, as we wanted to show.

Let $\{v_1,\ldots,v_n\}$ be a basis of $K_{\alpha_0}[L^p]$ over $K_{\alpha_0}[T^p]$ con-

sisting of elements $v_i \in L^p$ $(i=1,\ldots,n)$. For each $k_{\alpha'} \in \mathcal{J}$

with $k_{\alpha'} \subset k_{\alpha_0}$ the v_i are also linearly independent over

$K_{\alpha'}[T^p]$ and thus form a basis of $K_{\alpha'}[L^p]$ over $K_{\alpha'}[T^p]$,

because $\nu_{\alpha'} = \nu_{\alpha_0}$. Each $v \in L^p$ has a unique representation

$$v = \lambda_1 v_1 + \ldots + \lambda_n v_n \quad (\lambda_i \in \cap K_{\alpha'}[T^p] = T^p).$$

Therefore $\nu_{\alpha_0} = [K_{\alpha_0}[L^p]:K_{\alpha_0}[T^p]] = [L^p:T^p]$, q.e.d.

Theorem 13.13 and the remark before 13.12 show that

there is also a common admissible subfield $k_o \subset k$ for finitely many semianalytic field extensions L/k.

In what follows k is a ring as in the beginning of § 13. Let S be a semianalytic k-algebra, and choose a power series algebra R over which S is essentially of finite type. With the universally finite derivation $\delta : R \to \tilde{\Omega}^1_{R/k}$ we can form $\Omega^1_{S/\delta}$. In case S is reduced this is the differential module $D_k(S)$ introduced above, however for the moment we allow S to have nilpotent elements. We try to determine $\mu_{\mathfrak{p}}(\Omega^1_{S/\delta})$ for $\mathfrak{p} \in \mathrm{Spec}(S)$.

13.14. Theorem. Let $\mathfrak{p} \in \mathrm{Spec}(S)$ be given such that the residue field $k(\mathfrak{p})$ has characteristic 0. Then the canonical sequence

$$0 \to \mathfrak{p}S_{\mathfrak{p}}/\mathfrak{p}^2 S_{\mathfrak{p}} \to \Omega^1_{S_{\mathfrak{p}}/\delta}/\mathfrak{p}\Omega^1_{S_{\mathfrak{p}}/\delta} \to D_k(k(\mathfrak{p})) \to 0$$

is exact.

Proof. By 4.17 we have an exact sequence

$$\mathfrak{p}S_{\mathfrak{p}}/\mathfrak{p}^2 S_{\mathfrak{p}} \xrightarrow{\alpha} \Omega^1_{S_{\mathfrak{p}}/\delta}/\mathfrak{p}\Omega^1_{S_{\mathfrak{p}}/\delta} \to \Omega^1_{k(\mathfrak{p})/\delta} \to 0$$

Since $\Omega^1_{k(\mathfrak{p})/\delta} = D_k(k(\mathfrak{p}))$, as was mentioned above, it suffices to show that α is injective. As the canonical map

$$\Omega^1_{S_{\mathfrak{p}}/\delta}/\mathfrak{p}\Omega^1_{S_{\mathfrak{p}}/\delta} \to \Omega^1_{S_{\mathfrak{p}}/\mathfrak{p}^2 S_{\mathfrak{p}}/\delta}/\mathfrak{p}\Omega^1_{S_{\mathfrak{p}}/\mathfrak{p}^2 S_{\mathfrak{p}}/\delta}$$

is bijective, we may assume that $\mathfrak{p}^2 = (0)$.

By the assumption about the characteristic of $k(\mathfrak{p})$, the image k' of k in $S_{\mathfrak{p}}$ is a field or a complete discrete valuation ring with $\mathfrak{p} \cap k' = (0)$. The image R' of R in $S_{\mathfrak{p}}$ is finite over some power series algebra

$P = k'[X_1,\ldots,X_d] \subset R'$. By the transitive law for universal extensions $\Omega^1_{S_\mathfrak{p}/\delta}$ is the universal extension of $\widetilde{\Omega}^1_{P/k'}$. We may therefore assume that k and R are contained in $S_\mathfrak{p}$. Since $\mathfrak{p}^2 = (0)$, we have $\mathfrak{p} \cap R = (0)$, and $K := Q(R) \subset S_\mathfrak{p}$, $K \subset k(\mathfrak{p})$.

By 6.5 the canonical sequence

$$0 \to \mathfrak{p}S_\mathfrak{p} \xrightarrow{\alpha'} \Omega^1_{S_\mathfrak{p}/K}/\mathfrak{p}\Omega^1_{S_\mathfrak{p}/K} \to \Omega^1_{k(\mathfrak{p})/K} \to 0$$

is exact. Let $d : S_\mathfrak{p} \to \Omega^1_{S_\mathfrak{p}/\delta}$ be the derivation belonging to $\Omega^1_{S_\mathfrak{p}/\delta}$ and $d' : S_\mathfrak{p} \to \Omega^1_{S_\mathfrak{p}/K}$ the universal derivation of $S_\mathfrak{p}/K$. By the universal property of $\Omega^1_{S_\mathfrak{p}/\delta}$ there is an $S_\mathfrak{p}$-linear map $h : \Omega^1_{S_\mathfrak{p}/\delta} \to \Omega^1_{S_\mathfrak{p}/K}$ with $h(dx)=d'x$ for all $x \in S_\mathfrak{p}$. We obtain a commutative diagram

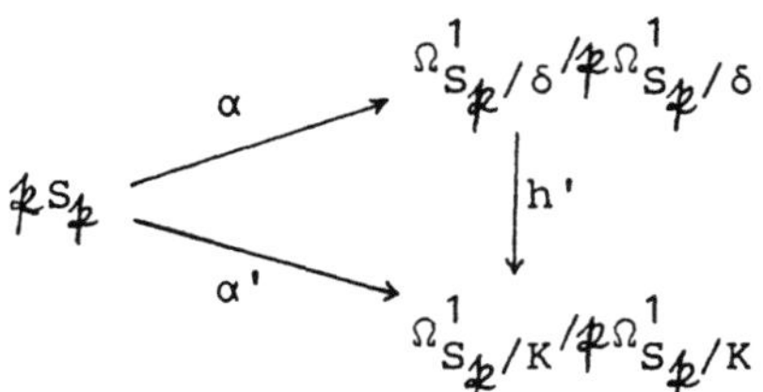

where h' is induced by h. Since α' is injective, so is α, q.e.d.

<u>13.15. Corollary.</u>   Under the assumptions of 13.14 let K be the quotient field of $A(k(\mathfrak{p}))$, the maximal analytic subalgebra of $k(\mathfrak{p})/k$. Then

$$\mu_\mathfrak{p}(\Omega^1_{S/\delta}) = \operatorname{edim} S_\mathfrak{p} + \dim_{k(\mathfrak{p})} D_k(k(\mathfrak{p})).$$

If the image of k in $k(\mathfrak{p})$ is a field, then

$$\mu_\mathfrak{p}(\Omega^1_{S/\delta}) = \operatorname{edim} S_\mathfrak{p} + \operatorname{aTrdeg}(K/k) + \operatorname{Trdeg}(k(\mathfrak{p})/K).$$

If the image of k in $k(\mathfrak{p})$ is a discrete valuation ring, then

$$\mu_\mathfrak{p}(\Omega^1_{S/\delta}) = \operatorname{edim} S_\mathfrak{p} + \dim A(k(\mathfrak{p}))-1 + \operatorname{Trdeg}(k(\mathfrak{p})/K).$$

The proof results from 13.14 by means of 13.6 and Nakayama's lemma.

We now consider the case that k is a field of characteristic $p > 0$ and $k_o \subset k$ a subfield with $k^p \subset k_o$ and $[k:k_o] < \infty$. Let $R = k[\![X_1,\ldots,X_d]\!]$ be a power series algebra over which S is essentially of finite type, let $\delta_o : R \to \tilde{\Omega}^1_{R/k_o}$ be the universally finite derivation of R/k_o and $\mathfrak{p} \in \mathrm{Spec}(S)$.

13.16. Theorem. a) The canonical sequence

$$0 \to \mathfrak{p} S_{\mathfrak{p}} / (\mathfrak{p}^2 + \eta) S_{\mathfrak{p}} \to \Omega^1_{S_{\mathfrak{p}}/\delta_o} / \mathfrak{p}\Omega^1_{S_{\mathfrak{p}}/\delta_o} \to D_{k_o}(k(\mathfrak{p})) \to 0$$

where $\eta := \mathfrak{p} \cap k_o[\![X_1^p,\ldots,X_d^p]\!][S^p]$, is exact. In particular,

$$\mu_{\mathfrak{p}}(\Omega^1_{S/\delta_o}) = \mathrm{edim}\, S_{\mathfrak{p}}/\eta S_{\mathfrak{p}} + \dim_{k(\mathfrak{p})} D_{k_o}(k(\mathfrak{p})).$$

b) If k_o is admissible for $k(\mathfrak{p})/k$, then

$$\mu_{\mathfrak{p}}(\Omega^1_{S/\delta_o}) = \mathrm{edim}\, S_{\mathfrak{p}} + a\mathrm{Trdeg}(K/k) + \mathrm{Trdeg}(k(\mathfrak{p})/K) + p\text{-}\deg(k/k_o)$$

where K is the maximal analytic subfield of $k(\mathfrak{p})/k$.

Proof. a) As in formula (2) above we have
$$\Omega^1_{S_{\mathfrak{p}}/\delta_o} = \Omega^1_{S_{\mathfrak{p}}/k_o[\![X_1^p,\ldots,X_d^p]\!]} \quad \text{and} \quad D_{k_o}(k(\mathfrak{p})) = \Omega^1_{k(\mathfrak{p})/k_o[\![X_1^p,\ldots,X_d^p]\!]}.$$
Therefore the sequence in a) is that of 6.7.

b) There is a noetherian normalization $k[\![t_1,\ldots,t_d]\!] \hookrightarrow R$ such that $\mathfrak{p} \cap k[\![t_1,\ldots,t_d]\!] = (t_{\delta+1},\ldots,t_d)$ for some $\delta \leq d$. Then $k(\mathfrak{p})$ is essentially of finite type over $k[\![t_1,\ldots,t_\delta]\!]$ and hence $\delta = a\mathrm{Trdeg}(K/k)$. $\tilde{\Omega}^1_{R/k_o}$ is the universal R-extension of $\tilde{\Omega}^1_{k[\![t_1,\ldots,t_d]\!]/k_o}$. By the transitive law for universal extensions we may assume that $R = k[\![t_1,\ldots,t_d]\!]$ from the beginning. The formula in b) follows from the formula in a), once it is shown that $\eta S_{\mathfrak{p}} \subset \mathfrak{p}^2 S_{\mathfrak{p}}$.

In order to prove this we put $L := k(\mathfrak{p})$,
$K' := k((t_1,\ldots,t_\delta))$, and $K_0 := k_0((t_1^p,\ldots,t_\delta^p))$. Since k_0
is admissible for L/k, the fields K_0 and L^p are linearly
disjoint over $(K')^p$. Consider the canonical commutative
diagram

$$k_0[\![t_1^p,\ldots,t_d^p]\!]\,(t_{\delta+1}^p,\ldots,t_d^p)\otimes_{R^p}(S_{\mathfrak{p}})^p \xrightarrow{\ \delta\ } k_0[\![t_1^p,\ldots,t_d^p]\!]\,(t_{\delta+1}^p,\ldots,t_d^p)[(S_{\mathfrak{p}})^p]$$

$$\Big\downarrow{\scriptstyle\alpha} \qquad\qquad\qquad\qquad\qquad\qquad\qquad \Big\downarrow{\scriptstyle\gamma}$$

$$K_0 \otimes_{(K')^p} L^p \xrightarrow{\qquad\qquad \beta \qquad\qquad} L$$

By assumption β is injective and α has kernel

$$(t_{\delta+1}^p,\ldots,t_d^p)\otimes(S_{\mathfrak{p}})^p + k_0[\![t_1^p,\ldots,t_d^p]\!]\,(t_{\delta+1}^p,\ldots,t_d^p)\otimes\mathfrak{p}^{(p)}(S_{\mathfrak{p}})^p$$

where $\mathfrak{p}^{(p)}$ is the set of all p-th power of the elements
of $\mathfrak{p}$. Moreover, $\ker\gamma$ is the extension ideal of $\mathfrak{a}$ and, on
the other hand, is the image of $\ker\alpha$ under δ. Thus we see
that $\mathfrak{a}S_{\mathfrak{p}} = (t_{\delta+1}^p,\ldots,t_d^p)S_{\mathfrak{p}} + \mathfrak{p}^{(p)}S_{\mathfrak{p}} \subset \mathfrak{p}^2 S_{\mathfrak{p}}$, q.e.d.

Now we have generalized most of §5 and §6 to our present
situation.

Exercises

1) Let k be a field of characteristic 0, $R = k[\![X]\!]$ the ring
   of formal power series in one variable X over k, and
   $S = R[Y]$ the polynomial ring in a variable Y over R.
   Show that the universally finite module of differentials
   of S/k does not exist. (Hint: $(1-XY)$ is a maximal ideal
   of S with residue class field $k((X))$, the quotient field
   of $k[\![X]\!]$).

2) Let k be a field, and let

$$L := k((X))[Y]/(Y^2) = k((X)) \oplus k((X))\varepsilon$$

where ε denotes the residue class of Y mod (Y^2). Put $R := k[\![X]\!]$ and $B := k[\![X]\!] \oplus k((X))\varepsilon$.

a) L is the full ring of quotients of B and a semianalytic k-algebra.

b) B is not noetherian, but integral over R.

c) There is a strictly increasing sequence

$$R_o \subset R_1 \subset R_2 \subset \dots$$

of analytic subalgebras R_i of L/k with $R_o = R$.

3) Under the assumptions of exercise 2) let k be a field of characteristic $p > 2$ with $[k:k^p] = \infty$. Put $K := k((X))$.

a) $p\text{-deg}(K/k) = \infty$.

b) There exists a non-trivial k-derivation $\delta : R \to K$ with $\delta X = 0$.

c) The mapping $\varphi : R \to L$ with $\varphi(r) = r + \delta r \cdot \varepsilon$ for all $r \in R$ is an injective k-algebra homomorphism.

d) Let $S := R[\varepsilon]$, $S' := R'[\varepsilon]$ where $R' := \varphi(R)$. Then $L = Q(S) = Q(S')$.

e) Let $D_k(L)$ denote the universal L-extension of $\tilde{\Omega}^1_{R/k}$ and $D'_k(L)$ the universal L-extension of $\tilde{\Omega}^1_{R'/k}$. Let $d : L \to D_k(L)$ and $d' : L \to D'_k(L)$ be the derivations corresponding to these differential modules. $D_k(L)$ and $D'_k(L)$ are isomorphic L-vector spaces, but there is no isomorphism that maps dz onto d'z for all $z \in L$.

§ 14. Regularity Criteria for Semianalytic Algebras

The purpose of this section is to generalize the results of § 7 to the case of (semi-)analytic algebras. As in § 13 let k be a complete noetherian local ring whose maximal ideal is generated by $p \cdot 1$, where p is the characteristic of the residue field of k. Let S be a semianalytic k-algebra, and let R be an analytic k-algebra such that S is essentially of finite type over R. Being a complete semi-local noetherian ring, R is universally catenary, hence for $\mathfrak{p}, \mathfrak{p}' \in \mathrm{Spec}(S)$ with $\mathfrak{p}' \subset \mathfrak{p}$ we have

$$(1) \qquad \dim S_{\mathfrak{p}} = \dim S_{\mathfrak{p}'} + \dim S_{\mathfrak{p}}/\mathfrak{p}'S_{\mathfrak{p}}.$$

Moreover, if S is a domain with quotient field L, if $R \subset S$ and $K := Q(R)$, then for any $\mathfrak{p} \in \mathrm{Spec}(S)$

$$(2) \qquad \dim S_{\mathfrak{p}} = \dim R_{\mathfrak{q}} + \mathrm{Trdeg}(L/K) - \mathrm{Trdeg}(k(\mathfrak{p})/k(\mathfrak{q}))$$

where $\mathfrak{q} := \mathfrak{p} \cap R$ (B.1). Using the formula $\dim R_{\mathfrak{q}} = \dim R - \dim R/\mathfrak{q}$ and observing that

$$\dim R = \dim A(L), \quad \dim R/\mathfrak{q} = \dim A(k(\mathfrak{p}))$$

formula (2) may be written as follows

$$(3) \quad \dim S_{\mathfrak{p}} = \dim A(L) + \mathrm{Trdeg}(L/K) - \dim A(k(\mathfrak{p})) - \mathrm{Trdeg}(k(\mathfrak{p})/k(\mathfrak{q})).$$

(3) can be applied, in particular, in the case R = A(S).

<u>14.1. Theorem</u> (Regularity criterion for characteristic 0). Let k be a field of characteristic 0 or a discrete valuation ring. Let δ denote the universally finite derivation of R/k. For $\mathfrak{p} \in \mathrm{Spec}(S)$ with $\mathfrak{p} \cap k = (0)$ the following conditions are equivalent:

a) $S_{\mathfrak{p}}$ is regular.

b) $\Omega^1_{S_{\mathfrak{p}}/\delta}$ is a free $S_{\mathfrak{p}}$-module.

Proof. Choose $\mathfrak{q} \in \mathrm{Spec}(S)$ with $\mathfrak{q} \subset \mathfrak{p}$ and $\dim S_{\mathfrak{p}} = \dim S_{\mathfrak{p}}/\mathfrak{q} S_{\mathfrak{p}}$. Clearly $\mathfrak{q} \cap k = (0)$. By 13.15 we have
$$\mu_{\mathfrak{p}}(\Omega^1_{S/\delta}) - \mu_{\mathfrak{q}}(\Omega^1_{S/\delta}) = \mathrm{edim}\, S_{\mathfrak{p}} - \mathrm{edim}\, S_{\mathfrak{q}} + \dim A(k(\mathfrak{p}))$$
$- \dim A(k(\mathfrak{q})) + \mathrm{Trdeg}(k(\mathfrak{p})/K_{\mathfrak{p}}) - \mathrm{Trdeg}(k(\mathfrak{q})/K_{\mathfrak{q}})$ where $K_{\mathfrak{p}}$
and $K_{\mathfrak{q}}$ are the quotient field of $A(k(\mathfrak{p}))$ and $A(k(\mathfrak{q}))$.
Applying (3) with $S/\mathfrak{q}$ replacing S and $S_{\mathfrak{q}}/\mathfrak{q} S_{\mathfrak{q}} = k(\mathfrak{q})$ replacing L, yields

(4) $.\mu_{\mathfrak{p}}(\Omega^1_{S/\delta}) - \mu_{\mathfrak{q}}(\Omega^1_{S/\delta}) = \mathrm{edim}\, S_{\mathfrak{p}} - \mathrm{edim}\, S_{\mathfrak{q}} - \dim S_{\mathfrak{p}}/\mathfrak{q} S_{\mathfrak{p}}.$

If $S_{\mathfrak{p}}$ is regular, then $\mathfrak{q} S_{\mathfrak{p}} = (0)$ and $S_{\mathfrak{q}}$ is the quotient
field of $S_{\mathfrak{p}}$. In this case we see from (4) that
$\mu_{\mathfrak{p}}(\Omega^1_{S/\delta}) = \mu_{\mathfrak{q}}(\Omega^1_{S/\delta})$ and from 7.3 that $\Omega^1_{S_{\mathfrak{p}}/\delta}$ is a free $S_{\mathfrak{p}}$-
module.

Conversely, if $\Omega^1_{S_{\mathfrak{p}}/\delta}$ is free, then we conclude, as in
the proof of 7.2, that $S_{\mathfrak{p}}$ is generically reduced, because
$\mathbb{Q} \subset S_{\mathfrak{p}}$, since $\mathfrak{p} \cap k = (0)$. We then have $\mathfrak{q} S_{\mathfrak{p}} = (0)$ and by (4)
$$0 = \mu_{\mathfrak{p}}(\Omega^1_{S/\delta}) - \mu_{\mathfrak{q}}(\Omega^1_{S/\delta}) = \mathrm{edim}\, S_{\mathfrak{p}} - \dim S_{\mathfrak{p}}$$
hence $S_{\mathfrak{p}}$ is regular.

Observe that $\Omega^1_{S_{\mathfrak{p}}/\delta} = D_k(S_{\mathfrak{p}})$ whenever $S_{\mathfrak{p}}$ is reduced.

<u>14.2. Corollary.</u>   Let $\mathfrak{p}' \in \mathrm{Spec}(S)$ with $\mathfrak{p}' \subset \mathfrak{p}$ be given.
If $S_{\mathfrak{p}}$ is regular, so is $S_{\mathfrak{p}'}$.

<u>14.3. Theorem</u> (Regularity criterion for characteristc $p > 0$).
Let k be a field of characteristic $p > 0$. For $\mathfrak{p} \in \mathrm{Spec}(S)$
let $S_{\mathfrak{p}}$ be reduced. Assume k_0 is a subfield of k that is ad-

missible for $k(\mathfrak{p})/k$ and $k(\mathfrak{q})/k$, where $\mathfrak{q} \subset \mathfrak{p}$ is a prime of S with dim $S_{\mathfrak{p}}$ = dim $S_{\mathfrak{p}}/\mathfrak{q}S_{\mathfrak{p}}$. Then the following conditions are equivalent:

a) $S_{\mathfrak{p}}$ is regular.

b) $D_{k_0}(S_{\mathfrak{p}})$ is a free $S_{\mathfrak{p}}$-module.

Proof. If $S_{\mathfrak{p}}$ is regular, there is just one prime $\mathfrak{q}$ of S with dim $S_{\mathfrak{p}}$ = dim $S_{\mathfrak{p}}/\mathfrak{q}S_{\mathfrak{p}}$ and $\mathfrak{q}S_{\mathfrak{p}}$ = (0). L := $S_{\mathfrak{q}}$ = $k(\mathfrak{q})$ is the quotient field of $S_{\mathfrak{p}}$. By 13.16 we have $\mu(D_{k_0}(S_{\mathfrak{p}}))$ = dim $S_{\mathfrak{p}}$ + dim $A(k(\mathfrak{p}))$ + Trdeg$(k(\mathfrak{p})/K_{\mathfrak{p}})$ + p-deg(k/k_0) where $K_{\mathfrak{p}}$ is the quotient field of $A(k(\mathfrak{p}))$. By (3) we conclude that $\mu(D_{k_0}(S_{\mathfrak{p}}))$ = dim $A(L)$ + Trdeg(L/K) + p-deg(k/k_0) where K is the quotient field of $A(L)$. Since k_0 is admissible for L = $k(\mathfrak{q})$, the last sum equals $\dim_L D_{k_0}(L)$, and $D_{k_0}(S_{\mathfrak{p}})$ is a free $S_{\mathfrak{p}}$-module by 7.3.

Conversely, if $D_{k_0}(S_{\mathfrak{p}})$ is a free $S_{\mathfrak{p}}$-module, then the same arguments as in the proof of 14.1 show that

$$O = \mu(D_{k_0}(S_{\mathfrak{p}})) - \mu(D_{k_0}(S_{\mathfrak{q}})) = \text{edim } S_{\mathfrak{p}} - \text{dim } S_{\mathfrak{p}}$$

for any $\mathfrak{q} \subset \mathfrak{p}$ with dim $S_{\mathfrak{p}}$ = dim $S_{\mathfrak{p}}/\mathfrak{q}S_{\mathfrak{p}}$ and any k_0 that is admissible for $k(\mathfrak{q})/k$, hence $S_{\mathfrak{p}}$ is regular.

14.4. Corollary. Let $\mathfrak{p}' \in$ Spec(S) with $\mathfrak{p}' \subset \mathfrak{p}$ be given. If $S_{\mathfrak{p}}$ is regular, so is $S_{\mathfrak{p}'}$.

We may choose k_0 to be admissible for $k(\mathfrak{p}')$ too. Then we can apply 14.3.

Suppose k is a discrete valuation ring with maximal ideal $\mathfrak{m}$. For $\mathfrak{p} \in$ Spec(S) with $\mathfrak{p} \cap k = \mathfrak{m}$ (case of mixed characteristics) a regularity criterion of the sort of 14.1 and 14.3 does not hold, as is easily seen if one tries to imitate the proof of these theorems.

We illustrate the regularity criteria by some applications.

14.5. Proposition. Let R be a complete noetherian local domain and $S := R[\![X_1,\ldots,X_n]\!]$. Then for any $\mathfrak{p} \in \mathrm{Spec}(S)$ with $\mathfrak{p} \cap R = (0)$ the local ring $S_{\mathfrak{p}}$ is regular.

Proof. By Cohen's structure theorem there is a Cohen subring $k \subset R$ having the same residue field as R. S is an analytic k-algebra. If k is a field of characteristic 0 or a discrete valuation ring, put $k_0 := k$. If k is a field of characteristic $p > 0$, choose k_0 to be admissible for $k(\mathfrak{p})/k$ and L/k where $L := Q(S)$. Let K be the quotient field of R. $\widetilde{\Omega}^1_{S/k_0}$ is the universally finite extension of $\widetilde{\Omega}^1_{R/k_0}$ (12.7) and

$$\widetilde{\Omega}^1_{S/k_0} = S \otimes_R \widetilde{\Omega}^1_{R/k_0} \oplus \bigoplus_{i=1}^{n} S\,dX_i.$$

Since $\mathfrak{p} \cap R = (0)$, we have $K \subset S_{\mathfrak{p}}$, and $D_{k_0}(S_{\mathfrak{p}}) = S_{\mathfrak{p}} \otimes_S \widetilde{\Omega}^1_{S/k_0} = S_{\mathfrak{p}} \otimes_R \widetilde{\Omega}^1_{R/k_0} \oplus \bigoplus_{i=1}^{n} S_{\mathfrak{p}}\,dX_i = S_{\mathfrak{p}} \otimes_K D_k(K) \oplus \bigoplus_{i=1}^{n} S_{\mathfrak{p}}\,dX_i$ is a fee $S_{\mathfrak{p}}$-module. By 14.1 and 14.3 the local ring $S_{\mathfrak{p}}$ is regular.

14.6. Corollary. The algebra $R[\![X_1,\ldots,X_n]\!]/R$ is regular.

Proof. It is wellknown that $R[\![X_1,\ldots,X_n]\!]/R$ is flat. For any $\mathfrak{p} \in \mathrm{Spec}(R[\![X_1,\ldots,X_n]\!])$ we have (with $\mathfrak{q} := \mathfrak{p} \cap R$)

$$R[\![X_1,\ldots,X_n]\!]_{\mathfrak{p}}/\mathfrak{q}R[\![X_1,\ldots,X_n]\!]_{\mathfrak{p}} = R/\mathfrak{q}[\![X_1,\ldots,X_n]\!]_{\bar{\mathfrak{p}}}$$

where $\bar{\mathfrak{p}}$ is the image of $\mathfrak{p}$ in $R/\mathfrak{q}[\![X_1,\ldots,X_n]\!]$. We can apply

14.5 to conclude that $R/\mathfrak{q}[X_1,\ldots,X_n]_{\overline{\mathfrak{p}}}$ is regular.

14.7. <u>Proposition.</u>   Let S be a semianalytic algebra over a field k. Then Reg(S) is open in Spec(S).

Proof.   a) Assume char.k = O, and let S be essentially of finite type over $R := k[X_1,\ldots,X_d] \subset S$. If δ is the universally finite derivation of R/k, then by 14.1 Reg(S) is the set of all $\mathfrak{p} \in$ Spec(S) for which $(\Omega^1_{S/\delta})_{\mathfrak{p}}$ is a free $S_{\mathfrak{p}}$-module, hence Reg(S) is open in Spec(S).

b) Assume now that char.k =: p > O. As in the proof of 7.7 we can reduce to the case of S being a domain. In this case we shall show that for any $\mathfrak{p} \in$ Reg(S) there is a neighbourhood of $\mathfrak{p}$ in Spec(S) that is contained in Reg(S).

Let S be essentially of finite type over a power series ring $R := k[X_1,\ldots,X_d]$ contained in S, and write $K := Q(R)$, $L := Q(S)$, $\mathfrak{q} := \mathfrak{p} \cap R$. Choose an admissible field k_o for $k(\mathfrak{p})/k$ and L/k. By 14.3, $D_{k_o}(S_{\mathfrak{p}})$ is a free $S_{\mathfrak{p}}$-module. Its rank equals the dimension of the L-vector space $\tilde{\Omega}^1_{L/k_o}$ which is aTrdeg(K/k) + Trdeg(L/K) + p-deg(k/k_o). There is an open neighbourhood U of $\mathfrak{p}$ in Spec(S) such that for all $\mathfrak{p}' \in$ U the $S_{\mathfrak{p}'}$-module $D_{k_o}(S_{\mathfrak{p}'})$ is free of the same rank. Formula (3) shows that aTrdeg(K/k) + Trdeg(L/K) + p-deg(k/k_o) =
dim $S_{\mathfrak{p}'}$ + dim A(k($\mathfrak{p}'$)) + Trdeg(k($\mathfrak{p}'$)/k($\mathfrak{q}'$)) + p-deg(k/k_o) =
dim $S_{\mathfrak{p}'}$ − dim $R_{\mathfrak{q}'}$ + Trdeg(k($\mathfrak{p}'$)/k($\mathfrak{q}'$)) + rank $\tilde{\Omega}^1_{R/k_o}$ where $\mathfrak{q}' := \mathfrak{p}' \cap R$. By 7.9, $S_{\mathfrak{p}'}$ is regular for all $\mathfrak{p}' \in$ U.

The statement of the proposition is true for any semianalytic algebra S (Matsumura [M_1], Thm.74), but the proof in the case of mixed characteristics is more complicated.

__14.8. Proposition.__ Let S be a semianalytic algebra over a field k of characteritic p > O. Then there exists a subfield $k_o \subset k$ with $k^p \subset k_o$, $[k:k_o] < \infty$ such that for all $\mathfrak{p} \in \mathrm{Reg}(S)$

$$\mu(D_{k_o}(S_\mathfrak{p})) = \dim S_\mathfrak{p} + \dim A(k(\mathfrak{p})) + \mathrm{Trdeg}(k(\mathfrak{p})/K_\mathfrak{p}) + p\text{-}\deg(k/k_o)$$

where $K_\mathfrak{p}$ is the quotient field of $A(k(\mathfrak{p}))$.

Proof. In this proof a field k_o, for which the formula in 14.8 holds, will be called admissible for $S_\mathfrak{p}$. Observe that by (3) this formula is equivalent with

$$\mu(D_{k_o}(S_\mathfrak{p})) = \dim A(L) + \mathrm{Trdeg}(L/K) + p\text{-}\deg(k/k_o)$$

where $L := Q(S_\mathfrak{p})$ and $K := Q(A(L))$.

a) We first show: If $k_o' \subset k_o$ is a subfield with $k^p \subset k_o'$, $[k_o:k_o'] < \infty$, and if k_o is admissible for $S_\mathfrak{p}$, so is k_o'.

Let $k[\![X_1,\ldots,X_d]\!] \subset S_\mathfrak{p}$ be a power series algebra over which $S_\mathfrak{p}$ is essentially of finite type. We have $D_{k_o}(S_\mathfrak{p}) = \Omega^1_{S_\mathfrak{p}/k_o}[\![X_1^p,\ldots,X_d^p]\!]$ and hence a canonical exact sequence (3.24)

$$S_\mathfrak{p} \otimes_{k_o} \Omega^1_{k_o/k_o'} \to D_{k_o'}(S_\mathfrak{p}) \to D_{k_o}(S_\mathfrak{p}) \to O \ .$$

Consequently $\mu(D_{k_o'}(S_\mathfrak{p})) \leq \mu(D_{k_o}(S_\mathfrak{p})) + p\text{-}\deg(k_o/k_o') = \dim A(L) + \mathrm{Trdeg}(L/K) + p\text{-}\deg(k/k_o')$. On the other hand $\mu(D_{k_o'}(S_\mathfrak{p})) \geq \dim_L D_{k_o'}(L) \geq \dim_K D_{k_o'}(K) + \mathrm{Trdeg}(L/K)$ by 13.6. $D_{k_o'}(K) = \Omega^1_{K/k_o'}((X_1^p,\ldots,X_d^p))$, being the universal K-extension of $\Omega^1_{k((X_1,\ldots,X_d))/k_o'((X_1^p,\ldots,X_d^p))}$, has at least dimension $d + p\text{-}\deg(k/k_o') = \dim A(L) + p\text{-}\deg(k/k_o')$. We thus obtain $\mu(D_{k_o'}(S_\mathfrak{p})) \geq \dim A(L) + \mathrm{Trdeg}(L/K) + p\text{-}\deg(k/k_o')$. Since the opposite inequality was already established, we have equality, and hence k_o' is admissible for $S_\mathfrak{p}$.

b) If k_0 is admissible for $S_{\mathfrak{p}}$, then $D_{k_0}(S_{\mathfrak{p}})$ is free of rank

$r := \dim A(L) + \mathrm{Trdeg}(L/K) + p\text{-deg}(k/k_0)$. Indeed, we always

have

$$\mu(D_{k_0}(S_{\mathfrak{p}})) \geq \dim_L D_{k_0}(L) \geq \dim A(L) + \mathrm{Trdeg}(L/K) + p\text{-deg}(k/k_0).$$

as was shown in a). For admissible k_0 we have equality, and

hence $D_{k_0}(S_{\mathfrak{p}})$ is free by 7.3.

There is an open neighbourhood $U \subset \mathrm{Reg}(S)$ of $\mathfrak{p}$ such that

$D_{k_0}(S_{\mathfrak{q}})$ is free of rank r for all $\mathfrak{q} \in U$. Hence k_0 is ad-

missible for all $S_{\mathfrak{q}}$ with $\mathfrak{q} \in U$.

c) For each intermediate field k_0 of k/k^p with $[k:k_0] < \infty$

define U_{k_0} as the set of all $\mathfrak{q} \in \mathrm{Reg}(S)$ such that k_0 is ad-

missible for $S_{\mathfrak{q}}$. By b) the sets U_{k_0} are open in $\mathrm{Spec}(S)$

and by a) we have $U_{k_0} \subset U_{k_0'}$ for $k_0' \subset k_0$.

Since $\mathrm{Spec}(S)$ is noetherian, there is a k_0 for which U_{k_0}

is maximal among all these sets. We shall show that

$U_{k_0} = \mathrm{Reg}(S)$. Indeed, if there is a $\mathfrak{p} \in \mathrm{Reg}(S) \smallsetminus U_{k_0}$, we

choose a subfield $k_0' \subset k_0$ with $k^p \subset k_0'$, $[k_0:k_0'] < \infty$, that

is admissible for $k(\mathfrak{p})/k$ (13.11). By 13.16b) this k_0' is

admissible for $S_{\mathfrak{p}}$ too. Since $U_{k_0} \subset U_{k_0'}$ and $\mathfrak{p} \notin U_{k_0}$, we

arrive at a contradiction, hence $U_{k_0} = \mathrm{Reg}(S)$.

We now shall discuss the notion corresponding to geo-

metric regularity in the category of analytic algebras.

Analytic algebras are semilocal rings. In the following

we call a homomorphism $\varphi : R \to S$ between semilocal rings a

local homomorphism, if it maps the radical of R into the

radical of S.

__14.9. Remark.__ Let A and B be analytic algebras over a

field k. Any k-homomorphism $\varphi : A \to B$ is local.

For $\mathfrak{M} \in \text{Max}(B)$ let $\varphi_{\mathfrak{M}} : A \to B/\mathfrak{M}$ be the composition of φ and the canonical map $B \to B/\mathfrak{M}$. $B/\mathfrak{M}$ is a finite extension field of k. Hence $\varphi_{\mathfrak{M}}(A)$ is a field, and ker $\varphi_{\mathfrak{M}} \in \text{Max}(A)$. In particular, rad(A) is mapped into $\mathfrak{M}$ for any $\mathfrak{M} \in \text{Max}(B)$, hence $\varphi(\text{rad}(A)) \subset \text{rad}(B)$.

Each analytic k-algebra A is a finite extension of a power series algebra $k[\![X_1, \ldots, X_d]\!] \subset A$. As in the local case we call $k[\![X_1, \ldots, X_d]\!] \hookrightarrow A$ a noetherian normalization of A.

<u>14.10. Definition.</u>   Let A be an analytic algebra over a field k and ℓ/k a field extension. A <u>constant field exten-sion</u> A_ℓ of A with ℓ is an analytic ℓ-algebra A_ℓ for which there is a local k-homomorphism $\alpha : A \to A_\ell$ satisfying the following universal property: If $\beta : A \to B$ is any local k-homomorphism into an analytic ℓ-algebra B, then there is exactly one ℓ-homomorphism $h : A_\ell \to B$ such that $\beta = h \circ \alpha$.

Of course, if A_ℓ exists, then it is unique up to a cano-nical ℓ-isomorphism. The <u>existence of a constant field ex-tension</u> is shown as follows: Given A, take a noetherian normalization $k[\![X_1, \ldots, X_d]\!] \hookrightarrow A$ and form

$$A_\ell := \ell[\![X_1, \ldots, X_d]\!] \otimes_{k[\![X_1, \ldots, X_d]\!]} A.$$

For α choose the canonical mapping of A into the tensor product.

Since A_ℓ is finite over $\ell[\![X_1, \ldots, X_d]\!]$, it is certainly an analytic ℓ-algebra, and the radical of A is mapped into the radical of A_ℓ by α. Given β as in definition 14.10, let $\beta' : k[\![X_1, \ldots, X_d]\!] \to B$ be the composition of $k[\![X_1, \ldots, X_d]\!] \to A$ with β. Then $\beta'(X_i) \in \text{Rad}(B)$ $(i=1, \ldots, d)$, and hence there is an ℓ-homomorphism $\beta'' : \ell[\![X_1, \ldots, X_d]\!] \to B$

with $\beta''(X_i) = \beta'(X_i)$ $(i=1,\ldots,d)$. β and β'' induce an ℓ-homomorphism

$$h : \ell[X_1,\ldots,X_d] \otimes_{k[X_1,\ldots,X_d]} A \to B$$

with $\beta = h \circ \alpha$.

If h' is an arbitrary ℓ-homomorphism with $\beta = h' \circ \alpha$, then $h'(X_i \otimes 1) = \beta'(X_i)$ $(i=1,\ldots,d)$ and $h'(1 \otimes a) = \beta(a)$ for all $a \in A$, hence $h' = h$.

For $A = k[X_1,\ldots,X_d]$ we have $A_\ell = \ell[X_1,\ldots,X_d]$. Observe that $\ell[X_1,\ldots,X_d] \neq \ell \otimes_k k[X_1,\ldots,X_d]$, if $[\ell:k] = \infty$ and $d > 0$.

14.11. Rules. Let A be an analytic k-algebra, $k[X_1,\ldots,X_d] \hookrightarrow A$ a noetherian normalization of A, and ℓ/k a field extension.

a) If $[\ell:k] < \infty$, then $A_\ell = \ell \otimes_k A$.

b) For any ideal I of A

$$(A/I)_\ell = A_\ell/IA_\ell.$$

c) A_ℓ is faithfully flat over A.

d) $\ell[X_1,\ldots,X_d] \to A_\ell$ is a noetherian normalization of A_ℓ.

e) $\dim A_\ell = \dim A$.

f) If A is equidimensional, so is A_ℓ.

g) For any $\mathfrak{q} \in \mathrm{Spec}(A)$ there is a $\mathfrak{p} \in \mathrm{Spec}(A_\ell)$ such that

$$\dim_{\mathfrak{p}} A_\ell \geq \dim_{\mathfrak{q}} A$$

h) $\tilde{\Omega}^1_{A_\ell/\ell} \cong A_\ell \otimes_A \tilde{\Omega}^1_{A/k}.$

Proof. a) If $[\ell:k] < \infty$, then $\ell[X_1,\ldots,X_d] = \ell \otimes_k k[X_1,\ldots,X_d]$ and $A_\ell = \ell[X_1,\ldots,X_d] \otimes_{k[X_1,\ldots,X_d]} A = \ell \otimes_k A$.

b) We can find a noetherian normalization of A such that $I \cap k[X_1,\ldots,X_d] = (X_{\delta+1},\ldots,X_d)$ for some $\delta \leq d$. Then

$k[\![X_1,\ldots,X_\delta]\!] \to A/I$ is a normalization of A/I and

$$(A/I)_\ell = \ell[\![X_1,\ldots,X_\delta]\!] \otimes_{k[\![X_1,\ldots,X_\delta]\!]} A/I =$$

$$\ell[\![X_1,\ldots,X_d]\!] \otimes_{k[\![X_1,\ldots,X_d]\!]} A/I =$$

$$\ell[\![X_1,\ldots,X_d]\!] \otimes_{k[\![X_1,\ldots,X_d]\!]} A/I \cdot (\ell[\![X_1,\ldots,X_d]\!] \otimes_{k[\![X_1,\ldots,X_d]\!]} A)$$

$$= A_\ell/IA_\ell.$$

c) It is known that $\ell[\![X_1,\ldots,X_d]\!]/k[\![X_1,\ldots,X_d]\!]$ is faithfully flat. The construction of A_ℓ shows that A_ℓ/A is faithfully flat too.

d) and e) follow, since A_ℓ is a finite ring extension of $\ell[\![X_1,\ldots,X_d]\!]$.

f) Let A be equidimensional, $\mathfrak{p} \in \mathrm{Min}(A_\ell)$, $\mathfrak{q} := \mathfrak{p} \cap A$. Then $\mathfrak{q} \in \mathrm{Min}(A)$. We wish to show that $\dim A_\ell/\mathfrak{p} = \dim A/\mathfrak{q}$, because then $\dim A_\ell/\mathfrak{p}$ is independent of the $\mathfrak{p} \in \mathrm{Min}(A)$ chosen. Let $\overline{\mathfrak{p}}$ be the image of $\mathfrak{p}$ in $(A/\mathfrak{q})_\ell = A_\ell/\mathfrak{q}A_\ell$. Since $A_\ell/\mathfrak{p} = (A/\mathfrak{q})_\ell/\overline{\mathfrak{p}}$ we may assume that A is a domain, hence $\mathfrak{p} \cap A = (0)$. It suffices to show that $\mathfrak{p} \cap \ell[\![X_1,\ldots,X_d]\!] = (0)$, because then $\dim A_\ell/\mathfrak{p} = d = \dim A$. Consider the composition

$$\ell[\![X_1,\ldots,X_d]\!] \to \ell[\![X_1,\ldots,X_d]\!] \otimes_{k[\![X_1,\ldots,X_d]\!]} k((X_1,\ldots,X_d))$$

$$\to A_\ell \otimes_{k[\![X_1,\ldots,X_d]\!]} k((X_1,\ldots,X_d)),$$ and let $\tilde{\mathfrak{p}}$ denote the extension ideal of $\mathfrak{p}$ in $S := A_\ell \otimes_{k[\![X_1,\ldots,X_d]\!]} k((X_1,\ldots,X_d))$. It is a minimal prime of that ring. It suffices to show that S is flat over $\ell[\![X_1,\ldots,X_d]\!]$, because then

$$\tilde{\mathfrak{p}} \cap \ell[\![X_1,\ldots,X_d]\!] = \mathfrak{p} \cap \ell[\![X_1,\ldots,X_d]\!]$$ is minimal in $\ell[\![X_1,\ldots,X_d]\!]$, that is, $\mathfrak{p} \cap \ell[\![X_1,\ldots,X_d]\!] = (0)$.

But $A_\ell \otimes_{k[\![X_1,\ldots,X_d]\!]} k((X_1,\ldots,X_d)) =$

$$(A \otimes_{k[\![X_1,\ldots,X_d]\!]} k((X_1,\ldots,X_d))) \otimes_{k[\![X_1,\ldots,X_d]\!]} \ell[\![X_1,\ldots,X_d]\!]$$

is flat over $k((X_1,\ldots,X_d)) \otimes_{k[\![X_1,\ldots,X_d]\!]} \ell[\![X_1,\ldots,X_d]\!]$ and this ring, being a localization of $\ell[X_1,\ldots,X_d]$, is flat over $\ell[X_1,\ldots,X_d]$.

g) Let $\mathscr{G}_0 \subset \mathscr{G}_1 \subset \ldots \subset \mathscr{G}_h = \mathscr{G} \subset \mathscr{G}_{h+1} \subset \ldots \subset \mathscr{G}_n$ be a chain of prime ideals in A with $n = \dim_{\mathscr{G}} A$. Since A_ℓ/A is faithfully flat, there is a $\mathscr{P}_n \in \mathrm{Spec}(A_\ell)$ with $\mathscr{P}_n \cap A = \mathscr{G}_n$. Suppose a chain $\mathscr{P}_i \subset \mathscr{P}_{i+1} \subset \ldots \subset \mathscr{P}_n$ of prime ideals of A_ℓ with $\mathscr{P}_j \cap A = \mathscr{G}_j$ $(j=1,\ldots,n)$ has already been constructed for some $i \leq n$. Then $(A_\ell)_{\mathscr{P}_i}$ is faithfully flat over $A_{\mathscr{G}_i}$ and we can find a $\mathscr{P}_{i-1} \in \mathrm{Spec}(A_\ell)$ with $\mathscr{P}_{i-1}(A_\ell)_{\mathscr{P}_i}$ lying over $\mathscr{G}_{i-1}A_{\mathscr{G}_i}$, hence $\mathscr{P}_{i-1} \subset \mathscr{P}_i$ and $\mathscr{P}_{i-1} \cap A = \mathscr{G}_{i-1}$. Thus we can construct a chain of prime ideals of A_ℓ lying of the given chain of A, and with $\mathscr{P} := \mathscr{P}_h$ the claim follows.

h) Write $R := k[X_1,\ldots,X_d]$, and let $A = R[Y_1,\ldots,Y_n]/(F_1,\ldots,F_m)$ $= R[y_1,\ldots,y_n]$ be a presentation of A/R by generators and relations. $\tilde{\Omega}^1_{A_\ell/\ell}$ is the universal extension of
$$\tilde{\Omega}^1_{R_\ell/\ell} = R_\ell dX_1 \oplus \ldots \oplus R_\ell dX_d = R_\ell \otimes_R \tilde{\Omega}^1_{R/k} \text{ to}$$
$A_\ell = R_\ell[Y_1,\ldots,Y_n]/(F_1,\ldots,F_m)R_\ell[Y_1,\ldots,Y_n]$, hence
$$\tilde{\Omega}^1_{A_\ell/\ell} = A_\ell \otimes_{R_\ell} \tilde{\Omega}^1_{R_\ell/\ell} \oplus \bigoplus_{k=1}^{n} A_\ell dY_k/U$$
where U is generated by $1 \otimes \delta F_j(y_1,\ldots,y_n) + \sum_{k=1}^{n} \dfrac{\partial F_j}{\partial Y_k} dY_k$ $(j=1,\ldots,m)$, with the usual notation.

Here $A_\ell \otimes_{R_\ell} \tilde{\Omega}^1_{R_\ell/\ell} = A_\ell \otimes_R \tilde{\Omega}^1_{R/k}$, and the $\delta F_j(y_1,\ldots,y_n)$ may be considered as elements of $\tilde{\Omega}^1_{R/k}$, since the F_j have coefficients in R $(j=1,\ldots,m)$. We obtain
$$\tilde{\Omega}^1_{A_\ell/\ell} = A_\ell \otimes_A (A \otimes_R \tilde{\Omega}^1_{R/k} \oplus \bigoplus_{i=1}^{n} AdY_i \big| <\{1 \otimes \delta F_j(y_1,\ldots,y_n) + \sum_{k=1}^{n} \dfrac{\partial F_j}{\partial Y_k} dY_k\} >) =$$
$$A_\ell \otimes_A \tilde{\Omega}^1_{A/k}.$$

<u>14.12. Definition.</u> An analytic k-algebra A is called <u>ab-solutely regular</u> at $\mathscr{G} \in \mathrm{Spec}(A)$, if for any field extension

ℓ/k and any $\mathfrak{p} \in \mathrm{Spec}(A_\ell)$ with $\mathfrak{p} \cap A = \mathfrak{q}$ the local ring $(A_\ell)_\mathfrak{p}$ is regular.

Of course, $A_\mathfrak{q}$ is then regular and even geometrically regular over k (14.11a). If char.k = 0, then A is absolutely regular at $\mathfrak{q} \in \mathrm{Spec}(A)$ if and only if $A_\mathfrak{q}$ is regular. In fact, for ℓ and $\mathfrak{p}$ as in 14.12 we have $(\widetilde{\Omega}^1_{A_\ell/\ell})_\mathfrak{p} = (A_\ell)_\mathfrak{p} \otimes_{A_\mathfrak{q}} (\widetilde{\Omega}^1_{A/k})_\mathfrak{q}$ by 14.11h), and with $A_\mathfrak{q}$ the ring $(A_\ell)_\mathfrak{p}$ is regulary too by 14.1.

For any field k the power series algebra $A = k[\![X_1, \ldots, X_d]\!]$ is absolutely regular at any $\mathfrak{q} \in \mathrm{Spec}(A)$, since $A_\ell = \ell[\![X_1, \ldots, X_d]\!]$.

<u>14.13. Theorem</u> (Criterion for absolute regularity).

Let A be an analytic k-algebra and $\mathfrak{q} \in \mathrm{Spec}(A)$. Then the following conditions are equivalent:

a) A is absolutely regular at $\mathfrak{q}$.

b) $(\widetilde{\Omega}^1_{A/k})_\mathfrak{q}$ is a free $A_\mathfrak{q}$-module of rank $\dim_\mathfrak{q} A$.

c) $\mu_\mathfrak{q}(\widetilde{\Omega}^1_{A/k}) \leq \dim_\mathfrak{q} A$.

Proof. a) $\to$ b). Let ℓ be the algebraic closure of k, and let $\mathfrak{p} \in \mathrm{Spec}(A_\ell)$ with $\mathfrak{p} \cap A = \mathfrak{q}$ be given. Then $(\widetilde{\Omega}^1_{A_\ell/\ell})_\mathfrak{p} = (A_\ell)_\mathfrak{p} \otimes_{A_\mathfrak{q}} (\widetilde{\Omega}^1_{A/k})_\mathfrak{q}$ is a free $(A_\ell)_\mathfrak{p}$-module (14.1 and 14.3), and since $(A_\ell)_\mathfrak{p}$ is faithfully flat over $A_\mathfrak{q}$ (14.11c), $(\widetilde{\Omega}^1_{A/k})_\mathfrak{q}$ is a free $A_\mathfrak{q}$-module of the same rank. This rank is $\dim(A_\ell)_\mathfrak{p} + \dim A_\ell/\mathfrak{p} = \dim_\mathfrak{p} A_\ell$ by 13.15 and 13.16.

As $A_\mathfrak{q}$ is regular, we may assume that A is a domain (replacing A by its image in $A_\mathfrak{q}$, if necessary). Then A_ℓ is equidimensional (14.11f), and hence $\dim_\mathfrak{p} A_\ell = \dim A_\ell = \dim A = \dim_\mathfrak{q} A$. Thus we have b).

c) $\to$ a). Under the assumption c) we first show that $A_{\mathfrak{q}}$ is regular. If char.$k = 0$, then $\mu_{\mathfrak{q}}(\tilde{\Omega}^1_{A/k}) = \text{edim } A_{\mathfrak{q}} + \dim A/\mathfrak{q}$ by 13.15, and c) implies edim $A_{\mathfrak{q}} = \dim A_{\mathfrak{q}}$. If char.$k =: p > 0$, let ℓ be the algebraic closure of k. Choose $\mathfrak{p} \in \text{Spec}(A_\ell)$ with $\mathfrak{p} \cap A = \mathfrak{q}$ and $\dim_{\mathfrak{q}} A \leq \dim_{\mathfrak{p}} A_\ell$ (14.11g)). Then $\mu_{\mathfrak{p}}(\tilde{\Omega}^1_{A_\ell/\ell}) = \mu_{\mathfrak{q}}(\tilde{\Omega}^1_{A/k}) \leq \dim_{\mathfrak{q}} A \leq \dim_{\mathfrak{p}} A_\ell$. On the other hand, $\mu_{\mathfrak{p}}(\tilde{\Omega}^1_{A_\ell/\ell}) \geq \text{edim}(A_\ell)_{\mathfrak{p}} + \dim A_\ell/\mathfrak{p}$ by 13.16b), and consequently $\text{edim}(A_\ell)_{\mathfrak{p}} = \dim(A_\ell)_{\mathfrak{p}}$. Since $(A_\ell)_{\mathfrak{p}}$ is faithfully flat over $A_{\mathfrak{q}}$, we conclude that $A_{\mathfrak{q}}$ is regular too.

Now let ℓ/k be an arbitrary field extension and $\mathfrak{p} \in \text{Spec}(A_\ell)$ with $\mathfrak{p} \cap A = \mathfrak{q}$. We may assume that A is a domain, hence $\dim_{\mathfrak{q}} A = \dim_{\mathfrak{p}} A_\ell$ (14.11e). From c) we obtain $\mu_{\mathfrak{p}}(\tilde{\Omega}^1_{A_\ell/\ell}) \leq \dim_{\mathfrak{p}} A_\ell$. As was shown above, this implies that $(A_\ell)_{\mathfrak{p}}$ is regular, q.e.d.

14.14. Corollary. The set of all $\mathfrak{q} \in \text{Spec } A$ at which A/k is absolutely regular is open in Spec A.

14.15. Corollary. If $A_{\mathfrak{q}}$ is regular and $k(\mathfrak{q})/k$ analytically separable, then A/k is absolutely regular at $\mathfrak{q}$. In particular, if k is a perfect field, A/k is absolutely regular at all $\mathfrak{q} \in \text{Spec } A$ if and only if A is regular.

Indeed, the condition 14.13c) is satisfied by 13.16b), k being admissible for $k(\mathfrak{q})/k$ by 13.12c).

14.16. Corollary. Let A be an analytic domain over k with $K := Q(A)$. A/k is absolutely regular at $\mathfrak{q} = (0)$ if and only if K/k is analytically separable.

This follows from 14.13 and 13.10.

14.17. <u>Corollary</u>.   Let A and A' be analytic k-algebras. For $\mathfrak{p} \in \mathrm{Spec}(A)$, $\mathfrak{p}' \in \mathrm{Spec}(A')$ assume there is a k-isomorphism $A_{\mathfrak{p}} \cong A'_{\mathfrak{p}'}$. Then A is absolutely regular at $\mathfrak{p}$ if and only if A' is so at $\mathfrak{p}'$.

In fact, $\dim_{\mathfrak{p}} A = \dim A_{\mathfrak{p}} + a\mathrm{Trdeg}(k(\mathfrak{p})/k) =$ $\dim A'_{\mathfrak{p}'} + a\mathrm{Trdeg}(k(\mathfrak{p}')/k) = \dim_{\mathfrak{p}'} A'$. Therefore condition 14.13c) depends only on the k-isomorphism class of $A_{\mathfrak{p}}$.

By 14.17 we are entitled to call $A_{\mathfrak{p}}$ <u>absolutely regular over k</u>, if A/k is absolutely regular at $\mathfrak{p}$.

We now wish to generalize the structure theorem 7.19.

14.18. <u>Theorem</u>.   Assume that A is a homomorphic image of a power series algebra over a field k, and that A is equidimensional. For $\mathfrak{p} \in \mathrm{Spec}(A)$ the following conditions are equivalent:

a) $A_{\mathfrak{p}}$ is absolutely regular over k.

b) There is a noetherian normalization $k[\![X_1,\ldots,X_d]\!] \hookrightarrow A$ such that $A_{\mathfrak{p}}$ is étale over $k[\![X_1,\ldots,X_d]\!]_{\mathfrak{q}}$ where $\mathfrak{q} := \mathfrak{p} \cap k[\![X_1,\ldots,X_d]\!]$.

If $\dim A_{\mathfrak{p}} > 0$, these conditions are also equivalent with:

b') There is a noetherian normalization $k[\![X_1,\ldots,X_d]\!] \hookrightarrow A$ such that $A_{\mathfrak{p}}$ is étale over $k[\![X_1,\ldots,X_d]\!]_{\mathfrak{q}}$ and $k(\mathfrak{p}) = k(\mathfrak{q})$ where $\mathfrak{q} := \mathfrak{p} \cap k[\![X_1,\ldots,X_d]\!]$.

Proof.   If condition b) (or the more strict condition b')) is fulfilled, then the exact sequence

$$A_{\mathfrak{p}} \otimes_{k[\![X_1,\ldots,X_d]\!]} \tilde{\Omega}^1_{k[\![X_1,\ldots,X_d]\!]/k} \to (\tilde{\Omega}^1_{A/k})_{\mathfrak{p}} \to \Omega^1_{A_{\mathfrak{p}}/k[\![X_1,\ldots,X_d]\!]_{\mathfrak{q}}} \to 0$$

shows (since $\Omega^1_{A_{\mathfrak{p}}/k[\![X_1,\ldots,X_d]\!]_{\mathfrak{q}}} = 0$ by 6.8) that $\mu_{\mathfrak{p}}(\tilde{\Omega}^1_{A/k}) \leq d = \dim_{\mathfrak{p}} A$ (A being equidimensional). By 14.13

we conclude that $A_{\mathfrak{p}}$ is absolutely regular over k.

Now assume a). If dim $A_{\mathfrak{p}} = 0$, then $A_{\mathfrak{p}}$ is an analytically separable analytic extension field of k by 14.16. As in the proof of 13.10 we can construct a system of parameters $\{X_1,\ldots,X_d\}$ of A such that $D_k(A_{\mathfrak{p}}) = A_{\mathfrak{p}}dX_1 \oplus \ldots \oplus A_{\mathfrak{p}}dX_d$. Since $\mathfrak{p}$ is a minimal prime of A and since dim $A/\mathfrak{p}$=dim A=d, we have $\mathfrak{q} := \mathfrak{p} \cap k[X_1,\ldots,X_d] = (0)$, and an injection $k((X_1,\ldots,X_d)) \hookrightarrow A_{\mathfrak{p}}$ with $\Omega^1_{A_{\mathfrak{p}}/k((X_1,\ldots,X_d))} = 0$. Then $A_{\mathfrak{p}}/k((X_1,\ldots,X_d))$ is separable (which is here the same as being étale).

In case dim $A_{\mathfrak{p}} > 0$ it remains to show that b') holds. We shall construct a parameter system $\{X_1,\ldots,X_d\}$ of A satisfying the following conditions:

α) $\{dX_1,\ldots,dX_d\}$ is a system of generators of $(\tilde{\Omega}^1_{A/k})_{\mathfrak{p}}$.

β) If $\xi_1,\ldots,\xi_d$ denote the images of $X_1,\ldots,X_d$ in $k(\mathfrak{p})$, then $k(\mathfrak{p})$ is the quotient field of $k[\xi_1,\ldots,\xi_d]$, which is, by definition, the image of $k[X_1,\ldots,X_d]$ in $k(\mathfrak{p})$.

Then $k[X_1,\ldots,X_d] \hookrightarrow A$ is a noetherian normalization as desired in b'): With $\mathfrak{q} := \mathfrak{p} \cap k[X_1,\ldots,X_d]$ we have $k(\mathfrak{p}) = k(\mathfrak{q})$ by β). From α) we obtain

$$\Omega^1_{A_{\mathfrak{p}}/k[X_1,\ldots,X_d]_{\mathfrak{q}}} = 0,$$

which implies that $A_{\mathfrak{p}}$ is unramified over $k[X_1,\ldots,X_d]_{\mathfrak{q}}$ (6.8). Since $A_{\mathfrak{p}}$ is regular, there is only one $\mathfrak{p}_0 \in \mathrm{Min}(A)$ with $\mathfrak{p}_0 \subset \mathfrak{p}$. $k[X_1,\ldots,X_d] \to A/\mathfrak{p}_0$ and $A/\mathfrak{p}_0 \to A_{\mathfrak{p}}$ are injective, hence $k[X_1,\ldots,X_d]_{\mathfrak{q}} \to A_{\mathfrak{p}}$ is injective too, and both rings have the same dimension. Since both are regular with the same residue field their completions are isomorphic and, in particular, $A_{\mathfrak{p}}$ is flat over $k[X_1,\ldots,X_d]_{\mathfrak{q}}$.

In order to construct a parameter system $\{X_1,\ldots,X_d\}$ of

A satisfying $\alpha)$ and $\beta)$ we start from an arbitrary noetherian normalization $k[X_1',\ldots,X_d'] \hookrightarrow A$ of A with $\mathfrak{g} \cap k[X_1',\ldots,X_d'] = (X_1',\ldots,X_h')$ for some $h \le d$. If ξ_i' denotes the image of X_i' in $k(\mathfrak{g})$, then $k[\xi_{h+1}',\ldots,\xi_d'] \hookrightarrow A/\mathfrak{g}$ is a noetherian normalization of $A/\mathfrak{g}$. Moreover $h > 0$, since dim $A_\mathfrak{g} > 0$.

If char.$k = 0$, then $\{d\xi_{h+1}',\ldots,d\xi_d'\}$ is a basis of $D_k(k(\mathfrak{g}))$. We write X_i for X_i' and ξ_i for ξ_i' $(i=h+1,\ldots,d)$ in this case. If char.$k =: p > 0$, we shall show that after changing the parameter system properly $\{d\xi_{h+1}',\ldots,d\xi_d'\}$ becomes part of a basis of $D_k(k(\mathfrak{g}))$. Start with an arbitrary basis $\{d\eta_1,\ldots,d\eta_m\}$ of $D_k(k(\mathfrak{g}))$ where the η_i are non-units of $A/\mathfrak{g}$ $(i=1,\ldots,m)$. Such elements can be found, since A is a homomorphic image of a power series algebra. Let $Y_1,\ldots,Y_m \in A$ a system of representatives of these elements. Consider $\overline{A} := A/(X_1',\ldots,\widehat{X_{h+1}'},\ldots,X_d')$ and denote images in $\overline{A}$ of elements or ideals of A by an overbar. Let $\mathfrak{g}_1,\ldots,\mathfrak{g}_\tau$ be the minimal prime ideals of $\overline{A}$ that contain $\overline{Y}_1$, and let $\mathfrak{g}_{\tau+1},\ldots,\mathfrak{g}_t$ be those that do not contain $\overline{Y}_1$. Since dim $\overline{A} = 1$, we can find a non-unit $a \in A$ such that $\overline{a} \in \bigcap_{i=\tau+1}^{t} \mathfrak{g}_i \smallsetminus \bigcup_{i=1}^{\tau} \mathfrak{g}_i$. Put $X_{h+1} := Y_1 + a^p$. Then $\overline{X}_{h+1}$ is not contained in any minimal prime ideal of $\overline{A}$, $\{X_1',\ldots,X_h',X_{h+1},X_{h+2}',\ldots,X_d'\}$ is a system of parameters of A, and if ξ_{h+1} denotes the image of X_{h+1} in $k(\mathfrak{g})$, then
$$d\xi_{h+1} = d\eta_1.$$

As $d-h \le m$ we can, by repeating this process, construct elements $X_{h+1},\ldots,X_d \in A$ such that $\{X_1',\ldots,X_h',X_{h+1},\ldots,X_d\}$ is a parameter system of A and $\{d\xi_{h+1},\ldots,d\xi_d\}$ is part of a basis of $D_k(k(\mathfrak{g}))$ where ξ_i denotes the image of X_i in $k(\mathfrak{g})$.

$k(\mathfrak{p})/k((\xi_{h+1},\ldots,\xi_d))$ is a finite field extension, and since $\dim_{k(\mathfrak{p})}\Omega^1_{k(\mathfrak{p})/k((\xi_{h+1},\ldots,\xi_d))} \leq h$, this field extension has p-degree $\leq h$. By 5.11 we can find elements $\xi_1,\ldots,\xi_h \in k(\mathfrak{p})$ which generate $k(\mathfrak{p})$ over $k((\xi_{h+1},\ldots,\xi_d))$. They can be chosen as non-units of $A/\mathfrak{p}$. In the separable case one element ξ_1 will do by the theorem of the primitive element; observe that $h > 0$. We may order the ξ_i in such a way that $\{d\xi_{\eta+1},\ldots,d\xi_h,d\xi_{h+1},\ldots,d\xi_d\}$ with $\eta \leq h$ becomes a basis of $D_k(k(\mathfrak{p}))$. In the case char.$k = 0$ we can, of course, also select non-units $\xi_1,\ldots,\xi_h \in A/\mathfrak{p}$ which generate $k(\mathfrak{p})$ over $k((\xi_{h+1},\ldots,\xi_d))$. In this case $\eta = h$.

We will finish the construction now for arbitrary characteristic of k. Let $z_1,\ldots,z_h \in A$ be arbitrary representatives of $\xi_1,\ldots,\xi_h$ in A. We shall show that having chosen such representatives properly the differentials $dz_1,\ldots,dz_h,dX_{h+1},\ldots,dX_d$ generate $(\tilde{\Omega}^1_{A/k})_\mathfrak{p}$. From the exact sequence

$$\mathfrak{p}A_\mathfrak{p}/\mathfrak{p}^2A_\mathfrak{p} \to (\tilde{\Omega}^1_{A/k})_\mathfrak{p}/\mathfrak{p}(\Omega^1_{A/k})_\mathfrak{p} \to D_k(k(\mathfrak{p})) \to 0$$

we see that there are elements $t_1,\ldots,t_\eta \in \mathfrak{p}$ such that $\{dt_1,\ldots,dt_\eta,dz_{\eta+1},\ldots,dz_h,dX_{h+1},\ldots,dX_d\}$ generates $(\tilde{\Omega}^1_{A/k})_\mathfrak{p}$. If $\eta = 0$, there is nothing to be done. Suppose $\eta > 0$. There is a relation

$$dz_1 = \sum_{i=1}^{\eta} a_i dt_i + \sum_{i=1}^{h-\eta} b_i dz_{\eta+i} + \sum_{i=1}^{d-h} c_i dX_{h+i} \quad (a_i,b_i,c_i \in A_\mathfrak{p}).$$

We may assume that a_1 is a unit, replacing z_1 by z_1+t_1, if necessary. Then in the above system of generators of $(\tilde{\Omega}^1_{A/k})_\mathfrak{p}$ the element dt_1 can be replaced by dz_1. Repeating the process we end up with a system of generators $\{dz_1,\ldots,dz_h,dX_{h+1},\ldots,dX_d\}$ of $(\tilde{\Omega}^1_{A/k})_\mathfrak{p}$ where $z_1,\ldots,z_h$ are representatives of $\xi_1, \ldots, \xi_h$.

A final change of $\{z_1, \ldots, z_h\}$ will lead to a system of parameters of A without destroying what was already achieved.

Let I be the inverse image of $\mathfrak{m}^2 A_{\mathfrak{m}}$ in A. We may assume that the original normalization $k[\![X_1', \ldots, X_d']\!] \hookrightarrow A$ was already constructed in such a way that $I \cap k[\![X_1', \ldots, X_d']\!] = (X_1', \ldots, X_h')$. Since $\mathfrak{m}^2 \subset I$, we also have $\mathfrak{m} \cap k[\![X_1', \ldots, X_d']\!] = (X_1', \ldots, X_h')$, which was our former condition. We pass to $\overline{A} := A/(X_2', \ldots, X_d')$ and denote images of ideals or elements of A in $\overline{A}$ by an overbar. We have $\dim \overline{A} = 1$ and $\overline{X_1'} \in \overline{I}$, therefore $\overline{I}$ is not contained in a minimal prime ideal of $\overline{A}$. We can find $a \in I$ such that $\overline{z}_1 + \overline{a}$ is a system of parameters of $\overline{A}$. Putting $X_1 := z_1 + a$ the set $\{X_1, z_2, \ldots, z_h, X_{h+1}, \ldots, X_d\}$ becomes a parameter system of A. Moreover $da \in \mathfrak{m}(\Omega^1_{A/k})_{\mathfrak{m}}$ and henceforth the differential dz_1 in the system of generators $\{dz_1, \ldots, dz_h, dX_{h+1}, \ldots, dX_d\}$ can be replaced by dX_1. Repeating this process we finally end up with a parameter system $\{X_1, \ldots, X_d\}$ of A having properties $\alpha)$ and $\beta)$, q.e.d.

The theorem is not true, if A is not a homomorphic image of a power series algebra (see exercise 3).

The results of the last two sections about analytic and semianalytic algebras also hold, with suitable modifications, if everywhere in the theory the algebra of formal power series $K[\![X_1, \ldots, X_n]\!]$ is replaced by the algebra of convergent power series $K\{X_1, \ldots, X_n\}$ over a valued field (see [BKKN], Dieudonné-Grothendieck [DG], Scheja [Sch] and Scheja-Storch [SS$_2$]).

Semianalytic algebras are excellent rings. In the case of convergent power series this was first established by Kiehl [Ki]. Scheja and Storch [SS$_2$] showed that any noetherian K-algebra A over a field K of characteristic 0, for which $\tilde{\Omega}^1_{A/K}$ exists, is excellent. For the general theory of excellent rings, which is strongly related to the theory of differential modules, the reader is referred to EGA, Matsumura [M$_1$], and the more recent treatment of Brezuleanu and Radu [BR$_2$]. Part of the theory developed in the last two sections also plays an important role in the proof of the "local Bertini theorems" of Flenner [Fl].

Some of the results of § 9 on complete intersections can easily be transferred to the "semianalytic" case. The Kähler differents $\vartheta^{(i)}$(S/K) of a reduced semianalytic algebrad S over a field K are defined to be the Fitting ideals of D_K(S). They have analogous properties as the Kähler differents of affine K-algebras.

Exercises

1) Let S be a semianalytic algebra over a field k. Given $\mathfrak{p}, \mathfrak{p}' \in$ Spec(S) with $\mathfrak{p}' \subset \mathfrak{p}$ such that $S_\mathfrak{p}$ is equidimensional show that

$$\text{edim } S_\mathfrak{p} - \text{dim } S_\mathfrak{p} \leq \text{edim } S_{\mathfrak{p}'} - \text{dim } S_{\mathfrak{p}'}.$$

2) Let $(X_i)_{i=1,2,..}$ be a countable family of indeterminates and $k := \mathbb{F}_p(X_1, X_2, ...)$ the rational function field in these indeterminates over $\mathbb{F}_p$. In $k[\![X,Y]\!]$ consider the power series

$$F := Y^p - \sum_{i=1}^{\infty} X_i X^{ip}. \quad \text{Let } A := k[\![X,Y]\!]/(F).$$

a) F is irreducible in $\ell[\![X,Y]\!]$ for any finite extension field ℓ of k.

b) $K := Q(A)$ is a separable extension field of k, but K/k is not analytically separable.

3) Let k be a field of characteristic $p > 0$ with $k \neq k^p$. Choose $\alpha \in k \smallsetminus k^p$.

a) The analytic k-algebra $A := k[\![X]\!][\sqrt[p]{\alpha+X}]$ is absolutely regular over k at its maximal ideal $\mathfrak{g}$.

b) Statement b) of theorem 14.18 is not true for A and $\mathfrak{g}$.

§ 15. Existence of p-Bases

Let S be a ring that contains a field of characteristic $p > 0$. If R is a subring of S such that $S \supset R \supset S^p$ and S is a finitely presented R-module, we say that a __Frobenius-sandwich__ $S \supset R \supset S^p$ is given. In this case S/R is also a finitely presented algebra (that is, there is a presentation $S = R[X_1,\dots,X_n]/I$ with a finitely generated ideal I of $R[X_1,\dots,X_n]$), and $\Omega^1_{S/R}$ is a finitely presented S-module.

We are going to discuss the question under which additional conditions S/R has locally p-bases. This means that we require for each $\mathfrak{p} \in \operatorname{Spec} S$ the existence of a p-basis of $S_{\mathfrak{p}}$ over $R_{\mathfrak{q}}$ ($\mathfrak{q} := \mathfrak{p} \cap R$). According to 6.18 this condition is equivalent with $\Omega^1_{S/R}$ being locally free (i.e. projective).

Observe that the canonical map $\operatorname{Spec} S \to \operatorname{Spec} R$ is a homeomorphism and that

$$(1) \qquad S_{\mathfrak{p}} = R_{\mathfrak{q}} \otimes_R S = S_{\mathfrak{q}}$$

for $\mathfrak{p} \in \operatorname{Spec} S$, $\mathfrak{q} := \mathfrak{p} \cap R$. Clearly $S_{\mathfrak{p}} \supset R_{\mathfrak{q}} \supset (S_{\mathfrak{p}})^p$ is also a Frobenius-sandwich.

__15.1. Remark.__ $S_{\mathfrak{p}}/R_{\mathfrak{q}}$ has a p-basis if and only if $S_{\mathfrak{p}}$ is a free $R_{\mathfrak{q}}$-module and $S_{\mathfrak{p}}/\mathfrak{q}S_{\mathfrak{p}}$ has a p-basis over $k(\mathfrak{q})$.

In fact, if $\{x_1,\dots,x_m\}$ is a system of representatives in $S_{\mathfrak{p}}$ for a p-basis of $S_{\mathfrak{p}}/\mathfrak{q}S_{\mathfrak{p}}$ over $k(\mathfrak{q})$, then $\{x_1,\dots,x_m\}$ is by Nakayama's lemma a p-generating set of $S_{\mathfrak{p}}/R_{\mathfrak{q}}$. If $S_{\mathfrak{p}}/R_{\mathfrak{q}}$ is free, then necessarily $[S_{\mathfrak{p}}:R_{\mathfrak{q}}] = p^m$, and the monomials $x_1^{\alpha_1}\dots x_m^{\alpha_m}$ ($0 \le \alpha_i < p$, $i=1,\dots,m$) form a basis of $S_{\mathfrak{p}}$ over $R_{\mathfrak{q}}$.

The question of existence of p-bases is connected with the

following notions.

15.2. Definition.   Let R be a ring, and let D be a subset
of $\mathrm{Der}_{\mathbb{Z}}(R)$.

a) An ideal I of R is called D-closed, if $dI \subset I$ for all
$d \in D$.

b) R is called D-simple, if $I = (0)$ and $I = R$ are the only
D-closed ideals of R.

c) An algebra S/R is called differentially simple, if S is
D-simple with $D = \mathrm{Der}_R(S)$.

   If R is D-simple with $D = \emptyset$, then R is a field. More generally

15.3. Lemma.   Let R be a D-simple ring. Then

a) $\bigcap_{n \in \mathbb{N}} I^n = (0)$ for any ideal $I \neq R$ of R.

b) If R is a local ring and $x \in R \setminus \{0\}$, there are derivations
$d_1, \ldots, d_m \in D$ such that $(d_1 \circ \ldots \circ d_m)(x)$ is a unit of R.

Proof.   a) Since $dI^{n+1} \subset I^n$ for all $d \in D$ and all $n \in \mathbb{N}$, the
ideal $\bigcap I^n$ is D-closed, hence $\bigcap I^n = (0)$.

b) The ideal J of R generated by all elements of the form
$(d_1 \circ \ldots \circ d_m)(x)$ with $d_1, \ldots, d_m \in D$, $m \geq 0$ is D-closed and
$\neq (0)$, hence $J = R$. Since R is local, the claim follows.

   The relation between differentially simple rings and p-
bases is given in the next theorem (Harper [Ha], Yuan [Y$_1$]).

15.4. Theorem.   Let R be a ring that contains a field of
characteristic $p > 0$ and is D-simple for some subset

$D \subset \mathrm{Der}_{\mathbb{Z}}(R)$. Put $K := \{x \in R \mid dx = 0 \text{ for all } d \in D\}$. Then

a) R is a local ring, and for any x of its maximal ideal $x^p = 0$.

b) K is a field, and there is a field $C \subset R$ that contains K and is isomorphic to $R/\mathfrak{m}$ where $\mathfrak{m}$ is the maximal ideal of R.

c) If $\dim_{R/\mathfrak{m}} \mathfrak{m}/\mathfrak{m}^2 < \infty$ and $\{x_1, \ldots, x_m\}$ is a system of representatives in $\mathfrak{m}$ of a basis of $\mathfrak{m}/\mathfrak{m}^2$, then the canonical C-homomorphism

$$\varphi : C[T_1, \ldots, T_m] \to R \qquad (T_i \mapsto x_i)$$

is an epimorphism with kernel $(T_1^p, \ldots, T_m^p)$, i.e.
$R \cong C[T_1, \ldots, T_m]/(T_1^p, \ldots, T_m^p)$, and $\{x_1, \ldots, x_m\}$ is a p-basis of R/C.

Proof. a) For any non-unit $x \in R$ the ideal (x^p) is D-closed, hence $x^p = 0$. Since all non-units of R are nilpotent, R is a local ring.

b) For $x \in K \smallsetminus \{0\}$ the ideal (x) of R is D-closed, hence $(x) = R$ and x is a unit of R. The quotient formula for derivations (1.9e) shows that $x^{-1} \in K$, hence K is a field. Any field C with $K \subset C \subset R$ which is maximal with this property is mapped isomorphically onto $R/\mathfrak{m}$ under the canonical residue map.

c) By 15.3a) R is separated in the $\mathfrak{m}$-topology. Any element of R has an expansion as a power series in $x_1, \ldots, x_m$ with coefficients in C. Since $x_i^p = 0$ $(i=1, \ldots, m)$, it is clear that φ is surjective and $(T_1^p, \ldots, T_m^p) \subset \ker \varphi$.

Proving equality is equivalent with showing that $x_1^{\alpha_1} \ldots x_m^{\alpha_m}$ $(0 \le \alpha_i < p,\ i=1, \ldots, m)$ are linearly independent over C. We need only to consider the case $m > 0$. It suffi-

234

ces to show, and this is the main point of the proof,

$$(2) \qquad (x_1 \cdot \ldots \cdot x_m)^{p-1} \neq 0.$$

For suppose (2) holds and there is a relation

$$\sum_{0 \leq \alpha_i < p} c_{\alpha_1 \ldots \alpha_m} x_1^{\alpha_1} \cdot \ldots \cdot x_m^{\alpha_m} \qquad (c_{\alpha_1 \ldots \alpha_m} \in C, \text{not all} = 0).$$

Then let $(\alpha_1, \ldots, \alpha_m)$ be the smallest index with respect to
the lexicographic ordering such that $c_{\alpha_1 \ldots \alpha_m} \neq 0$. Multiplying
the relation by $x_1^{p-1-\alpha_1} \cdot \ldots \cdot x_m^{p-1-\alpha_m}$ leads to $(x_1 \cdot \ldots \cdot x_m)^{p-1} = 0$,
a contradiction.

For $a \in R$ let $c_0(a)$ denote the constant term of the expansion of a as a power series in $x_1, \ldots, x_m$ with coefficients
in C, and let $c_i(a)$ for each $i \in \{1, \ldots, m\}$ denote the coefficient of x_i in that expansion. Since $\{x_1, \ldots, x_m\}$ is a minimal system of generators of $\mathcal{M}$, these coefficients are uniquely determined by a, and the following rules hold:

$$c_i(a+a') = c_i(a) + c_i(a') \quad \text{for } a, a' \in R, i=0, \ldots, m$$

$$(3)$$

$$c_i(a_1 \cdot \ldots \cdot a_n) = \sum_{j=1}^{n} c_i(a_j) \cdot \prod_{k \neq j} c_0(a_k) \quad \text{for } a_1, \ldots, a_n \in R, i=1, \ldots, m.$$

(2) is proved indirectly. Suppose $(x_1 \cdot \ldots \cdot x_m)^{p-1} = 0$.
Choose $(\ell_1, \ldots, \ell_m) \in \mathbb{N}^m$ $(0 \leq \ell_i < p, i=1, \ldots, m)$ of smallest
"degree" $\delta := \ell_1 + \ldots + \ell_m$ such that $x_1^{\ell_1} \cdot \ldots \cdot x_m^{\ell_m} = 0$. We may
assume that $\ell_i > 0$ for $i=1, \ldots, s$ and $\ell_i = 0$ for $i=s+1, \ldots, m$.
Then

$$z_i := x_1^{\ell_1} \cdot \ldots \cdot x_i^{\ell_i - 1} \cdot \ldots \cdot x_s^{\ell_s} \neq 0 \quad (i=1, \ldots, s).$$

By 15.3b) we may further assume that, after an appropriate
ordering of the x_i $(i=1, \ldots, s)$, for certain derivations
$d_1, \ldots, d_r \in D$ the element $(d_1 \circ \ldots \circ d_r)(z_1)$ is a unit of R,
while for $\rho < r$ and any set of derivations $\delta_1, \ldots, \delta_\rho \in D$

we have

$$(\delta_1 \circ \ldots \circ \delta_\rho)(z_i) \in \mathcal{M} \qquad (i=1,\ldots,s).$$

By the generalized Leibniz rule (1.13) there is a relation

$$(4) \qquad 0 = (d_1 \circ \ldots \circ d_r)(x_1^{\ell_1} \cdot \ldots \cdot x_s^{\ell_s}) = \sum_\Delta \prod_{(i,j)\in A} \Delta_{ij}(x_i).$$

Here $A := \{(i,j) \in \mathbb{N}\times\mathbb{N} \mid 1 \leq i \leq s, \ 1 \leq j \leq \ell_i\}$. The summation is over the set of all mappings $\Delta : \{1,\ldots,r\} \to A$, and for $(i,j) \in A$

$$\Delta_{ij}(x_i) = \begin{cases} (d_{\rho_1} \circ \ldots \circ d_{\rho_t})(x_i), & \text{if } \Delta^{-1}(i,j)=\{\rho_1,\ldots,\rho_t\}, \rho_1<\ldots<\rho_t, t>0 \\[2ex] x_i & \text{if } \Delta^{-1}(i,j) = \emptyset. \end{cases}$$

The rules (3), applied to (4), give us a relation

$$0 = c_1\left(\sum_\Delta \prod_{(i,j)\in A} \Delta_{ij}(x_i)\right) =$$

$$(5) \qquad = \sum_\Delta \sum_{(i,j)\in A} c_1(\Delta_{ij}(x_i)) \cdot \prod_{\substack{(k,\ell)\in A \\ (k,\ell)\neq(i,j)}} c_0(\Delta_{k\ell}(x_k))$$

$$= \sum_{(i,j)\in A} \sum_\Delta c_1(\Delta_{ij}(x_i)) \cdot \prod_{(k,\ell)\neq(i,j)} c_0(\Delta_{k\ell}(x_k)).$$

By studying this expression term by term we shall show that it is also $\neq 0$, which is the desired contradiction.

For fixed $(i,j) \in A$ with $i > 1$ consider first all mappings Δ with $\Delta^{-1}(i,j) = \emptyset$. For these we have $c_1(\Delta_{ij}(x_i)) = c_1(x_i) = 0$. Now consider a fixed non-empty subset $B \subset \{1,\ldots,r\}$ and take for fixed $(i,j) \in A$ all mappings Δ with $\Delta^{-1}(i,j) = B$. Then

$$\sum_{\Delta^{-1}(i,j)=B} \prod_{(k,\ell)\neq(i,j)} \Delta_{k\ell}(x_k)$$

is the expansion by Leibniz' rule of $(\delta_1 \circ \ldots \circ \delta_\rho)(z_i)$ for certain $\delta_1,\ldots,\delta_\rho \in D$, where $\rho < r$. We have

$(\delta_1 \ldots \delta_\rho)(z_i) \in \mathfrak{m}$ by assumption, hence

$$0 = c_1(\Delta_{ij}(x_i)) \cdot c_0 \Big(\sum_{\Delta^{-1}(i,j) \in B} \prod_{(k,\ell) \neq (i,j)} \Delta_{k\ell}(x_k) \Big) =$$

$$\sum_{\Delta^{-1}(i,j)=B} c_1(\Delta_{ij}(x_i)) \prod_{(k,\ell) \neq (i,j)} c_0(\Delta_{k\ell}(x_k)).$$

The remaining part of the sum in (5) is

$$\sum_{j=1}^{\ell_1} \sum_{\Delta^{-1}(1,j)=\emptyset} c_1(x_1) \cdot \prod_{(k,\ell) \neq (1,j)} c_0(\Delta_{k,\ell}(x_k)) = \ell_1 \cdot c_0((d_1 \circ .. \circ d_r)(z_1)).$$

Since, by construction, $(d_1 \circ \ldots \circ d_r)(z_1)$ is a unit of R and

$\ell_1 < p$ this expression is $\neq 0$, q.e.d.

We now obtain the following criterion for the existence

of a p-basis (Yuan [Y$_2$]):

<u>15.5. Theorem.</u>   Let $S \supset R \supset S^p$ be a Frobenius-sandwich, and

let $\mathfrak{g} := \mathrm{Der}_R(S)$. Then the following assertions are equi-

valent:

a) S/R has locally p-bases.

b) S is a projective R-module, and the ring of endomorphisms

$\mathrm{Hom}_R(S,S)$ is generated as an R-algebra by $\mathfrak{g}$:

$$\mathrm{Hom}_R(S,S) = R[\mathfrak{g}].$$

c) S is a projective R-module, and $S_{\mathfrak{g}}/\mathfrak{g}S_{\mathfrak{g}}$ is a differentially

simple $k(\mathfrak{g})$-algebra for each $\mathfrak{g} \in \mathrm{Spec}\ R$.

If the conditions a)-c) are satisfied, then for any

$\mathfrak{p} \in \mathrm{Spec}\ S$, $\mathfrak{g} := \mathfrak{p} \cap R$, the canonical sequence (6.4)

$$0 \to \mathfrak{p}S_{\mathfrak{p}}/(\mathfrak{g}+\mathfrak{p}^2)S_{\mathfrak{p}} \to k(\mathfrak{p}) \otimes_{S_{\mathfrak{p}}} \Omega^1_{S_{\mathfrak{p}}/R_{\mathfrak{g}}} \to \Omega^1_{k(\mathfrak{p})/k(\mathfrak{g})} \to 0$$

is exact and

$$\text{p-deg}(S_{\mathfrak{p}}/R_{\mathfrak{g}}) = \text{edim}\ S_{\mathfrak{p}}/\mathfrak{g}S_{\mathfrak{p}} + \text{p-deg}(k(\mathfrak{p})/k(\mathfrak{g})).$$

Proof. The theorem is of local nature, since $\mathrm{Hom}_{R_\wp}(S_\wp, S_\wp) = \mathrm{Hom}_R(S,S)_\wp$ and $\mathrm{Der}_{R_\wp}(S_\wp) = \mathrm{Der}_R(S)_\wp$ for each $\wp \in \mathrm{Spec}\, R$, the R-module S being finitely presented. We therefore can assume that S (and hence also R) is a local ring.

a) $\to$ b) If $\{x_1, \ldots, x_m\}$ is a p-basis of S/R, then $\Omega^1_{S/R} = Sdx_1 \oplus \ldots \oplus Sdx_m$, and there are derivations $d_i \in \mathcal{Y}$ with $d_i x_j = \delta_{ij}$ $(i,j=1,\ldots,m)$. We order the m-tuples $(\alpha_1, \ldots, \alpha_m) \in \mathbb{N}^m$ $(0 \le \alpha_i < p)$ lexicographically. We have

$$(d_1^{\beta_1} \circ \ldots \circ d_m^{\beta_m})(x_1^{\gamma_1} \cdot \ldots \cdot x_m^{\gamma_m}) = \begin{cases} 0 & \text{if } (\gamma_1, \ldots, \gamma_m) < (\beta_1, \ldots, \beta_m) \\ \\ \beta_1! \cdot \ldots \cdot \beta_m! & \text{if } (\gamma_1, \ldots, \gamma_m) = (\beta_1, \ldots, \beta_m) \end{cases}$$

for any pair of such m-tuples. Since $\beta_1! \cdot \ldots \cdot \beta_m!$ is a unit of R, it follows easily that

$$x_1^{\alpha_1} \cdot \ldots \cdot x_m^{\alpha_m} \cdot d_1^{\beta_1} \circ \ldots \circ d_m^{\beta_m} \quad (0 \le \alpha_i < p, 0 \le \beta_j < p)$$

is a basis of the R-module $\mathrm{Hom}_R(S,S)$.

b) $\to$ c) Let $\mathfrak{m}$ be the maximal ideal of R, and let $\bar{R} := R/\mathfrak{m}$, $\bar{S} := S/\mathfrak{m}S$. Since S is a free R-module, we have $\mathrm{Hom}_{\bar{R}}(\bar{S}, \bar{S}) = \mathrm{Hom}_R(S,S)/\mathfrak{m}\,\mathrm{Hom}_R(S,S)$. Any derivation $d \in \mathcal{Y}$ induces a derivation $\bar{d} \in \mathrm{Der}_{\bar{R}}(\bar{S}) =: \bar{\mathcal{Y}}$. From $\mathrm{Hom}_R(S,S) = R[\mathcal{Y}]$ we therefore obtain $\mathrm{Hom}_{\bar{R}}(\bar{S}, \bar{S}) = \bar{R}[\bar{\mathcal{Y}}]$. Since $\bar{R}$ is a field, any $x \in \bar{S} \setminus \{0\}$ can be mapped by an $\bar{R}$-linear endomorphism of $\bar{S}$ onto any given $y \in \bar{S}$. One easily concludes that $\bar{S}/\bar{R}$ is differentially simple.

c) $\to$ a) If $\bar{S}/\bar{R}$ is differentially simple, then by 15.4

$$\bar{S} = C[T_1, \ldots, T_m]/(T_1^p, \ldots, T_m^p)$$

with a field C such that $C^p \subset \bar{R} \subset C \subset \bar{S}$. The images of the T_i in $\bar{S}$ together with a p-basis of $C/\bar{R}$ form a p-basis of $\bar{S}/\bar{R}$. By 15.1, S/R has a p-basis.

Obviously its length is edim $\overline{S}$ + p-deg$(C/\overline{R})$, and C is isomorphic to the residue field of $\overline{S}$. Let $\mathcal{M}$ be the maximal ideal of S. Comparing dimensions of the vector spaces of the exact sequence

$$\mathcal{M}/\mathcal{M}S+\mathcal{M}^2 \to S/\mathcal{M} \otimes_S \Omega^1_{S/R} \to \Omega^1_{S/\mathcal{M}/\overline{R}} \to 0$$

we see that the first mapping is injective, q.e.d.

As an application of 15.5 we have

15.6. Proposition (Kimura-Niitsuma $[KN_1]$). Suppose $S \supset T \supset S^p$ is a Frobenius-sandwich, and R is a subring of S with $S \supset R \supset T$. Assume S is a projective R-module, and R is a finite projective T-module. If S/T has locally p-bases, so has R/T.

Proof. We can assume that R,S and T are local rings. Let $\mathcal{M}$ be the maximal ideal of T, let $\overline{S} := S/\mathcal{M}S$, $\overline{R} := R/\mathcal{M}R$, and $K := T/\mathcal{M}$. $\overline{S}$ is a free $\overline{R}$-module, and $\overline{S}/K$ is differentially simple by 15.5. We are going to show that $\overline{R}/K$ is differentially simple too. Then applying 15.5 again, we obtain that R/T has a p-basis.

Let $\{\omega_1,\ldots,\omega_r\}$ be a basis of $\overline{S}/\overline{R}$. For any $x \in \overline{S}$ and any K-derivation d of $\overline{S}$ we write

$dx = d_1x\cdot\omega_1 +\ldots+ d_rx\cdot\omega_r$ with certain $d_ix \in \overline{R}$.

Clearly the mappings $d_i : \overline{R} \to \overline{R}$ $(x \mapsto d_ix)$ are K-derivations of $\overline{R}$ $(i=1,\ldots,r)$. Let I be an ideal of $\overline{R}$, $I \neq (0)$, $I \neq \overline{R}$. Since $\overline{S}/K$ is differentially simple, there is a derivation $d \in \mathrm{Der}_K(\overline{S})$ with $d(I\overline{S}) \not\subset I\overline{S}$. This implies that $d_iI \not\subset I$ for at least one $i \in \{1,\ldots,r\}$. Hence $\overline{R}/K$ is differentially simple.

This proposition is the key to the proof of the following theorem of Kimura-Niitsuma [KN_1].

15.7. Theorem.   Let $S \supset R \supset S^p$ be a Frobenius-sandwich where S is a regular noetherian ring. Then the following assertions are equivalent:

a) S/R has locally p-bases.

b) R is a regular noetherian ring.

c) $\Omega^1_{S/R}$ is a projective S-module.

Proof   (Matsumura [M_2]). We have mentioned already that a) and c) are equivalent. If we assume a), then S/R is projective and R is noetherian, since S is. The homological characterization of regularity (Matsumura [M_1], Thm. 51) implies that with S the ring R is regular as well.

For the proof of b) $\to$ a) we may assume that R and S are regular local rings. Then S is a free R-module (Matsumura [M_1], Thm. 46). The existence of a p-basis of S/R is proved by induction on the rank [S:R].

Passing to the quotient fields of R and S one can see that for any $x \in S \setminus R$ the R-module R[x] has the basis $\{1, x, \ldots, x^{p-1}\}$. If $\{x_2, \ldots, x_m\}$ is a p-basis of S/R[x], then $\{x, x_2, \ldots, x_m\}$ is a p-basis of S/R. By the induction hypothesis it suffices therefore to find $x \in S \setminus R$ such that R[x] is a regular local ring.

Let $\mathfrak{m}_R$ and $\mathfrak{m}_S$ be the maximal ideals of R and S.

α) If $S/\mathfrak{m}_S \neq R/\mathfrak{m}_R$, we choose $x \in S$ whose residue in $S/\mathfrak{m}_S$ is not in $R/\mathfrak{m}_R$. Then $R[x]/\mathfrak{m}_R R[x]$ is a field, since $x^p \in R$, and henceforth $R[x]$ is a regular local ring.

β) If $\mathfrak{m}_S^{[p]} := \{x^p \mid x \in \mathfrak{m}_S\}$ is not contained in $\mathfrak{m}_R^2$, we choose $x \in \mathfrak{m}_S$ with $x^p \notin \mathfrak{m}_R^2$. Then we can find $y_2,\ldots,y_d \in \mathfrak{m}_R$ such that $\{x^p, y_2, \ldots, y_d\}$ is a regular system of parameters of R, where $d := \dim R = \dim S$. Since

$$R[x]/(x,y_2,\ldots,y_d) \cong R/(x^p, y_2, \ldots, y_d)$$

is a field, $R[x]$ is again a regular local ring.

γ) By α) and β) we can assume that R and S have the same residue field and that $\mathfrak{m}_S^{[p]} \subset \mathfrak{m}_R^2$. We shall show that in this case $R = S$, which finishes the proof.

Since the two conditions are preserved under completion, we may assume that R and S are complete, hence are power series algebras over a field K. We then have the following situation:
$$S = K[\![X_1, \ldots, X_d]\!] \supset R = K[\![Y_1, \ldots, Y_d]\!] \supset K[\![X_1^p, \ldots, X_d^p]\!] \supset S^p.$$
Put $T := K[\![X_1^p, \ldots, X_d^p]\!]$, and let $\mathfrak{m}_T$ be the maximal ideal of T. S/R and R/T are free, and S/T has a p-basis of length d.

By 15.6 the algebra R/T has a p-basis and its length is $\operatorname{edim} \bar{R}$, where $\bar{R} := R/\mathfrak{m}_T R$. But $\mathfrak{m}_T R \subset \mathfrak{m}_R^2$ by assumption, hence $\operatorname{edim} \bar{R} = d$. The equation $[S:T] = p^d = [R:T]$ implies that $R = S$, q.e.d.

Next we point out a relation between p-bases and complete intersections.

<u>15.8. Theorem.</u> Let $S \supset R \supset S^p$ be a Frobenius-sandwich of local rings. Then the following conditions are equivalent:

a) S/R has a p-basis.

b) S is a free R-module whose rank $[S:R]$ is a power of p, and $S/\mathfrak{m}_R S$ is a strict complete intersection (C.25).

Proof. Put $\overline{S} := S/\mathfrak{m}_R S$ and $K := R/\mathfrak{m}_R$.

a) $\to$ b) If S/R has a p-basis, then $\overline{S}/K$ is differentially simple and by 15.4

$$\overline{S} = C[X_1,\ldots,X_d]/(X_1^p,\ldots,X_d^p) = C[\![X_1,\ldots,X_d]\!]/(X_1^p,\ldots,X_d^p)$$

with a field C. This is a presentation of $\overline{S}$ as a strict complete intersection.

b) $\to$ a). Let $\mathfrak{m}$ be the maximal ideal of $\overline{S}$, and let $n := \operatorname{edim} \overline{S}$. $\overline{S}$ has a presentation

$$\overline{S} = C[\![X_1,\ldots,X_n]\!]/I$$

where C is a field containing K and

$$(X_1^p,\ldots,X_n^p) \subset I \subset (X_1,\ldots,X_n)^2.$$

Since $\overline{S}$ is a strict complete intersection, we know from C.26 that I can be generated by a super-regular sequence $\{F_1,\ldots,F_n\}$ and

$$\operatorname{gr}_{\mathfrak{m}}(\overline{S}) = C[X_1,\ldots,X_n]/(\varphi_1,\ldots,\varphi_n)$$

where φ_i is the leading form of F_i (i=1,...,n). By C.29 we have

$$[S:R] = \dim_K \overline{S} = [C:K]\cdot\dim_C \overline{S} = [C:K]\cdot\dim_C(\operatorname{gr}_{\mathfrak{m}}\overline{S}) = [C:K]\cdot d_1\cdot\ldots\cdot d_n$$

where $d_i := \deg \varphi_i$ (i=1,...,n).

By assumption $[S:R]$ is a power of p, hence $d_1\cdot\ldots\cdot d_n$ is a power of p. On the other hand, $d_i \geq 2$ for i=1,...,n, because $I \subset (X_1,\ldots,X_n)^2$, and $\dim_C \overline{S} \leq p^n$, because $(X_1^p,\ldots,X_n^p) \subset I$. Necessarily we must have $d_1 = ..= d_n = p$, which implies that $I = (F_1,\ldots,F_n) = (X_1^p,\ldots,X_n^p)$. This shows that $\overline{S}$ has a p-basis over C and over K. Hence S has a p-basis over R, q.e.d.

The previous existence theorems for p-bases can be applied

to "inseparable Galois theory". Consider a Frobenius-sand-
wich $S \supset R \supset S^p$ where S/R has locally p-bases. Let $\mathfrak{Z}$ denote
the set of all intermediate rings T of R and S such that S
is a finitely presented T-module and S/T has locally p-bases.

A p-Lie-subalgebra of $Der_R(S)$ is an S-submodule of $Der_R(S)$
which is closed with respect to the Lie-product (§ 1) and the
formation of p-th powers of derivations (In characteristic p
these are derivations, § 1, exercise 5). Let $\mathfrak{U}$ be the
set of all p-Lie-subalgebras of $Der_R(S)$ which as S-modules
are direct summands of $Der_R(S)$.

15.9. Theorem.   For any $T \in \mathfrak{Z}$ the S-module $\mathfrak{y} := Der_T(S)$ is
an element of $\mathfrak{U}$, and for each $\mathfrak{y} \in \mathfrak{U}$ the ring $T := \bigcap_{d \in \mathfrak{y}} \ker d$
belongs to $\mathfrak{Z}$. The mappings

$$\mathfrak{Z} \to \mathfrak{U} \qquad\qquad (T \mapsto Der_T(S))$$

and

$$\mathfrak{U} \to \mathfrak{Z} \qquad\qquad (\mathfrak{y} \to \bigcap_{d \in \mathfrak{y}} \ker d)$$

are inverse to each other.

For the proof we refer to Yuan $[Y_2]$.

15.10. Corollary.   Let $S \supset R \supset S^p$ be a Frobenius-sandwich
where R and S are regular noetherian rings. Then the regular
noetherian rings T with $S \supset T \supset R$ are in natural one-to-one
correspondence with the p-Lie-subalgebras of $Der_R(S)$ which
as S-modules are direct summands of $Der_R(S)$.

In fact, by 15.7 the set $\mathfrak{Z}$ above is just the set of regular intermediate rings of S/R.

We close this section with some references. Theorem 15.7
was first claimed in [KK] as a corollary of Satz 5, which,

unfortunately, is false. Its correct version is given above as theorem 15.8. However, 15.7 is not an immediate consequence of 15.8.

15.7 was first proved for 2-dimensional regular local rings by Rudakov-Shafarevich [RS] with global methods, and by Ganong [G], who used Hamburger-Noether expansions. The proof of Kimura-Niitsuma [KN_1] in the general case was given a simplification by Matsumura [M_2]. Its the proof we used above.

Ganong [G] also proves the existence of a p-basis for Frobenius-sandwiches $S \supset R \supset S^p$ where R and S are polynomial rings of 2 variables over an (algebraically closed) field of characteristic $p > 0$. It seems that for polynomial rings in more than two variables a proof of the corresponding result is still missing. Kimura-Niitsuma [KN_2] have proved that for an arbitrary regular local ring R of prime characteristic p the existence of a p-basis of R/R^p is equivalent with Ω^1_{R/R^p} being free.

Informations about differentially closed ideals and differentially simple rings in characteristic O are contained in the papers [Se_1] and [Se_2] by Seidenberg.

Exercises

1) Let $(R, \mathcal{m})$ and S be local rings, and let $R \to S$ be a local inclusion such that $S^p \subset R$. If S/R has a p-basis (which need not be finite), then $\bar{S} := S/\mathcal{m}S$ is differentially simple over $R/\mathcal{m}$.

2) In 15.4c) the condition that m/m^2 is a finite-dimensional vector space may be replaced by the condition that m is a nilpotent ideal.

3) In the situation of exercise 1) let R' be a ring with $R \subset R' \subset S$ such that R'/R is finite and free and S/R' is free. If S/R has a p-basis, so has R'/R.

4) Let K be a field of characteristic $p > 0$, and let $S := K[\![X_1,\ldots,X_d]\!]$, $R := K[\![X_1^p,\ldots,X_d^p]\!]$. For each regular local ring T with $S \supset T \supset R$ there is a regular system of parameters $\{Y_1,\ldots,Y_d\}$ of S and an $r \in \{0,\ldots,d\}$ such that
$$T = K[\![Y_1^p,\ldots,Y_r^p,Y_{r+1},\ldots,Y_d]\!].$$

§ 16. Traces of Differential Forms

For a finite locally free algebra S/R the canonical trace $\sigma_{S/R} : S \to R$ and the canonical norm $n_{S/R} : S \to R$ are defined (F.3). It would be very usefull if for any pair $(S/R,\Omega)$ consisting of an algebra S/R as above and a differential algebra Ω of R a trace mapping

$$\sigma^{\Omega}_{S/R} : \Omega_S \to \Omega$$

where Ω_S is the universal S-extension of Ω, could be constructed such that the following conditions ("trace axioms") are satisfied:

Tr1) (Linearity) If we consider Ω_S as a left Ω-module via the canonical map $\Omega \to \Omega_S$, then $\sigma^{\Omega}_{S/R}$ is Ω-linear and homogeneous of degree 0 (i.e. Ω^p_S is mapped into Ω^p for each $p \in \mathbb{N}$).

Tr2) (Relation to the canonical trace) The restriction $\sigma^{\Omega}_{S/R}|_{\Omega^0_S} : S \to R$ to the elements of degree 0 is the canonical trace $\sigma_{S/R}$.

Tr3) (Base change) If $\alpha : R \to R'$ is a ring homomorphism, if $S' := R' \otimes_R S$, and if $\Omega \to \Omega'$ is an α-homomorphism of Ω into a differential algebra Ω' of R', then the following diagram commutes

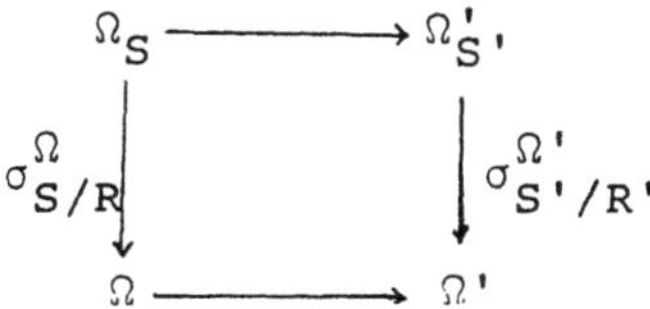

Tr4) (Direct products) If $S = S_1 \times \ldots \times S_h$ is a direct product of algebras S_i/R $(i=1,\ldots,h)$, then for each $\omega = (\omega_1,\ldots,\omega_h) \in \Omega_{S_1} \times \ldots \times \Omega_{S_h} = \Omega_S$ we have

$$\sigma^{\Omega}_{S/R}(\omega) = \sum_{i=1}^{h} \sigma^{\Omega}_{S_i/R}(\omega_i).$$

Tr5) (Transitivity) If S/R and T/S are finite locally free algebras, then

$$\sigma^{\Omega}_{T/R} = \sigma^{\Omega}_{S/R} \circ \sigma^{\Omega_S}_{T/S}.$$

Tr6) (Differentiation) Let d be the differentiation of Ω, and let d' be the differentiation of Ω_S. Then

$$d \circ \sigma^{\Omega}_{S/R} = \sigma^{\Omega}_{S/R} \circ d'.$$

Tr7) (Logarithmic derivative) If a $\in$ S is a unit, then

$$\sigma^{\Omega}_{S/R}\left(\frac{da}{a}\right) = \frac{dn_{S/R}(a)}{n_{S/R}(a)}.$$

In this section we shall construct a trace $\sigma^{\Omega}_{S/R}$ for all pairs (S/R,Ω) where

a) R is a noetherian ring,

b) S/R is finite and locally a complete intersection (C.3),

c) Ω is an arbitrary differential algebra of R.

It will be shown that the system of traces $\{\sigma^{\Omega}_{S/R}\}$ for these pairs obeys the axioms Tr1)-Tr7), and is, moreover, uniquely determined by the axioms Tr1)-Tr4). It is under-stood here that in Tr3) we allow only base changes with noetherian rings R'. For other situations in which traces of differential forms have been constructed, see the re-ferences at the end of this section.

We start with the question of uniqueness.

<u>16.1. Theorem.</u>   Let C be the class of all pairs (S/R,Ω) where R is noetherian, S/R is finite and locally a complete intersection, and Ω is a differential algebra of R. Then

there is at most one system of traces $\sigma = \{\sigma^{\Omega}_{S/R}\}_{(S/R,\Omega)\in C}$ satisfying Tr1)-Tr4).

Proof. Let $\tilde{\sigma} = \{\tilde{\sigma}^{\Omega}_{S/R}\}$ be another family of this kind. Starting with the case of étale extensions and passing step by step to more general situations we shall show that $\tilde{\sigma}^{\Omega}_{S/R} = \sigma^{\Omega}_{S/R}$ for all $(S/R,\Omega) \in C$. In what follows we always assume that $(S/R,\Omega) \in C$.

a) If S/R is étale, then $\Omega_S = S \otimes_R \Omega$, as follows easily from 6.16a). Hence by Tr1) and Tr2)

$$\sigma^{\Omega}_{S/R} = \sigma_{S/R} \otimes \mathrm{id}_{\Omega}$$

and we see that in this case the trace is already uniquely determined by Tr1) and Tr2). This is true, in particular, if S/R is a finite separable field extension (and only this case will be used in the rest of the proof).

b) Suppose now that R and S are complete regular local rings of characteristic O where R is a power series algebra over a Cohen ring and the residue field of R is finitely generated over its prime field. Then the universally finite differential algebra $\tilde{\Omega}_{R/\mathbf{Z}}$ exists (12.9). Put $\Omega := \tilde{\Omega}_{R/\mathbf{Z}}$, then $\Omega_S = \tilde{\Omega}_{S/\mathbf{Z}}$ (11.9). Since R is a power series algebra over a Cohen ring, Ω is a free R-module (12.16).

L/K with $L := Q(S)$ and $K := Q(R)$ is a separable field extension. Let $\Omega_K = K \otimes_R \Omega$ be the universal K-extension of Ω. By Tr3) there is a commutative diagram

$$
\begin{array}{ccc}
\Omega_S & \longrightarrow & \Omega_L \\
\sigma^{\Omega}_{S/R} \downarrow & & \downarrow \sigma^{\Omega_K}_{L/K} \\
\Omega & \longrightarrow & \Omega_K
\end{array}
$$

and an analogous diagram for the traces $\tilde{\sigma}$. Since $\tilde{\sigma}_{L/K}^{\Omega_K} = \sigma_{L/K}^{\Omega_K}$ by a) and since $\Omega \to \Omega_K$ is injective, we conclude that $\tilde{\sigma}_{S/R}^{\Omega} = \sigma_{S/R}^{\Omega}$.

c) Now let R and S be arbitrary complete noetherian local rings.

<u>16.2. Lemma.</u>   There are complete regular local rings R_O and S_O of characteristic zero where R_O is a power series algebra over a Cohen ring, and there is a flat local homomorphism $R_O \to S_O$ such that

α) S_O/R_O is finite.

β) R is a homomorphic image of R_O, and

$$ S = R \otimes_{R_O} S_O. $$

Suppose we have proved the lemma. If we assume in addition that the residue field of R is finitely generated over its prime field, then $\tilde{\Omega}_{R/\mathbf{Z}}$ and $\tilde{\Omega}_{R_O/\mathbf{Z}}$ exist (12.9). Clearly S_O/R_O is locally a complete intersection. By Tr3) there is a commutative diagram

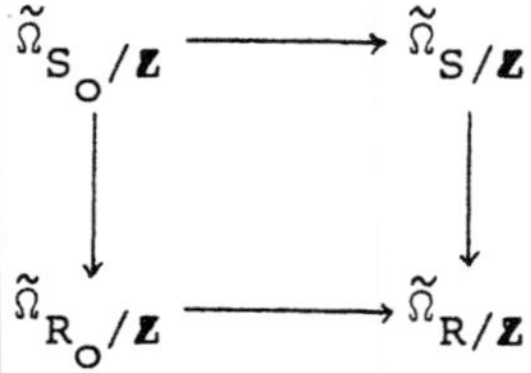

where the horizontal maps are surjective and the vertical maps are the traces from the system σ or $\tilde{\sigma}$. By b) we know already that $\sigma\,{}_{S_O/R_O}^{\tilde{\Omega}_{R_O/\mathbf{Z}}} = \tilde{\sigma}\,{}_{S_O/R_O}^{\tilde{\Omega}_{R_O/\mathbf{Z}}}$, and we conclude that $\sigma\,{}_{S/R}^{\tilde{\Omega}_{R/\mathbf{Z}}} = \tilde{\sigma}\,{}_{S/R}^{\tilde{\Omega}_{R/\mathbf{Z}}}$.

Lemma 16.2 will be shown at the end of the uniqueness proof.

d) Let R be an arbitrary complete noetherian local ring whose residue field is finitely generated over its prime field, and let $\Omega := \tilde{\Omega}_{R/\mathbb{Z}}$. Then $S = S_1 \times \dots \times S_h$ where S_i is a complete local ring and S_i/R is locally a complete intersection ($i=1,\dots,h$) by C.19a). By d) we have $\tilde{\sigma}^{\Omega}_{S_i/R} = \sigma^{\Omega}_{S_i/R}$ ($i=1,\dots,h$), hence $\tilde{\sigma}^{\Omega}_{S/R} = \sigma^{\Omega}_{S/R}$ by Tr4).

e) Let R be a local ring that is essentially of finite type over $\mathbb{Z}$, and let $\hat{R}$ be its completion. Then $\tilde{\Omega}_{\hat{R}/\mathbb{Z}}$ exists and $\tilde{\Omega}_{\hat{R}/\mathbb{Z}} = \hat{R} \otimes_R \Omega_{R/\mathbb{Z}}$ (12.10). Putting $\hat{S} := \hat{R} \otimes_R S$ we obtain a commutative diagram

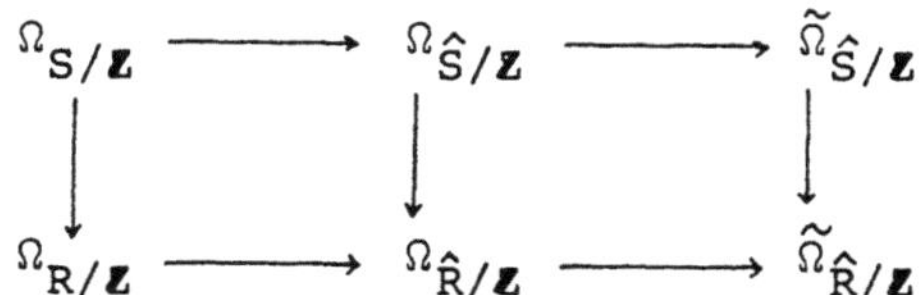

where the vertical maps are the traces from σ or $\tilde{\sigma}$, and the composed map $\Omega_{R/\mathbb{Z}} \to \Omega_{\hat{R}/\mathbb{Z}} \to \tilde{\Omega}_{\hat{R}/\mathbb{Z}}$ is injective (being the canonical map $\Omega_{R/\mathbb{Z}} \to \hat{R} \otimes_R \Omega_{R/\mathbb{Z}}$). Since the traces σ and $\tilde{\sigma}$ on the right side of the diagram coincide by d), so do the traces on the left.

f) Let R be an arbitrary noetherian local ring and let $\Omega := \Omega_{R/\mathbb{Z}}$.

__16.3. Lemma.__   Given $\omega \in \Omega_S$ there exists a local ring R_0 that is essentially of finite type over $\mathbb{Z}$, and there exists a finite algebra S_0/R_0 such that

α) S_0/R_0 is locally a complete intersection.

β) There is a local homomorphism $R_0 \to R$ such that $S = R \otimes_{R_0} S_0$.

γ) ω is in the image of $\Omega_{S_0/\mathbb{Z}}$ in $\Omega_{S/\mathbb{Z}}$.

  The proof of this lemma will be given after the proof of 16.2.

In the situation of the lemma we have a commutative diagram

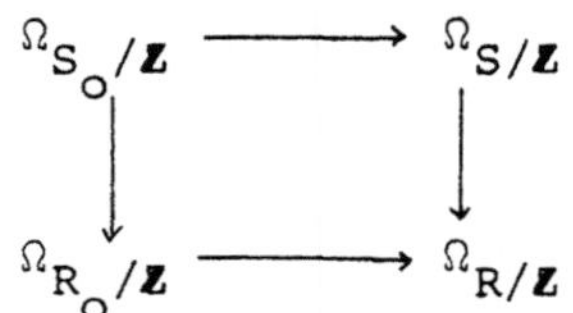

$$\begin{array}{ccc} \Omega_{S_0/\mathbb{Z}} & \longrightarrow & \Omega_{S/\mathbb{Z}} \\ \downarrow & & \downarrow \\ \Omega_{R_0/\mathbb{Z}} & \longrightarrow & \Omega_{R/\mathbb{Z}} \end{array}$$

where the vertical arrows are the traces. We obtain
$\tilde{\sigma}_{S/R}^{\,\Omega_{R/\mathbb{Z}}}(\omega) = \sigma_{S/R}^{\,\Omega_{R/\mathbb{Z}}}(\omega)$, since ω is in the image of $\Omega_{S_0/\mathbb{Z}} \to \Omega_{S/\mathbb{Z}}$
and the traces $\tilde{\sigma}$ and σ coincide on $\Omega_{S_0/\mathbb{Z}}$ by e).

g) In the general case for each $\mathcal{M} \in \mathrm{Max}(R)$ the algebra $S_\mathcal{M}/R_\mathcal{M}$
is finite and locally a complete intersection. We have
$$\tilde{\sigma}_{S_\mathcal{M}/R_\mathcal{M}}^{\,\Omega_{R_\mathcal{M}/\mathbb{Z}}} = \sigma_{S_\mathcal{M}/R_\mathcal{M}}^{\,\Omega_{R_\mathcal{M}/\mathbb{Z}}}$$
by f) and hence $\tilde{\sigma}_{S/R}^{\,\Omega_{R/\mathbb{Z}}} = \sigma_{S/R}^{\,\Omega_{R/\mathbb{Z}}}$ by Tr3) and the local-global
principle. Finally, if Ω is an arbitrary differential alge-
bra of R, then Ω is a homomorphic image of $\Omega_{R/\mathbb{Z}}$ and we can
apply Tr3) once more to obtain $\tilde{\sigma}_{S/R}^{\,\Omega} = \sigma_{S/R}^{\,\Omega}$, q.e.d.

Proof of 16.2.

By C.14 the algebra S/R has a presentation

$$S = R[X_1,\dots,X_n]/(t_1,\dots,t_n)$$

as a complete intersection. Write

$$t_i = \sum_{\nu_1+\dots+\nu_n \leq N} \rho^{(i)}_{\nu_1\dots\nu_n} X_1^{\nu_1}\dots X_n^{\nu_n} \qquad (\rho^{(i)}_{\nu_1\dots\nu_n} \in R,\ i=1,\dots,n).$$

By Cohen's structure theorem R is a homomorphic image of a
power series algebra R_2 over a Cohen ring of characteristic O.
Choose indeterminates $Y^{(i)}_{\nu_1\dots\nu_n}$ $(i=1,\dots,n; \nu_1+\dots+\nu_n \leq N)$ corresponding to the coefficients $\rho^{(i)}_{\nu_1\dots\nu_n}$ of the t_i, and put
$R_1 := R_2[\{Y^{(i)}_{\nu_1\dots\nu_n}\}]$. There is an epimorphism $R_1 \to R$ extending

the epimorphism $R_2 \to R$ to R_1 and sending $Y^{(i)}_{\nu_1 \dots \nu_n}$ to $\rho^{(i)}_{\nu_1 \dots \nu_n}$ $(i=1,\dots,n; \nu_1 + \dots + \nu_n \leq N)$. Let I denote its kernel, let

$$T_i := \Sigma\, Y^{(i)}_{\nu_1 \dots \nu_n} X_1^{\nu_1} \dots X_n^{\nu_n} \qquad (i=1,\dots,n)$$

and let

$$S_1 := R_1[X_1,\dots,X_n]/(T_1,\dots,T_n).$$

Then $S_1 \cong R_2[\{Y^{(i)}_{\nu_1 \dots \nu_n}\}_{(\nu_1,\dots,\nu_n)\neq(0,\dots,0)}, X_1,\dots,X_n]$

is a polynomial ring over R_2 and, in particular, a regular domain. Moreover,

$$R = R_1/I \text{ and } S = R \otimes_{R_1} S_1 = S_1/IS_1.$$

Let R_0 be the completion of R_1 at the preimage of the maximal ideal of R, and let S_0 denote the completion of S_1 at the preimage of the maximal ideal of S. Since R and S are complete, we have

$$R = R_0/IR_0 \text{ and } S = S_0/IS_0 = R \otimes_{R_0} S_0.$$

R_0 and S_0 are complete regular local rings of characteristic 0, having the same dimension, and R_0 is a power series algebra over a Cohen ring. R_0 has the same residue field as R, and S_0 has the same residue field as S. Since S/R is finite, it is clear that S_0/R_0 is quasifinite and hence finite. S_0/R_0 is flat, because both rings are regular local rings of the same dimension. We now have shown all assertions of the lemma.

Proof of 16.3.

By C.14 the algebra S/R has a presentation

$$S = R[X_1,\dots,X_n]/(t_1,\dots,t_n) = R[x_1,\dots,x_n]$$

as a complete intersection. Let $\{y_1,\dots,y_m\}$ be a basis of the free R-module S consisting of monomials in $x_1,\dots,x_n$:

$$y_i = x^{\alpha_i} \qquad (i=1,\ldots,m).$$

In S there are equations

$$(1) \qquad x_j y_i = \Sigma\, r_{jk}^{(i)} y_k \quad (r_{jk}^{(i)} \in R,\ j=1,\ldots,n,\ i=1,\ldots,m).$$

and therefore there are equations in $R[X_1,\ldots,X_n]$ of the form

$$(2) \qquad X_j X^{\alpha_i} - \Sigma\, r_{jk}^{(i)} X^{\alpha_k} = \sum_{s=1}^{n} A_s^{ji} t_s \quad (A_s^{ji} \in R[X]).$$

Let R_1 be a subring of R which is of finite type over the prime ring of R and contains the following elements:

α) all coefficients of the polynomials t_s (s=1,\ldots,n),

β) all coefficients $r_{jk}^{(i)}$ of the equations (1),

γ) all coefficients of all polynomials A_s^{ji} in the equations (2).

We put $S_1 := R_1[X_1,\ldots,X_n]/(t_1,\ldots,t_n)$ and denote the image of X_i in S_1 again by x_i (i=1,\ldots,n). Then

$$S = R \otimes_{R_1} S_1 \, .$$

Since the equations (2) hold in S_1, the equations (1) also hold in S_1, and we can conclude that S_1 is generated as an R_1-module by the elements x^{α_i} (i=1,\ldots,m). Since these elements are linearly independent over R, they are linearly independent over R_1, hence S_1/R_1 is free.

Let $\mathscr{m}$ be the maximal ideal of R, let $\mathscr{m}_1 := \mathscr{m} \cap R_1$, $R_0 := (R_1)_{\mathscr{m}_1}$, and let $S_0 := (S_1)_{\mathscr{m}_1}$. Then $S = R \otimes_{R_0} S_0$, $S_0 = R_0[X_1,\ldots,X_n]/(t_1,\ldots,t_n)$, and S_0 is a free R_0-module of rank m.

It has to be shown that $\{t_1,\ldots,t_n\}$ is a quasiregular sequence of $R_0[X_1,\ldots,X_n]$. Each maximal ideal $\mathscr{M}$ of $R_0[X_1,\ldots,X_n]$ that contains $t_1,\ldots,t_n$ lies over the maximal ideal $\mathcal{M}_o$ of R_o, because S_o/R_o is finite. Let $\bar{t}_1,\ldots,\bar{t}_n$ be the images of $t_1,\ldots,t_n$ in $R_o/\mathcal{M}_o[X_1,\ldots,X_n]$. These elements form a regular sequence of $R_o/\mathcal{M}_o[X_1,\ldots,X_n]$, and we conclude with B.23 that $\{t_1,\ldots,t_n\}$ is a regular sequence in $R_o[X_1,\ldots,X_n]_{\mathcal{M}}$ for any $\mathcal{M}$ as above.

Finally we observe that ω is a finite sum of forms $s_o ds_1 \ldots ds_p$ ($s_i \in S$) and each s_i is a linear combination of $y_1,\ldots,y_m$ with coefficients in R. We can assume that R_1 contains all these coefficients. Then ω is contained in the image of $\Omega_{S_o/\mathbb{Z}} \to \Omega_{S/\mathbb{Z}}$, q.e.d.

We now turn to the construction of the trace and assume at first that S/R has a presentation

$$(3) \qquad S = R[X_1,\ldots,X_n]/(t_1,\ldots,t_n) = R[x_1,\ldots,x_n]$$

as a complete intersection. Let $\tau_t^x : S \to R$ be the trace associated with it (F.20). Put $P := R[X_1,\ldots,X_n]$. We then have an exact sequence

$$0 \to \langle \overline{dt}_1,\ldots,\overline{dt}_n \rangle \to S \otimes_P \Omega_P^1 \to \Omega_S^1 \to 0$$

where $\overline{dt}_i := 1 \otimes dt_i \in S \otimes_P \Omega_P^1$ $(i=1,\ldots,n)$. Moreover,

$$(4) \qquad S \otimes_P \Omega_P^{n+p} = \bigoplus_{i=0}^{n} \bigoplus_{1 \le \nu_1 < \ldots < \nu_i \le n} (S \otimes_R \Omega^{n+p-i})\overline{dX}_{\nu_1}\ldots\overline{dX}_{\nu_i}.$$

Given $\omega \in \Omega_S^p$, let $\omega' \in S \otimes_P (\Omega_P)^p$ be a preimage of ω. On account of (4) we can write

$$\omega'\overline{dt}_1\ldots\overline{dt}_n = \sum_{i=0}^{n} \omega_i' \quad \text{with } \omega_i' \in \bigoplus_{1 \le \nu_1 < \ldots < \nu_i \le n} (S \otimes_R \Omega^{n+p-i})\overline{dX}_{\nu_1}\ldots\overline{dX}_{\nu_i}.$$

Here ω_n' is of the form

$$\omega_n' = \eta\,\overline{dX}_1\ldots\overline{dX}_n \qquad (\eta \in S \otimes_R \Omega^p).$$

16.4. <u>Definition.</u> $\sigma^{\Omega}_{S/R}(\omega) := (\tau^{X}_{t} \otimes id_{\Omega})(\eta)$.   *)

We must show that $\sigma^{\Omega}_{S/R}(\omega)$ does not depend on the choice

of the preimage ω' of ω, nor on the special presentation (3).

To this end we consider more generally a presentation

(5) $S = R[\![Y_1,\ldots,Y_s]\!][Z_1,\ldots,Z_t]_N / (u_1,\ldots,u_{s+t})$

where $N \subset R[\![Y]\!][Z]$ is multiplicatively closed and $\{u_1,\ldots,u_{s+t}\}$

is a quasiregular sequence of $Q := R[\![Y]\!][Z]_N$. By F.19 there

is a trace

$$\tau^{Y-y,Z-z}_u : S \to R$$

associated with (5). Here $\{Y-y, Z-z\}$ is the system

$\{Y_1-y_1,\ldots,Y_s-y_s,Z_1-z_1,\ldots,Z_t-z_t\}$ of $S[\![Y]\!][Z]_N = S \otimes_R R[\![Y]\!][Z]_N$

where y_i and z_j are the images of Y_i and Z_j in S.

Let Ω^* be the universally finite $R[\![Y]\!]$-extension of Ω.

As above there is an exact sequence

$$0 \to \langle \overline{du}_1,\ldots,\overline{du}_{s+t}\rangle \to S \otimes_Q \Omega^*_Q \to \Omega_S \to 0$$

and, if ω^* is a preimage of ω in $S \otimes_Q (\Omega^*_Q)^P$, then there is

in $S \otimes_Q \Omega^*_Q$ an equation

$$\omega^* \overline{du}_1 \ldots \overline{du}_{s+t} = \eta^* \overline{dY}_1 \ldots \overline{dY}_s \overline{dZ}_1 \ldots \overline{dZ}_t + \ldots$$

with $\eta^* \in S \otimes_R \Omega$. We claim

16.5. <u>Proposition.</u> $\sigma^{\Omega}_{S/R}(\omega) = (\tau^{Y-y,Z-z}_u \otimes id_{\Omega})(\eta^*)$.

Proof.   Let $\tilde{Q} := P \otimes_R Q = R[\![Y_1,\ldots,Y_s]\!][Z_1,\ldots,Z_t,X_1,\ldots,X_n]_N$

---

*) This construction of the trace was suggested to the
author by J.Lipman (1976).

Then $S = \tilde{Q}/(u_1,\ldots,u_{s+t},X_1-f_1,\ldots,X_n-f_n)$

$\qquad = \tilde{Q}/(t_1,\ldots,t_n,Y_1-g_1,\ldots,Y_s-g_s,Z_1-h_1,\ldots,Z_t-h_t)$

with $f_i \in Q$, $g_i,h_i \in P$. ω' and ω^* are two preimages of ω in $S \otimes \Omega_{\tilde{Q}}^*$ and $\omega'-\omega^*$ is contained in the ideal generated by $\overline{du}_1,\ldots,\overline{du}_{s+t},\overline{d(X_1-f_1)},\ldots,\overline{d(X_n-f_n)}$. Therefore

$$\omega^*\overline{du}_1..\overline{du}_{s+t}\overline{d(X_1-f_1)}..\overline{d(X_n-f_n)}=\omega'\overline{du}_1..\overline{du}_{s+t}\overline{d(X_1-f_1)}..\overline{d(X_n-f_n)}.$$

Moreover, we have

$(6)\; \omega^*\overline{du}_1..\overline{du}_{s+t}\overline{d(X_1-f_1)}..\overline{d(X_n-f_n)}=\eta^*\overline{dY}_1..\overline{dY}_s\overline{dZ}_1..\overline{dZ}_t\overline{dX}_1..\overline{dX}_n+..$

and

$(7)\quad \omega'\overline{dt}_1..\overline{dt}_n\overline{d(Y_1-g_1)}..\overline{d(Y_s-g_s)}\overline{d(Z_1-h_1)}..\overline{d(Z_t-h_t)} \;=$

$\qquad \eta\overline{dX}_1..\overline{dX}_n\overline{dY}_1..\overline{dY}_s\overline{dZ}_1..\overline{dZ}_t \;+\ldots$

Here the dots stand for terms of lower order in dX,dY,dZ.
With the notation introduced in connection with F.26

$$\overline{du}_1..\overline{du}_{s+t}\overline{d(X_1-f_1)}..\overline{d(X_n-f_n)} \;=$$

$$\Delta_{t,Y-g,Z-h}^{u,X-f}\cdot\overline{dt}_1..\overline{dt}_n\overline{d(Y_1-g_1)}..\overline{d(Y_s-g_s)}\cdot\overline{d(Z_1-h_1)}..\overline{d(Z_t-h_t)}.$$

Comparing (6) and (7) we obtain

$$\eta^* = (-1)^{n(s+t)}\Delta_{t,Y-g,Z-h}^{u,X-f}\cdot\eta = \Delta_{t,Y-g,Z-h}^{X-f,u}\cdot\eta$$

and F.29 implies that

$$(\tau_u^{Y-y,Z-z}\otimes id_\Omega)(\eta^*) = (\tau_u^{Y-y,Z-z}\otimes id_\Omega)(\Delta_{t,Y-g,Z-h}^{X-f,u}\cdot\eta) \;=$$

$$= (\tau_t^x\otimes id_\Omega)(\eta) = \sigma_{S/R}^\Omega(\omega).$$

Next we shall study the properties of the trace and verify, in particular, that it satisfies the axioms Tr1)-Tr7). It follows readily from the definition 16.4 that $\sigma_{S/R}^\Omega$ is Ω-linear and homogeneous of degree 0 (Tr1)). Given a zero-form $s \in S = \Omega_S^0 = S \otimes_P \Omega_P^0$ we have

$$s\overline{dt}_1\ldots\overline{dt}_n = s\,\frac{\partial(t_1,\ldots,t_n)}{\partial(x_1,\ldots,x_n)}\,\overline{dX}_1\ldots\overline{dX}_n +\ldots$$

and F.23 implies

$$\sigma^{\Omega}_{S/R}(s) = \tau^{x}_{t}\left(\frac{\partial(t_1,\ldots,t_n)}{\partial(x_1,\ldots,x_n)} \cdot s\right) = \sigma_{S/R}(s)$$

which is Tr2).

16.6. <u>Example.</u>   Suppose $S = R[x] = R[X]/(t)$ with a monic

polynomial $t = r_o + r_1 X + \ldots + r_{n-1} X^{n-1} + X^n$ ($r_i \in R$). Each

$\omega \in \Omega_S$ is of the form $\omega = \omega_o + \omega_1 dx$ with ω_o, ω_1 in the image

of $S \otimes_R \Omega \to \Omega_S$. We have

$$\sigma^{\Omega}_{S/R}(\omega) = \sigma^{\Omega}_{S/R}(\omega_o) + \sigma^{\Omega}_{S/R}(\omega_1 dx).$$

We wish to describe $\sigma^{\Omega}_{S/R}(\omega_1 dx)$ more closely. Let

$\omega'_1 = 1 \otimes \eta_o + x \otimes \eta_1 + \ldots + x^{n-1} \otimes \eta_{n-1}$ ($\eta_i \in \Omega$) be a preimage of

ω_1 in $S \otimes_R \Omega$. Then $\omega'_1 \overline{dxdt} =$

$$-(1 \otimes \eta_o + x \otimes \eta_1 + \ldots + x^{n-1} \otimes \eta_{n-1})(1 \otimes dr_o + x \otimes dr_1 + \ldots + x^{n-1} \otimes dr_{n-1})\overline{dX} + \ldots$$

hence $\sigma^{\Omega}_{S/R}(\omega_1 dx) =$

$$(-\tau^{x}_{t} \otimes id_{\Omega})((1 \otimes \eta_o + \ldots + x^{n-1} \otimes \eta_{n-1})(1 \otimes dr_o + \ldots + x^{n-1} \otimes dr_{n-1})).$$

If $r_1 = \ldots = r_{n-1} = 0$, this reduces to

$$\sigma^{\Omega}_{S/R}(\omega_1 dx) = -\eta_{n-1} dr_o.$$

From this formula we see, for example, that for an insepara-

ble field extension S/R the trace of a form of degree >0

may very well be different from zero, although the trace of

a 0-form (element of S) is always zero.

Let us now establish Tr3). Under the assumptions of Tr3)

suppose R' is noetherian. (3) induces a presentation

$$S' = R'[X_1,\ldots,X_n]/(t'_1,\ldots,t'_n)$$

of S'/R' as a complete intersection where t'_i is the image of

t_i in $R'[X]$ (C.18). There is a canonical commutative diagram

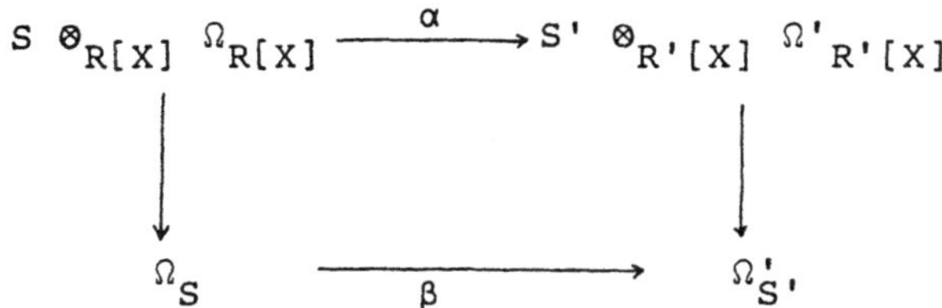

and since τ_t^X is compatible with base change (F.27) the following diagram commutes too

$$(8) \qquad \begin{array}{ccc} S \otimes_R \Omega & \xrightarrow{\;\gamma\;} & S' \otimes_{R'} \Omega' \\ \tau_t^X \otimes id_\Omega \downarrow & & \downarrow \tau_{t'}^X \otimes id_{\Omega'} \\ \Omega & \longrightarrow & \Omega' \end{array}$$

Now, given $\omega \in \Omega_S$, if $\omega' \in S \otimes \Omega_{R[X]}$ is a preimage of ω, then $\alpha(\omega')$ is a preimage of $\beta(\omega)$ in $S' \otimes \Omega'_{R'[X]}$. To the relation

$$\omega'\overline{dt}_1\ldots\overline{dt}_n = \eta\overline{dX}_1\ldots\overline{dX}_n + \ldots \qquad (\eta \in S \otimes_R \Omega)$$

corresponds in $S' \otimes \Omega'_{R'[X]}$ the relation

$$\alpha(\omega')\overline{dt}'_1\ldots\overline{dt}'_n = \gamma(\eta)\overline{dX}_1\ldots\overline{dX}_n + \ldots$$

and (8) shows that $\sigma_{S'/R'}^{\Omega'}(\beta(\omega))$ is the image of $\sigma_{S/R}^{\Omega}(\omega)$ in Ω'.

Tr3) implies, in particular, that the trace is compatible with localization. Given any element $(S/R,\Omega) \in C$, and given $\mathcal{M} \in \mathrm{Max}(R)$, there is an $f \in R\setminus\mathcal{M}$ such that S_f/R_f has a presentation (3) as a complete intersection (C.15). The traces $\sigma_{S_f/R_f}^{\Omega_f}$, where $\mathcal{M}$ runs over $\mathrm{Max}(R)$, define a unique map $\sigma_{S/R}^{\Omega} : \Omega_S \to \Omega$ with $(\sigma_{S/R}^{\Omega})_f = \sigma_{S_f/R_f}^{\Omega_f}$ for all f (A.8b)). Thus we have defined $\sigma_{S/R}^{\Omega}$ for any $(S/R,\Omega) \in C$. It is clear that this trace also satisfies Tr1)-Tr3), which have already been established for the algebras S_f/R_f. In order to prove Tr4)-Tr7), it is again enough to consider algebras that have a presentation (3).

Under the assumptions of Tr4) let $1 = e_1 + \ldots + e_h$ be the decomposition of 1 into idempotents corresponding to the product decomposition $S = S_1 \times \ldots \times S_h$. Then $S_i = S_{e_i}$, and (3) induces a presentation

$$S_i = R[X_1, \ldots, X_n, X]/(t_1, \ldots, t_n, E_i X - 1)$$

of S_i/R as a complete intersection where E_i is a preimage of e_i in $R[X_1, \ldots, X_n]$ (see C.19 and its proof). With $P := R[X_1, \ldots, X_n]$ and $Q := R[X_1, \ldots, X_n, X]$ there is a canonical commutative diagram

$$
\begin{array}{ccc}
S \otimes_P \Omega_P & \longrightarrow & \Omega_S \\
\alpha \downarrow & & \downarrow \\
S_i \otimes_Q \Omega_Q & \longrightarrow & \Omega_{S_i}
\end{array}
$$

Given $\omega = (\omega_1, \ldots, \omega_h) \in \Omega_{S_1} \times \ldots \times \Omega_{S_h} = \Omega_S$, let $\omega' \in S \otimes_P \Omega_P$ be a preimage of ω. Then $\alpha(\omega')$ is a preimage of $\omega_i = e_i \omega$ in $S_i \otimes_Q \Omega_Q$. From the equation

$$\omega' \overline{dt}_1 \ldots \overline{dt}_n = \eta \overline{dX}_1 \ldots \overline{dX}_n + \ldots \qquad (\eta \in S \otimes_R \Omega)$$

in $S \otimes_P \Omega_P$ we obtain the equation

$$\alpha(\omega') \overline{dt}_1 \ldots \overline{dt}_n d\overline{(E_i X - 1)} = e_i \alpha(\eta) \overline{dX}_1 \ldots \overline{dX}_n \overline{dX} + \ldots$$

in $S_i \otimes_Q \Omega_Q$, and $e_i \alpha(\eta)$ is the projection of η in $S_i \otimes_R \Omega$. Using F.30 we get $\sigma_{S/R}^{\Omega}(\omega) =$

$$(\tau_t^x \otimes id)(\omega) = \sum_{i=1}^{h} (\tau_{t_1, \ldots, t_n, E_i X - 1}^{X_1, \ldots, X_n, e_i} \otimes id)(e_i \eta) = \sum_{i=1}^{h} \sigma_{S_i/R}^{\Omega}(\omega_i).$$

In order to prove Tr5) we may assume that R is local. Then S is semilocal, and both T/S and S/R are complete intersections. Let

$$T = S[Y_1, \ldots, Y_m]/(u_1, \ldots, u_m) = S[y_1, \ldots, y_m]$$

be a presentation of T/S as a complete intersection, and let $\{U_1,\ldots,U_m\}$ be a system of representatives of the u_i in $R[X_1,\ldots,X_n,Y_1,\ldots,Y_m]$. Then

$$T = R[X_1,\ldots,X_n,Y_1,\ldots,Y_m]/(t_1,\ldots,t_n,U_1,\ldots,U_m)$$

is a presentation of T/R as a complete intersection (C.17).

For $\omega \in \Omega_T^p$ we choose a preimage $\tilde{\omega}$ in $\Omega_{R[X,Y]}^p$ and write

$$(9) \quad \tilde{\omega}dU_1..dU_m = \eta\, dY_1..dY_m + \text{ terms of lower order in } dY_1,..,dY_m$$

where

$$\eta = \sum_{\nu_1<..<\nu_r} \eta_{\nu_1..\nu_r} dX_{\nu_1}..dX_{\nu_r} \qquad (\eta_{\nu_1..\nu_r} \in R[X,Y] \otimes_R \Omega^{p-r})$$

In $\Omega_{R[X]}$ there is a relation

$$dt_1..dt_n = \sum_{\mu_1<..<\mu_s} \delta_{\mu_1..\mu_s} dX_{\mu_1}..dX_{\mu_s} \qquad (\delta_{\mu_1..\mu_s} \in R[X] \otimes_R \Omega^{n-s})$$

and therefore the following relation holds in $\Omega_{R[X,Y]}$:

$$\tilde{\omega}dt_1..dt_n dU_1..dU_m = (-1)^{mn}\tilde{\omega}dU_1..dU_m dt_1..dt_n =$$

$$(\sum_{r+s=n} (-1)^r \eta_{\nu_1..\nu_r} \delta_{\mu_1..\mu_s} dX_{\nu_1}..dX_{\nu_r} dX_{\mu_1}..dX_{\mu_s}) dY_1..dY_m +...=$$

$$(\sum_{r+s=n} (-1)^{r+\varepsilon(\nu,\mu)} \eta_{\nu_1..\nu_r} \delta_{\mu_1..\mu_s}) dX_1..dX_n dY_1..dY_m +...$$

where $\{\nu_1,\ldots,\nu_r,\mu_1,\ldots,\mu_s\} = \{1,\ldots,n\}$ and where $\varepsilon(\nu,\mu)$ is defined by

$$dX_{\nu_1}..dX_{\nu_r} dX_{\mu_1}..dX_{\mu_s} = (-1)^{\varepsilon(\nu,\mu)} dX_1..dX_n.$$

Let $\overline{\eta_{\nu_1..\nu_r}}$ and $\overline{\delta_{\mu_1..\mu_s}}$ denote the images of $\eta_{\nu_1..\nu_r}$ and $\delta_{\mu_1..\mu_s}$ in $T \otimes_R \Omega$. Observe that $\overline{\delta_{\mu_1..\mu_s}} \in S \otimes_R \Omega \subset T \otimes_R \Omega$.

Using the transitive law F.28 the definition of the trace yields the formula

$$(10) \quad \sigma_{T/R}^{\Omega}(\omega) = (\tau_{t,U}^{X,Y} \otimes \mathrm{id}_\Omega)(\Sigma(-1)^{r+\varepsilon(\nu,\mu)}\overline{\eta}_{\nu_1..\nu_r}\overline{\delta}_{\mu_1..\mu_s}) =$$

$$(\tau_t^X \otimes \mathrm{id}_\Omega)(\Sigma(-1)^{r+\varepsilon(\nu,\mu)} (\tau_u^Y \otimes \mathrm{id}_\Omega)(\overline{\eta}_{\nu_1..\nu_r}) \cdot \overline{\delta}_{\mu_1..\mu_s}).$$

On the other hand, if ω^* denotes the image of $\tilde{\omega}$ in $\Omega_{S[Y]}$, formula (9) leads to the following relation in $\Omega_{S[Y]}$:

$$\omega^* du_1 \ldots du_m = \eta^* dY_1 \ldots dY_m + \ldots$$

where

$$\eta^* = \Sigma \eta^*_{\nu_1 \ldots \nu_r} dx_{\nu_1} \ldots dx_{\nu_r} \qquad (\eta^*_{\nu_1 \ldots \nu_r} \in S[Y] \otimes_R \Omega^{p-r}).$$

By the definition of the trace

$$\sigma^{\Omega_S}_{T/S}(\omega) = \Sigma (\tau^Y_u \otimes \mathrm{id}_{\Omega_S})(\overline{\eta^*_{\nu_1 \ldots \nu_r}}) dx_{\nu_1} \ldots dx_{\nu_r}$$

where $\overline{\eta^*_{\nu_1 \ldots \nu_r}} \in T \otimes_S \Omega_S^{p-r}$ is the image of $\eta^*_{\nu_1 \ldots \nu_r}$.

$\overline{\eta^*_{\nu_1 \ldots \nu_r}}$ is also the image of $\overline{\eta}_{\nu_1 \ldots \nu_r}$ under the map

$T \otimes_R \Omega \to T \otimes_S \Omega_S$, and $(\tau^Y_u \otimes \mathrm{id}_{\Omega_S})(\overline{\eta^*_{\nu_1 \ldots \nu_r}})$ is the image of

$(\tau^Y_u \otimes \mathrm{id}_\Omega)(\overline{\eta}_{\nu_1 \ldots \nu_r})$ under $S \otimes_R \Omega \to \Omega_S$.

A preimage $\theta_{\nu_1 \ldots \nu_r}$ of $(\tau^Y_u \otimes \mathrm{id}_\Omega)(\overline{\eta}_{\nu_1 \ldots \nu_r})$ in

$R[X] \otimes_R \Omega^{p-r}$ is also one of $(\tau^Y_u \otimes \mathrm{id}_{\Omega_S})(\overline{\eta^*_{\nu_1 \ldots \nu_r}})$, and

$$\omega' := \Sigma \, \theta_{\nu_1 \ldots \nu_r} dX_{\nu_1} \ldots dX_{\nu_r}$$

is a representative of $\sigma^{\Omega_S}_{T/S}(\omega)$ in $\Omega_{R[X]}$. Applying now the definition of $\sigma^{\Omega}_{S/R}$ to the relation

$$\omega' dt_1 \ldots dt_n = \Sigma (-1)^r \theta_{\nu_1 \ldots \nu_r} \delta_{\mu_1 \ldots \mu_s} dX_{\nu_1} \ldots dX_{\nu_r} dX_{\mu_1} \ldots dX_{\mu_s} + \ldots =$$

$$(\Sigma (-1)^{r+\varepsilon(\nu,\mu)} \theta_{\nu_1 \ldots \nu_r} \delta_{\mu_1 \ldots \mu_s}) dX_1 \ldots dX_n + \ldots$$

we obtain

$$\sigma^{\Omega}_{S/R}(\sigma^{\Omega_S}_{T/S}(\omega)) = (\tau^X_t \otimes \mathrm{id}_\Omega)(\Sigma (-1)^{r+\varepsilon(\nu,\mu)} \overline{\theta}_{\nu_1 \ldots \nu_r} \overline{\delta}_{\mu_1 \ldots \mu_s})$$

where the bar denotes the image in $S \otimes_R \Omega$. Since

$$\overline{\theta}_{\nu_1 \ldots \nu_r} = (\tau^Y_u \otimes \mathrm{id}_\Omega)(\overline{\eta}_{\nu_1 \ldots \nu_r})$$ we end up with the same

expression as in (10), hence $\sigma^{\Omega}_{S/R} \circ \sigma^{\Omega_S}_{T/S} = \sigma^{\Omega}_{T/R}$.

We prove Tr6) and Tr7) simultaneously by the method

employed in the proof of theorem 16.1. We may assume R is
local. By Tr3) it suffices to consider the case $\Omega = \Omega_{R/\mathbb{Z}}$.
Applying lemma 16.3 and using Tr3) again, we reduce to
the case that R is essentially of finite type over $\mathbb{Z}$. We
then pass to the completion $\hat{R}$ of R and consider the diagram

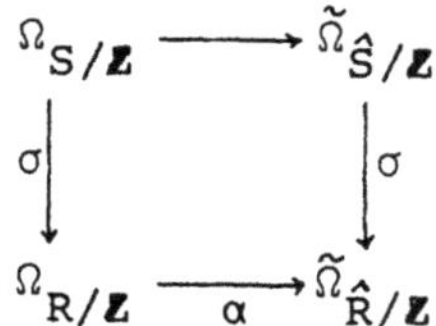

where $\hat{S} := \hat{R} \otimes_R S$. Since α is injective it is enough to
prove Tr6) and Tr7) for $\hat{S}/\hat{R}$, hence we may assume R is a
complete local ring. In this case $S = S_1 \times \ldots \times S_h$ is a
direct product of complete local rings S_i, and if
$a = (a_1, \ldots, a_h)$ is a unit of S, then a_i is a unit of S_i
$(i=1,\ldots,h)$. Since $n_{S/R}(a) = \prod_{i=1}^{h} n_{S_i/R}(a_i)$, and since the trace
and the differentiation are compatible with direct products,
we reduce to the case that S is a complete local ring too,
and still $\Omega = \tilde{\Omega}_{R/\mathbb{Z}}$. An application of lemma 16.2 show that it
is enough to consider extensions S/R of complete regular
local rings of characteristic 0 where R is a power series alge-
bra over a Cohen ring. Then $\Omega = \tilde{\Omega}_{R/\mathbb{Z}}$ is a free R-module, and
passing to the quotient fields, we may assume that S/R is a
separable field extension. Let N/R be the smallest Galois extension containing S/R and let $T := N \otimes_R S$. Then T is a direct
product of copies of N, and Ω_T is a direct product of copies of
Ω_N. Since $\Omega \to \Omega_N$ is injective and the diagram

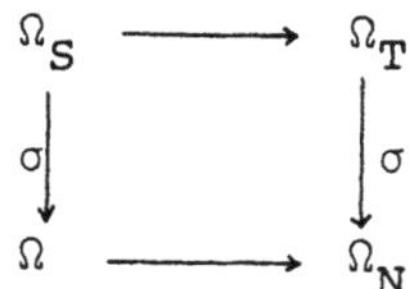

is commutative (Tr3), it suffices to consider T/N, and using Tr4) once more, we are led to consider N/N, in which case the statements of Tr6) and Tr7) are trivial.

This completes the proof of the claim that the trace defined in 16.4 satisfies the axioms Tr1)-Tr7).

16.7. Corollary. $\sigma^{\Omega}_{S/R} : \Omega_S \to \Omega$ induces a map

$$H_{DR}(\sigma^{\Omega}_{S/R}) : H_{DR}(\Omega_S) \to H_{DR}(\Omega)$$

of the deRham cohomologies, which is $H_{DR}(\Omega)$-linear.

We shall show now that, under a special assumption about Ω, the trace $\sigma^{\Omega}_{S/R}$ is a basis of the Ω_S-module $\mathrm{Hom}_{\Omega}(\Omega_S, \Omega)$.

16.8. Theorem. Given $(S/R, \Omega) \in C$ assume that Ω is an exterior differential algebra where Ω^1 is a projective R-module of rank r and Ω^1_S is a projective S-module of rank r. Then the canonical map

$$\Omega_S \longrightarrow \mathrm{Hom}_{\Omega}(\Omega_S, \Omega)$$

$$\eta \longmapsto (\omega \mapsto \sigma^{\Omega}_{S/R}(\omega\eta))$$

is bijective. In other words:

$$\mathrm{Hom}_{\Omega}(\Omega_S, \Omega) = \Omega_S \cdot \sigma^{\Omega}_{S/R}.$$

For $p \in \mathbb{N}$ let $\mathrm{Hom}_{\Omega}(\Omega_S, \Omega)^p$ be the R-module of homogeneous Ω-linear maps $\ell : \Omega_S \to \Omega$ of degree p (i.e. $\ell(\Omega^q_S) \subset \Omega^{p+q}$).

16.9. Lemma. Let S/R be an algebra, and let Ω be an exterior differential algebra of R where Ω^1 is a projective R-module of rank r. Then the R-linear map

$$\gamma \; : \; \mathrm{Hom}_{\Omega}(\Omega_S, \Omega)^p \rightarrow \mathrm{Hom}_R(\Omega_S^{r-p}, \Omega^r)$$

$$\ell \longmapsto \ell\big|_{\Omega_S^{r-p}}$$

is bijective.

Proof. Suppose $\ell \in \ker \gamma$. Clearly $\ell(\Omega_S^q) = 0$ for $q \geq r - p$. For $\omega \in \Omega_S^q$ with $q < r - p$ and any $\eta \in \Omega^{r-(p+q)}$ we have $\eta \cdot \ell(\omega) = \ell(\omega\eta) = 0$. However, the multiplication in Ω defines a non-degenerate bilinear map $\Omega^{r-(p+q)} \times \Omega^{p+q} \rightarrow \Omega^r$. Hence $\ell(\omega) = 0$, and we see that γ is injective.

For $\lambda \in \mathrm{Hom}_R(\Omega_S^{r-p}, \Omega^r)$ and $\omega \in \Omega_S^q$ with $q \leq r - p$ define $\ell(\omega)$ to be the unique element of Ω^{p+q} with

$$\eta \cdot \ell(\omega) = \lambda(\omega\eta) \quad \text{for all } \eta \in \Omega^{r-(p+q)}.$$

Put $\ell(\Omega_S^q) = 0$ for $q > r - p$. This defines an element $\ell \in \mathrm{Hom}_{\Omega}(\Omega_S, \Omega)^p$ with $\gamma(\ell) = \lambda$.

Proof of 16.8.

By the lemma it suffices to show that, for all $p \in \mathbb{N}$,

$$\sigma \; : \; \Omega_S^p \rightarrow \mathrm{Hom}_R(\Omega_S^{r-p}, \Omega^r)$$

$$\eta \longmapsto (\omega \rightarrow \sigma_{S/R}^{\Omega}(\omega\eta))$$

is bijective. We may assume R is local. Then $\Omega^r = R \cdot \omega_o$ with a basis element $\omega_o \in \Omega^r$, and S/R has a presentation (3) as a complete intersection. Let $\tau_t^x : S \rightarrow R$ be the trace associated with (3). Given a basis $\{y_1, \ldots, y_m\}$ of S/R, let $\{y_1', \ldots, y_m'\}$ be the dual basis of S/R with respect to τ_t^x (F.21), i.e. $\tau_t^x(y_i \cdot y_j') = \delta_{ij}$.

Put $P := R[X_1, \ldots, X_n]$ and $I := (t_1, \ldots, t_n)$. From the exact sequence of (free) S-modules

$$0 \to I/I^2 \to S \otimes_P \Omega_P^1 \to \Omega_S^1 \to 0$$

we derive a canonical isomorphism of S-modules

$$\varphi : \Omega_S^r \xrightarrow{\sim} \operatorname{Hom}_S(\Lambda^n I/I^2, S \otimes_P \Omega_P^{r+n})$$

as follows: Let $\overline{t_i}$ denote the image of t_i in I/I^2. For $\omega \in \Omega_S^r$ let $\omega' \in S \otimes_P \Omega_P^r$ be a preimage of ω. Then $\varphi(\omega)$ is the S-linear map sending $\overline{t_1} \wedge \ldots \wedge \overline{t_n}$ to $\omega' \overline{dt_1} \ldots \overline{dt_n}$, the bars denoting the images in $S \otimes_P \Omega_P^1$. This map does not depend on the choice of the preimage ω', nor on the generators $t_1, \ldots, t_n$ of I.

Now let $\omega_j \in \Omega_S^r$ be the form with

$$\varphi(\omega_j)(\overline{t_1} \wedge \ldots \wedge \overline{t_n}) = y_j' \otimes \omega_o \overline{dX_1} \ldots \overline{dX_n}$$

that is, with a preimage ω_j' of ω_j in $S \otimes_P \Omega_P^r$ we have

$$\omega_j' \overline{dt_1} \ldots \overline{dt_n} = y_j' \otimes \omega_o \overline{dX_1} \ldots \overline{dX_n}.$$

The definition of the trace yields

$$(11) \qquad \sigma_{S/R}^{\Omega}(y_i \omega_j) = \tau_t^x(y_i y_j') \cdot \omega_o = \delta_{ij} \omega_o \qquad (i,j = 1, \ldots, m).$$

Clearly $\{\omega_1, \ldots, \omega_m\}$ is a basis of Ω_S^r as an R-module. If $\{\beta_1, \ldots, \beta_r\}$ is a basis of the S-module Ω_S^1, then $\{y_i \beta_{i_1} \ldots \beta_{i_p} \mid 1 \leq i_1 < \ldots < i_p \leq r, i = 1, \ldots, m\}$ is a basis of the R-module Ω_S^p. Write $\omega_j = z_j \beta_1 \ldots \beta_r$ ($z_j \in S, j = 1, \ldots, m$). Then $\{z_j \beta_{j_1} \ldots \beta_{j_{r-p}} \mid 1 \leq j_1 < \ldots < j_{r-p} \leq r, j = 1, \ldots, m\}$ is a basis of the R-module Ω_S^{r-p}, and by (11) the R-bilinear map

$$\Omega_S^p \times \Omega_S^{r-p} \to \Omega^r$$

sending $(y_i \beta_{i_1} \ldots \beta_{i_p}, z_j \beta_{j_1} \ldots \beta_{j_{r-p}})$ to

$$\sigma_{S/R}^{\Omega}(y_i z_j \beta_{i_1} \ldots \beta_{i_p} \beta_{j_1} \ldots \beta_{j_{r-p}}) \text{ is non-degenerate. This}$$

proves that σ is bijective, q.e.d.

Following Angéniol ([An],7.1.2) one can construct a system
of traces $\{\sigma^{\Omega}_{S/R}\}$ in the following situation:

I) S/R is a finite projective algebra with $\mathbb{Q} \subset R$, and Ω is an
arbitrary differential algebra of R. Lipman [Lip$_4$] has given
an approach to these traces via Hochschild homology. Lipman's
method also allows the construction of traces under one of
the following assumptions.

II) R is a direct product of fields, S/R a finite algebra,
and Ω a differential algebra of R.

III) S/R is a finite projective algebra where 2 is not a zerodivisor of R, and Ω is an exterior differential algebra of R
for which Ω^1 is a projective R-module of finite rank.

All these traces have properties similar to the traces
considered in this chapter, and it can be shown that any two
of them agree, whenever both are defined. However, it is not
possible to construct a family of traces $\{\sigma^{\Omega}_{S/R}\}$ for <u>all</u> finite free algebras S/R, and <u>all</u> differential algebras Ω of R,
such that the trace axioms Tr1)-Tr4) are satisfied, see
exercises 1) and 2) below. There remains the problem of
finding a common construction for traces of differential
forms covering all cases in which the traces are known to
exist.

The notion of trace is closely related to the notion of residue and to duality theory. In the next section this will be
illustrated in the one-dimensional case. For residues and
duality in higher dimensions we refer to [Lip$_3$], [Lip$_4$] and
their bibliographies.

Exercises

1) In the polynomial algebra $\mathbf{Z}[X_1,\ldots,X_9]$ consider the ideal I generated by the 2×2-minors of the matrix

$$\begin{bmatrix} X_1 & X_2 & X_3 \\ X_4 & X_5 & X_6 \\ X_7 & X_8 & X_9 \end{bmatrix}.$$

It is known that

$$R = \mathbf{Z}[X_1,\ldots,X_9]/I = \mathbf{Z}[x_1,\ldots,x_9]$$

is a domain. In $K := Q(R)$ we have

$$x_5 = \frac{x_2 x_4}{x_1}, \quad x_6 = \frac{x_3 x_4}{x_1}, \quad x_8 = \frac{x_2 x_7}{x_1}, \quad x_9 = \frac{x_3 x_7}{x_1}$$

hence $K = \mathbb{Q}(x_1,x_2,x_3,x_4,x_7)$.

a) Show that x_1,x_2,x_3,x_4,x_7 are algebraically independent over $\mathbb{Q}$.

b) The 84 forms $dx_i dx_j dx_k \in \Omega_{R/\mathbf{Z}}$ with $1 \le i < j < k \le 9$ are linearly independent over $\mathbf{Z}$. Show that their images under the canonical map $\Omega_{R/\mathbf{Z}} \to \Omega_{K/\mathbb{Q}}$ are also linearly independent over $\mathbf{Z}$.

(Hint: R is a graded ring where $\deg x_i = i$ $(i=1,\ldots,9)$. Extend this grading to $\Omega_{R/\mathbf{Z}}$ ($\S 3$, exercise 1). It suffices to consider the forms $dx_i dx_j dx_k$ with fixed degree $i+j+k$ separately. Now calculate....?)

2) Let R be the ring of exercise 1). On the free R-module

$$S := R \oplus RY_1 \oplus RY_2 \oplus RY_3$$

define a (commutative) multiplication with the multiplication table

$$Y_1^2 = -x_4x_8 + (x_5+x_7)Y_1 + x_4Y_2$$

$$Y_1Y_2 = x_1Y_3$$

$$Y_1Y_3 = x_3x_4 + x_7Y_3$$

$$Y_2^2 = x_1x_3 + x_8Y_2 - x_2Y_3$$

$$Y_2Y_3 = -x_2x_6 + x_3Y_1$$

$$Y_3^2 = -x_6x_8 + x_9Y_1 + x_6Y_2$$

a) Show that S is an associative R-algebra.

b) Let $K := Q(R)$ and $L := K \otimes_R S$. Show that L/K is a complete intersection. (In fact: $L = K \oplus KY_1 \oplus KY_1^2 \oplus KY_1^3$).

c) Let $\sigma_{L/K} : \Omega_{L/\mathbb{Q}} \to \Omega_{K/\mathbb{Q}}$ be the trace. Show that

$$\sigma_{L/K}(dY_1dY_2dY_3) =$$

$$\frac{1}{x_1^2}((x_1x_7-x_2x_4)dx_1dx_3dx_7 - x_3x_4dx_1dx_2dx_7 - x_2x_3dx_1dx_4dx_7).$$

d) Let $\varphi : \Omega_{R/\mathbb{Z}} \to \Omega_{K/\mathbb{Q}}$ be the canonical map. Show that the 3-form

$$-dx_1dx_5dx_9 - dx_1dx_6dx_8 - 2dx_1dx_7dx_9 + dx_2dx_4dx_9 - dx_2dx_6dx_7$$

$$+dx_3dx_4dx_8 - dx_3dx_5dx_7$$

has image $2\sigma_{L/K}(dY_1dY_2dY_3)$ under φ.

e) Using the statement of exercise 1b) conclude that $\sigma_{L/K}(dY_1dY_2dY_3)$ is not in the image of φ. Hence there cannot exist a map $\sigma_{S/R} : \Omega_{S/\mathbb{Z}} \to \Omega_{R/\mathbb{Z}}$ such that the following diagram commutes:

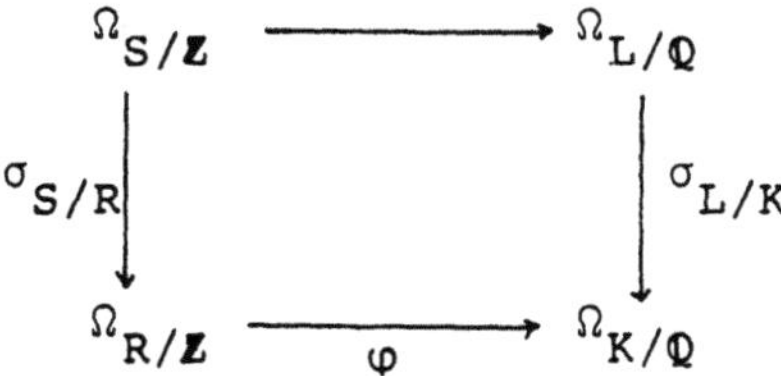

Anton Kliegl helped me to find this example.

3) Let R be a discrete valuation ring with quotient field K
   and let L be a finite extension field of K. Suppose the
   integral closure S of R in L is a discrete valuation ring
   and is finite over R (which is the case f.i. if R is complete). Let Ω be an exterior differential algebra of R
   where Ω^1 is free of rank r, and suppose that Ω_S^1 is a free
   S-module too. Define the order $\nu_S(\omega)$ of $\omega \in \Omega_L$ as follows:
   Choose an S-basis of Ω_S, express ω as a linear combination
   with coefficients in L of the basis elements, and take the
   minimum of the orders of the coefficients as the order of
   ω. $\nu_R(\eta)$ for $\eta \in \Omega_K$ is defined in the same way. Show that
   for each $\omega \in \Omega_L$

$$\nu_R(\sigma_{L/K}^{\Omega}(\omega)) \geq \frac{\nu_S(\omega)+1}{e} - 1$$

   where e is the ramification index of S/R.

4) Let F be the class of all pairs $(L/K,\Omega)$ where L/K is a
   finite field extension and Ω a differential algebra of K.
   Show that there is at most one (and hence exactly one)
   system of traces $\{\sigma_{L/K}^{\Omega}\}$ for $(L/K,\Omega) \in F$ such that the trace
   axioms Tr1),Tr2),Tr5),Tr6), and Tr7), are satisfied.

5) Let S/K be a finite-dimensional algebra over a field K, and
   let $S = S_1 \times \ldots \times S_t$ be its canonical decomposition into local
   algebras S_i/K (i=1,..,t). Write L_i for the residue field of
   S_i. For a differential algebra Ω of K and for $\omega \in \Omega_S$ let
   ω_i denote the canonical image of ω in Ω_{L_i} (i=1,..,t).
   Define the trace of ω by the formula

$$\sigma_{S/K}^{\Omega}(\omega) := \sum_{i=1}^{t} \frac{[S_i:K]}{[L_i:K]} \cdot \sigma_{L_i/K}^{\Omega}(\omega_i)$$

   with the traces $\sigma_{L_i/K}^{\Omega}$ from exercise 4). Show that the trace
   axioms (if properly modified) are satisfied for the system
   $\{\sigma_{S/K}^{\Omega}\}$ with $(S/K,\Omega)$ as above.

§ 17. Residues in Algebraic Function Fields of One Variable

Let L/K be an algebraic function field with $\mathrm{Trdeg}(L/K)=1$. The set X of all discrete valuation rings R with $K \subset R$ and $Q(R) = L$ is called the "non-singular model" of L/K, or the "abstract Riemann surface" of L/K. It is well-known that X together with L is also the set of local rings of a non-singular projective algebraic curve defined over K.

We shall use the trace introduced in §16 to construct for each $R \in X$ a canonical residue map

$$\mathrm{Res}_R : \Omega^1_{L/K} \to K$$

generalizing the classical residue of differentials on compact Riemann surfaces.

Let $\mathfrak{m}$ be the maximal ideal of $R \in X$, and let $\mathfrak{k} := R/\mathfrak{m}$ be the residue field of R. It is known that R/K is essentially of finite type and $[\mathfrak{k}:K] < \infty$.

I) Suppose $\mathfrak{k}/K$ is separable.

In this case there is a canonical isomorphism
$\mathfrak{m}/\mathfrak{m}^2 \xrightarrow{\sim} \Omega^1_{R/K}/\mathfrak{m}\Omega^1_{R/K}$ (6.5a)), hence $\Omega^1_{R/K} = Rdt, \Omega^1_{L/K} = Ldt$
where t is a regular parameter of R, i.e. $\mathfrak{m} = (t)$. Let $\hat{R}$ denote the completion of R, and let $\hat{L} := Q(\hat{R})$. Since $\mathfrak{k}/K$ is separable, there is by Cohen's structure theorem an iso-morphism $\hat{R} \cong \mathfrak{k}[\![t]\!]$ of K-algebras. $\tilde{\Omega}^1_{\hat{R}/K}$ denotes the univer-sally finite differential module of $\hat{R}/K$, and $D_K(\hat{L}) := \hat{L} \otimes_{\hat{R}} \tilde{\Omega}^1_{\hat{R}/K}$. By 12.10

$$\tilde{\Omega}^1_{\hat{R}/K} = \hat{R} \otimes_R \Omega^1_{R/K} = \hat{R}dt, \text{ hence } D_K(\hat{L}) = \hat{L} \otimes_R \Omega^1_{R/K} = \hat{L}dt \text{ and}$$

$D_K(\hat{L}) = \hat{L} \otimes_L \Omega^1_{L/K}$. Any $\omega \in D_K(\hat{L})$ has a "Laurent expansion"

$$(1) \qquad \omega = \left(\sum_{\nu \geq \nu_0} a_\nu t^\nu \right) dt \qquad (a_\nu \in \mathfrak{k}).$$

If $a_{\nu_o} \neq 0$, we put $\nu_{\hat{R}}(\omega) := \nu_o$ and call ν_o <u>the order of ω</u> <u>at $\hat{R}$</u>. It is easily seen that this number does not depend on the regular parameter t chosen. For $\omega \in \Omega^1_{L/K}$ we write $\nu_R(\omega)$ for $\nu_{\hat{R}}(\omega)$ and call this number <u>the order of ω at R.</u>

<u>17.1. Proposition.</u> The coefficient a_{-1} in the expansion (1) is independent of the choice of the regular parameter t of R.

Proof. a) If $\nu_{\hat{R}}(\omega) \geq 0$, then this coefficient vanishes for any choice of the regular parameter t.

b) Suppose $\nu_{\hat{R}}(\omega) = -1$, and let $t = \varepsilon \cdot \tau$ with a unit ε of $\hat{R}$. If d denotes the differentiation $d : \hat{L} \to D_K(\hat{L})$, then

$$(2) \qquad \frac{dt}{t} = \frac{d\varepsilon}{\varepsilon} + \frac{d\tau}{\tau}$$

and $\nu_{\hat{R}}(\frac{d\varepsilon}{\varepsilon}) \geq 0$. It is therefore clear that in the Laurent expansion of ω with respect to τ the same coefficient a_{-1} as in (1) will appear.

c) Suppose now that $\omega = df$ with some $f \in \hat{L}$. If we write

$$f = \sum_{\nu \geq n_o} b_\nu t^\nu \quad (b_\nu \in \hat{k}), \text{ then}$$

$$(3) \qquad df = f'dt = (\sum_{\nu \geq n_o} \nu b_\nu t^{\nu-1})dt$$

since the universally finite derivation $\hat{R} \to \tilde{\Omega}^1_{\hat{R}/K}$ is given by the formal differentiation of power series. The coefficient of t^{-1} in the expansion (3) vanishes for any choice of the regular parameter t. So we are done, if $\omega = df$.

d) If Char.k $= 0$ and ω is given as in (1), then

$$\omega = a_{-1} \frac{dt}{t} + df \text{ with } f := \sum_{\nu \geq \nu_o, \nu \neq -1} \frac{1}{\nu+1} a_\nu t^{\nu+1}$$

and the claim follows from c) by using (2).

e) Suppose now that Char.$K =: p > 0$. Choose $e \in \mathbb{N}$ such that $p^e > |\nu_0|$. Write $\hat{R}_e := k[\![R^{p^e}]\!]$, $\hat{L}_e := Q(\hat{R}_e)$ and $\tau := t^{p^e}$. Then $\hat{R}_e = k[\![t^{p^e}]\!] = k[\![\tau]\!]$ and $\hat{L} = \hat{L}_e[X]/(X^{p^e}-\tau) = \hat{L}_e \oplus \hat{L}_e t \oplus \ldots \oplus \hat{L}_e t^{p^e-1}$. We may write

$$\omega = (b_0 + b_1 t + \ldots + b_{p^e-1} t^{p^e-1})\frac{dt}{\tau}$$

with $b_i \in \hat{R}_e$ ($i=0,\ldots,p^e-1$), since $p^e > |\nu_0|$. Observe that $b_{p^e-1} \equiv a_{-1} \bmod(\tau)$.

For the finite field extension $\hat{L}/\hat{L}_e$ the trace

$$\sigma_{\hat{L}/\hat{L}_e} : D_K(\hat{L}) \rightarrow D_K(\hat{L}_e)$$

is defined (§ 16). Using 16.6 we see that

$$\sigma_{\hat{L}/\hat{L}_e}(\omega) = a_{-1}\frac{d\tau}{\tau} + \omega_0$$

with some $\omega_0 \in \hat{\Omega}^1_{\hat{R}_e/K}$. Applying b) to the differential $\sigma_{\hat{L}/\hat{L}_e}(\omega)$, which is of order -1 with respect to $\hat{R}_e$, we conclude that a_{-1} is indeed independent of t.

17.2. Definition. We write $\mathrm{Res}_{\hat{R}}\omega := \sigma_{\hat{k}/K}(a_{-1})$ and call this element of K the <u>residue of ω at $\hat{R}$</u>. If $\omega \in \Omega^1_{L/K}$, we put $\mathrm{Res}_R\omega := \mathrm{Res}_{\hat{R}}\omega$ (Residue of ω at R).

17.3. Remarks.

a) $\mathrm{Res}_R : \Omega^1_{L/K} \rightarrow K$ is K-linear.

b) $\mathrm{Res}_R\omega = 0$ for all $\omega \in \Omega^1_{R/K}$, and $\mathrm{Res}_R(df) = 0$ for $f \in L$.

c) If Char.$K =: p > 0$, if $L_e := K(L^{p^e})$, and if $R_e := R \cap L_e$, then

$$\mathrm{Res}_R\omega = \mathrm{Res}_{R_e}(\sigma_{K/K_e}(\omega))$$

for each $\omega \in \Omega^1_{L/K}$.

These facts follow from the definition 17.2 and the proof of 17.1.

II) Suppose now that Char.$K =: p > 0$ and k/K is arbitrary.

Put $L_i := K(L^{p^i})$ for $i \in \mathbb{N}$. To the tower of fields

(4) $\qquad L \supset L_1 \supset L_2 \supset \ldots$

corresponds the tower of discrete valuation rings

(5) $\qquad R \supset R_1 \supset R_2 \supset \ldots \qquad$ with $R_i := R \cap L_i$

and the tower of their residue fields

(6) $\qquad k \supset k_1 \supset k_2 \supset \ldots$

Here $k_i \supset K[k^{p^i}]$, and hence k/k_i is purely inseparable for all $i \in \mathbb{N}$. Since $[k:K] < \infty$, there is a smallest number $e \in \mathbb{N}$ such that $k_e = k_{e+1} = \ldots$, and k_e is the separable closure of K in k.

If $\omega \in \Omega^1_{L/K}$, then definition 17.2 can be applied to $\sigma_{L/L_e}(\omega) \in \Omega^1_{L_e/K}$, since k_e/K is separable.

17.4. Definition. $\quad \mathrm{Res}_R\omega := \mathrm{Res}_{R_e}(\sigma_{L/L_e}(\omega))$
is the __residue of ω at R__.

Thus $\mathrm{Res}_R\omega$ is defined for arbitrary residue field extensions k/K. For $\omega \in D_K(\hat{L})$ we define $\mathrm{Res}_{\hat{R}}\omega$ in the same manner as above, using the field tower $\hat{L} \supset \hat{L}_1 \supset \hat{L}_2 \supset \ldots$ where $\hat{L}_i := Q(\hat{R}_i)$ is the quotient field of the completion $\hat{R}_i$ of R_i ($i \in \mathbb{N}$). For an entirely different construction of the residue map, see Tate [T$_2$].

17.5. Remarks. $\quad$ For $\omega \in \Omega^1_{L/K}$ we have

a) $\mathrm{Res}_R\omega = \mathrm{Res}_{R_i}(\sigma_{L/L_i}(\omega))$ for all $i \in \mathbb{N}$.

b) $\mathrm{Res}_R : \Omega^1_{L/K} \to K$ is a K-linear map.

c) If ω is in the image of the canonical map $\Omega^1_{R/K} \to \Omega^1_{L/K}$, then $\mathrm{Res}_R\omega = 0$.

d) $\mathrm{Res}_R(df) = O$ for each $f \in L$.

e) For each $\omega \in \Omega^1_{L/K}$ an $i \in \mathbb{N}$ can be chosen such that
$$v_{R_i}(\sigma_{L/L_i}(\omega)) \geq -1.$$

Proof. a) follows from the transitivity of the trace and

b) is clear.

c) Since R/R_e is a complete intersection, the trace
$\sigma_{R/R_e} : \Omega^1_{R/K} \to \Omega^1_{R_e/K}$ is defined and the diagram

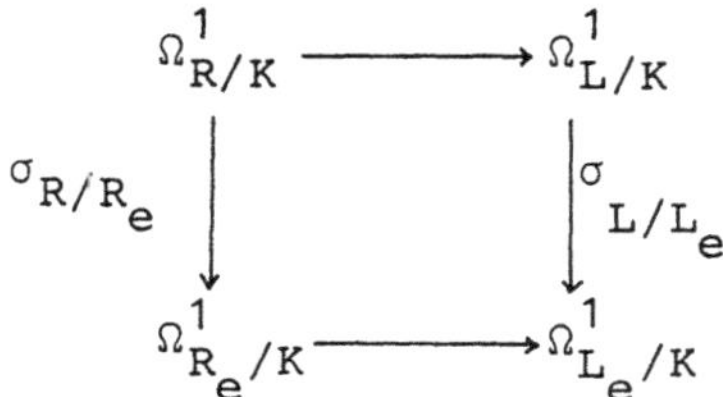

is commutative (Tr3)). Hence $\sigma_{L/L_e}(\omega) = \sigma_{R/R_e}(\omega)$, and 17.3b)
can be applied.

d) We have $\sigma_{L/L_e}(df) = d(\sigma_{L/L_e}(f))$ by Tr6). The claim
follows from the second statement of 17.3b).

e) follows, if we apply part e) of the proof of 17.1 to the
differential $\sigma_{L/L_e}(\omega)$.

The next result leads to the residue theorem in algebraic function fields of one variable:

<u>17.6. Theorem</u> (Trace formula for residues).
Let L/K be an algebraic function field of one variable,
let X be its non-singular model, let L' be an intermediate
field of L/K with $\mathrm{Trdeg}\ L'/K = 1$, and let X' be the non-
singular model of L'/K. For $R' \in X'$ suppose $R_1, \ldots, R_h$ are
the discrete valuation rings in X containing R'. Then for

each $\omega \in \Omega^1_{L/K}$

$$\mathrm{Res}_{R'}(\sigma_{L/L'}(\omega)) = \sum_{j=1}^{h} \mathrm{Res}_{R_j}\omega.$$

Proof. a) It suffices to prove the trace formula under the assumptions that the residue field k_j of R_j is separable over K for j=1,...,h. For, if Char.K =: $p > 0$, let

$L_i := K(L^{p^i})$, $R_{ij} := R_j \cap L_i$, $L'_i := K(L'^{p^i})$, and $R'_i := R' \cap L'_i$

(i $\in \mathbb{N}$,j=1,...,h). For sufficiently large i the residue fields of the R_{ij} (j=1,...,h) are separable over K. Moreover,

$\mathrm{Res}_{R'}(\sigma_{L/L'}(\omega))=\mathrm{Res}_{R'_i}(\sigma_{L'/L'_i}(\sigma_{L/L'}(\omega)))=\mathrm{Res}_{R'_i}(\sigma_{L_i/L'_i}(\sigma_{L/L_i}(\omega)))$

and

$$\mathrm{Res}_{R_j}(\omega) = \mathrm{Res}_{R_{ij}}(\sigma_{L/L_i}(\omega))$$

by 17.5a) and the transitivity of the trace. If the trace formula holds for $\sigma_{L/L_i}(\omega)$ and L_i/L'_i,it also holds for ω and L/L'.

b) Suppose now that k_j/K is separable (j=1,...,h), and let R be the integral closure of R' in L, let $\hat{R}$ be the completion of R with respect to the $\mathcal{M}'$-topology where $\mathcal{M}'$ is the maximal ideal of R',and let $\hat{L} := Q(\hat{R})$.

If $\hat{R}_j$ and $\hat{R}'$ denote the completions of the local rings R_j and R', if $\hat{L}_j := Q(\hat{R}_j)$ (j=1,...,h),and $\hat{L}' := Q(\hat{R}')$, then

$$\hat{R} = \hat{R}' \otimes_{R'} R = \hat{R}_1 \times \ldots \times \hat{R}_h$$
$$\hat{L} = \hat{L}' \otimes_{L'} L = \hat{L}_1 \times \ldots \times \hat{L}_h .$$

By Tr3) and Tr4) we have

$$\sigma_{L/L'}(\omega) = \sigma_{\hat{L}/\hat{L}'}(\omega) = \sum_{j=1}^{h} \sigma_{\hat{L}_j/\hat{L}'}(\omega)$$

hence

$$\text{Res}_{R'}(\sigma_{L/L'}(\omega)) = \sum_{j=1}^{h} \text{Res}_{\hat{R}'}(\sigma_{\hat{L}_j/\hat{L}'}(\omega)).$$

Therefore it remains to be shown that

(7) $$\text{Res}_{R_j}\omega = \text{Res}_{\hat{R}_j}\omega = \text{Res}_{\hat{R}'}(\sigma_{\hat{L}_j/\hat{L}'}(\omega))$$

in other words, that the trace formula holds in the complete case.

c) It suffices now to consider the following situation:

$$\hat{R} = k[\![t]\!] \ , \ \hat{R}' = k'[\![\tau]\!] \ , \ K \subset k' \subset k$$

where k' is the residue field of R', t is a regular parameter of $\hat{R}$, and τ an element of the maximal ideal of $\hat{R}$.

$S := k[\![\tau]\!]$ is an intermediate ring of $\hat{R}/\hat{R}'$. Put $Z := Q(S)$, and let $\eta \in D_K(Z)$. Write

$$\eta = (\sum_{\nu \geq \nu_o} a_\nu \tau^\nu) d\tau \qquad (a_\nu \in k).$$

Then $\sigma_{Z/\hat{L}'}(\eta) = (\sum_{\nu \geq \nu_o} \sigma_{k/k'}(a_\nu) \cdot \tau^\nu) d\tau$, and hence

$$\text{Res}_S(\eta) = \text{Res}_{\hat{R}'}(\sigma_{Z/\hat{L}'}(\eta)).$$

Substitute $\sigma_{\hat{L}/Z}(\omega)$ for η, and assume that (7) holds for $\hat{L}/Z$. Then we see that (7) also holds for $\hat{L}/\hat{L}'$. In other words, it suffices to consider the case $k' = k$.

d) Let $\bar{k}$ be an algebraic closure of k, and let $\hat{R}_{\bar{k}} := \bar{k}[\![t]\!]$ and $\hat{R}'_{\bar{k}} := \bar{k}[\![\tau]\!]$ be the constant field extensions of $\hat{R}$ and $\hat{R}'$ with $\bar{k}$. Since $\hat{R}/\hat{R}'$ is finite, we have

$$\hat{R}_{\bar{k}} = \hat{R}'_{\bar{k}} \otimes_{\hat{R}'} \hat{R} .$$

Let $L_1 := Q(\hat{R}_{\bar{k}})$, $L_2 := (\hat{R}'_{\bar{k}})$, and consider $\omega = (\sum_{\nu \geq \nu_o} a_\nu t^\nu) dt$ ($a_\nu \in k$) as an element of $D_{\bar{k}}(L_1)$. Then $\text{Res}_{\hat{R}_{\bar{k}}}\omega = a_{-1}$, hence

$$\text{Res}_{\hat{R}}\omega = \sigma_{k/K}(\text{Res}_{\hat{R}_{\bar{k}}}\omega).$$

Suppose we have already shown that

$$(8) \qquad \operatorname{Res}_{\hat{R}\overline{/\hat{k}}} \omega = \operatorname{Res}_{\hat{R}'\overline{/\hat{k}}} (\sigma_{L_1/L_2}(\omega)).$$

Since the trace is compatible with base change, we have $\sigma_{L_1/L_2}(\omega) = \sigma_{\hat{L}/\hat{L}'}(\omega)$. If we apply $\sigma_{\hat{k}/K}$ to both sides of (8), we obtain the formula (7) for $\hat{L}/\hat{L}'$. This shows that we may assume that $\hat{k}$ is even algebraically closed.

e) Suppose Char.$K = 0$ and $\hat{k} = K$ is algebraically closed. We may assume $\tau = t^m$ for some $m > 0$, since in $\hat{R} = K[\![t]\!]$ each unit has an m-th root. We then have

$$\omega = (\sum_{\nu \geq \nu_o} a_\nu t^\nu) dt = \frac{1}{mt^{m-1}} (\sum_{\nu \geq \nu_o} a_\nu t^\nu) d\tau \qquad (a_\nu \in K)$$

and by Tr1), Tr2) and F.24

$$\sigma_{\hat{L}/\hat{L}'}(\omega) = \sigma_{\hat{L}/\hat{L}'}(\frac{1}{mt^{m-1}}(\sum_{\nu \geq \nu_o} a_\nu t^\nu)) \cdot d\tau = (\sum_{\nu \equiv -1 \bmod m} a_\nu \tau^{\frac{\nu+1}{m}-1}) d\tau.$$

Now $\operatorname{Res}_{\hat{R}'}(\sigma_{\hat{L}/\hat{L}'}(\omega)) = a_{-1} = \operatorname{Res}_{\hat{R}}(\omega)$.

f) Finally assume that Char.$K =: p > 0$ and $\hat{k} = K$ is algebraically closed. By 17.5e) we may assume that $\nu_{\hat{R}}(\omega) \geq -1$. If even $\nu_{\hat{R}}(\omega) \geq 0$, then

$$\nu_{\hat{R}'}(\sigma_{\hat{L}/\hat{L}'}(\omega)) = \nu_{\hat{R}'}(\sigma_{\hat{R}/\hat{R}'}(\omega)) \geq 0$$

and (8) is trivial. It remains to consider only the special differential $\omega = \frac{dt}{t}$.

It is known that t satisfies an Eisenstein equation over $\hat{R}'$:

$$t^m - r_1 t^{m-1} - \dots - r_m = 0 \qquad (r_k \in \hat{R}')$$

where $\nu_{\hat{R}'}(r_k) > 0$ ($k=1,\dots,m-1$) and $\nu_{\hat{R}'}(r_m) = 1$, if $\nu_{\hat{R}'}$ is the order function on $\hat{R}'$. We may choose τ to be r_m. Then by Tr7)

$$\sigma_{\hat{L}/\hat{L}'}(\frac{dt}{t}) = \frac{d\tau}{\tau}$$

and

$$\operatorname{Res}_{\hat{R}'}\left(\sigma_{\hat{L}/\hat{L}'},\left(\frac{dt}{t}\right)\right) = 1 = \operatorname{Res}_{\hat{R}}\left(\frac{dt}{t}\right), \quad \text{q.e.d.}$$

17.7. Corollary (Residue theorem). For each $\omega \in \Omega^1_{L/K}$

$$\sum_{R \in X} \operatorname{Res}_R \omega = 0.$$

Proof. On account of the trace formula we may assume that $L = K(T)$ is a rational function field of one variable T over K. In this case the residue theorem can be proved by direct computation, using the decomposition of rational functions into partial fractions (see Lang [L], Chap. I, Thm. 8).

Certain applications of residues (i.e. the duality theorem) only remain valid for inseparable function fields if we modify the residue map.

Let L/K be as in the beginning, and assume Char.$K =: p > 0$. By 6.24 there exists a subfield $K_O \subset K$ which is admissible for all $R \in X$, since the $R \in X$ are localizations of finitely many affine K-algebras $A \subset L$, as is well-known. If $\dim_K \Omega^1_{K/K_O} =: r$, then Ω^1_{R/K_O} is a free R-module of rank $r+1$ for alle $R \in X$, hence

$$\Omega^{r+1}_{R/K_O} = \Lambda^{r+1} \Omega^1_{R/K_O} = R \cdot \omega_O$$

with a basis element $\omega_O \in \Omega^{r+1}_{R/K_O}$. Let ν_R denote the normed discrete valuation belonging to R.

17.8. Definition. For $\omega = f \cdot \omega_O \in \Omega^{r+1}_{L/K_O}$ $(f \in L)$

$$\nu_R(\omega) := \nu_R(f)$$

is called the <u>order of ω at R</u>.

Clearly this number does not depend on the special choice of ω_o. The modified residue will be defined for the $\omega \in \Omega_{L/K_o}^{r+1}$. Consider the field towers (4) and (6) and choose $e \in \mathbb{N}$ such that we have $\hat{k}_e = \hat{k}_{e+1} = \ldots$. Then, since $\hat{k}_e/K$ is separable, the canonical sequence

$$O \to R_e \otimes_K \Omega_{K/K_o}^1 \to \Omega_{R_e/K_o}^1 \to \Omega_{R_e/K}^1 \to O$$

is split-exact, and there are canonical isomorphisms

$$\Omega_{R_e/K_o}^{r+1} \cong \Omega_{R_e/K}^1 \otimes_K \Omega_{K/K_o}^r$$

and

$$\Omega_{L_e/K_o}^{r+1} \cong \Omega_{L_e/K}^1 \otimes_K \Omega_{K/K_o}^r.$$

Let $\alpha \in \Omega_{K/K_o}^r$ be a basis element of Ω_{K/K_o}^r. For $\omega \in \Omega_{L/K_o}^{r+1}$ the trace $\sigma_{L/L_e}(\omega)$ is mapped under the above isomorphism to an element

$$\eta \otimes \alpha \in \Omega_{L_e/K}^1 \otimes_K \Omega_{K/K_o}^r \quad \text{with } \eta \in \Omega_{L_e/K}^1.$$

<u>17.9. Definition.</u> $\mathrm{Res}_R \omega = \mathrm{Res}_{R_e}(\eta) \cdot \alpha$
is the <u>residue of ω at R</u>. Here $\mathrm{Res}_{R_e}(\eta)$ is the residue introduced in 17.2.

It is clear that $\mathrm{Res}_R(\omega)$ does not depend on the choice of α and is an element of Ω_{K/K_o}^r. In case K is a perfect field, we have $K_o = K$. Then the new residue coincides with the previous one. The trace formula and the residue theorem immediately generalize to the new situation, if we choose K_o to be admissible for all discrete valuation rings under consideration.

In the rest of this section let K_o be admissible for all $R \in X$, if Char.$K > 0$, and put $K_o := K$, if Char.$K = 0$. We first show that the residue map given in 17.9 is not trivial:

<u>17.10. Lemma.</u>   For any $R \in X$ there exists a form $\omega \in \Omega^{r+1}_{L/K_o}$ with $\nu_R(\omega) = -1$ and $\mathrm{Res}_R \omega \neq 0$.

Proof.   If the residue field $\mathcal{R}$ of R is separable over K, this statement follows from the definitions 17.9 and 17.2 and from the fact that $\sigma_{\mathcal{R}/K}$ is not trivial.

If Char.$K =: p > 0$, let L_e be chosen as in the construction of the residue. Then it suffices to prove: (*) For any $\omega' \in \Omega^{r+1}_{L_e/K_o}$ with $\nu_{R_e}(\omega') = -1$ there exists an $\omega \in \Omega^{r+1}_{L/K_o}$ with $\nu_R(\omega) = -1$ and $\sigma_{L/L_e}(\omega) = \omega'$.

Since L/L_e is purely inseparable, there exists a chain $L = Z_o \supset Z_1 \supset \ldots \supset Z_s = L_e$ of fields where $[Z_i : Z_{i+1}] = p$ for $i=0,\ldots,s-1$. Let $S_i := R \cap Z_i$ be the discrete valuation ring of Z_i/K belonging to R, and let ℓ_i denote its residue field $(i=0,\ldots,s)$.

a) If $\ell_i = \ell_{i+1}$, then $S_i = S_{i+1}[\xi_i]$ with a regular parameter ξ_i of S_i, and $\tau_i := \xi_i^p$ is a regular parameter of S_{i+1}.

b) If $[\ell_i : \ell_{i+1}] = p$, then $S_i = S_{i+1}[\xi_i]$ with a unit ξ_i of S_i whose image $\overline{\xi_i}$ in ℓ_i is a primitive element of ℓ_i/ℓ_{i+1}. In this case $\tau_i := \xi_i^p$ is a unit of S_{i+1}, and the maximal ideal of S_i is generated by the maximal ideal of S_{i+1}.

In any case we have

$$\Omega^1_{S_i/K_o} = S_i \otimes_{S_{i+1}} (\Omega^1_{S_{i+1}/K_o}/\langle d\tau_i \rangle) \oplus S_i d\xi_i$$

and we conclude that $\mu(\Omega^1_{S_i/K_o}) \geq \mu(\Omega^1_{S_{i+1}/K_o})$. But Ω^1_{R/K_o} and

$\Omega^1_{R_e/K_o}$ are free modules of rank r+1. Hence all the $\Omega^1_{S_i/K_o}$ are free of rank r+1.

In order to prove (*) it suffices now to consider an inseparable extension L/L' of degree p. With R' := R ∩ L' we can assume that

$$R = R'[\xi] \quad , \quad \eta := \xi^p \in R'$$

$$\Omega^1_{R/K_o} = R \otimes_{R'} (\Omega^1_{R'/K_o}/\langle d\eta\rangle) \oplus Rd\xi$$

where $d\eta$ belongs to a basis of the free R'-module Ω^1_{R'/K_o}. If $\{d\eta, d\alpha_1, \ldots, d\alpha_r\}$ is a basis of Ω^1_{R'/K_o} $(\alpha_i \in R')$, then $\{d\xi, d\alpha_1, \ldots, d\alpha_r\}$ is a basis of Ω^1_{R/K_o}. Moreover, by Tr7)

$$\sigma_{L/L'} (\frac{d\xi}{\xi} d\alpha_1 \ldots d\alpha_r) = \frac{d\eta}{\eta} d\alpha_1 \ldots d\alpha_r.$$

We now write $\omega' = \varphi \frac{d\eta}{\eta} d\alpha_1 \ldots d\alpha_r$ $(\varphi \in L')$. If, as in case a), ξ is a regular parameter of R, then η is a regular parameter of R' and $v_{R'}(\varphi) = 0$. If, as in case b), ξ is a unit of R, then η is a unit of R' and $v_{R'}(\varphi) = v_R(\varphi) = -1$. In both cases

$$\omega := \varphi \frac{d\xi}{\xi} d\alpha_1 \ldots d\alpha_r \in \Omega^{r+1}_{L/K_o}$$

is a form with $v_R(\omega) = -1$ and $\sigma_{L/L'}(\omega) = \omega'$.

We now investigate the relation between the Kähler differentials of L/K and the Chevalley differentials, introduced in [Ch].

A <u>divisor</u> D of L/K is an element of the free abelian group on X:

$$D = \sum_{R \in X} n_R \cdot R \qquad (n_R \in \mathbf{Z}).$$

Put $v_R(D) := n_R$ for $R \in X$.

The elements $(f_R)_{R \in X} \in \prod_{R \in X} L$ $(f_R \in L)$ for which $v_R(f_R) < 0$ holds only for finitely many $R \in X$ are called <u>repartitions</u> of L/K. They form a subring P of $\prod_{R \in X} L$, and there is a canonical injection $L \to P$, identifying $f \in L$ with the family $(f_R)_{R \in X}$ for which $f_R = f$ for all $R \in X$. For each divisor D of L/K define

$$P_D := \{(f_R) \in P \mid v_R(f_R) \geq v_R(D) \text{ for all } R \in X\}$$

and let J_D be the vector space of K-linear forms $\lambda : P \to K$ that vanish on $L + P_D$:

$$J_D \cong \mathrm{Hom}_K(P/L+P_D, K).$$

If D' is another divisor of L/K such that $v_R(D) \geq v_R(D')$ for all $R \in X$, then $P_D \subset P_{D'}$ and $J_{D'} \subset J_D$.

$J := \bigcup_D J_D$ is an L-vector space in which for $f \in L$ and $\lambda \in J$ the product $f \cdot \lambda$ is defined by

$$(f \cdot \lambda)((f_R))_{R \in X} = \lambda((ff_R)_{R \in X}).$$

Using the Riemann-Roch theorem one can show that $\dim_L J = 1$ ([Ch], Chap. II, Thm. 5). J is called vector space of <u>Chevalley differentials</u> of L/K. For its cohomological interpretation, see Serre [S], Chap. II, Prop. 3. Serre's duality theorem gives a relation between Chevalley differentials and Kähler differentials.

For each divisor D of L/K define

$$\Omega_D := \{\omega \in \Omega^{r+1}_{L/K_o} \mid v_R(\omega) \geq v_R(D) \text{ for all } R \in X\}.$$

<u>17.11. Duality theorem.</u>   The canonical map

$$P \times \Omega^{r+1}_{L/K_o} \to \Omega^r_{K/K_o} \qquad ((f_R), \omega) \mapsto \sum_{R \in X} \mathrm{Res}_R(f_R \omega))$$

induces for each divisor D of L/K a non-degenerate bilinear map

$$P/L+P_D \times \Omega_{-D} \to \Omega^r_{K/K_o}.$$

In particular, there are canonical isomorphisms of K-vector spaces

$$P/K+P_D \cong \mathrm{Hom}_K(\Omega_{-D}, \Omega^r_{K/K_O})$$

and

$$\Omega_{-D} \cong \mathrm{Hom}_K(P/L+P_D, \Omega^r_{K/K_O}).$$

Proof (see [S] and [N]). For $(f_R) \in P$ and $\omega \in \Omega^{r+1}_{L/K_O}$ put

$$<(f_R),\omega> := \sum_{R \in X} \mathrm{Res}_R(f_R\omega).$$

This "scalar product" is well-defined, because $\nu_R(f_R) < 0$ and $\nu_R(\omega) < 0$ only hold for finitely many $R \in X$, and hence only finitely many residues of the sum do not vanish. The scalar product is K-linear and has the following properties:

α) $<f \cdot (f_R),\omega> = <(f_R), f\omega>$ for all $f \in L$.

β) $<f,\omega> = 0$ for all $f \in L$ and all $\omega \in \Omega^{r+1}_{L/K_O}$.

This is the statement of the residue theorem.

γ) For all $(f_R) \in P_D$ and all $\omega \in \Omega_{-D}$ we have $<(f_R),\omega> = 0$.

In fact, $\nu_R(f_R\omega) = \nu_R(f_R) + \nu_R(\omega) \geq \nu_R(D) + \nu_R(-D) = 0$, and hence $\mathrm{Res}_R(f_R\omega) = 0$ for all $R \in X$.

δ) If $<(f_R),\omega> = 0$ for some $\omega \in \Omega^{r+1}_{L/K_O}$ and all $(f_R) \in P$, then $\omega = 0$.

Indeed, if $\omega \neq 0$, then ω is a basis element of Ω^{r+1}_{L/K_O}. By 17.10 there exists for each $R \in X$ an $f \in L$ such that $\mathrm{Res}_R(f\omega) \neq 0$. Now choose the repartion which is f at R and 0 everywhere else.

β) and γ) imply that the scalar product induces a K-bilinear map

$$P/L+P_D \times \Omega_{-D} \to \Omega^r_{K/K_O}.$$

$$i_D \; : \; \Omega_{-D} \to \mathrm{Hom}_K(P/L+P_D, \Omega^r_{K/K_o})$$

is injective. Put $J'_D := \mathrm{Hom}_K(P/L+P_D, \Omega^r_{K/K_o})$ and $J' := \bigcup_D J'_D$. Clearly $J'_D \cong J_D$, and J' is an L-vector space of dimension 1. The mappings i_D define an injection

$$i \; : \; \Omega^{r+1}_{L/K_o} \to J'$$

which is L-linear by $\alpha)$. Since both vector spaces have dimension 1, it must be an isomorphism.

It remains to show that $i^{-1}(J'_D) = \Omega_{-D}$ or, in other words, that for a given $\omega \in \Omega^{r+1}_{L/K_o}$ the condition:

$$<(f_R), \omega> = 0 \text{ for all } (f_R) \in L+P_D$$

implies that $\omega \in \Omega_{-D}$.

Suppose there were an $R \in X$ such that $v_R(\omega) \leq -v_R(D) - 1$. Choose $\omega' \in \Omega^{r+1}_{L/K_o}$ with $v_R(\omega') = -1$ and $\mathrm{Res}_R(\omega') \neq 0$ (17.10). Then $\omega' = f \cdot \omega$ for some $f \in L$ and $v_R(f) = v_R(\omega') - v_R(\omega) \geq v_R(D)$. Let $(f_{R'})_{R' \in X}$ be the repartion with $f_R = f$ and $f_{R'} = 0$ for $R' \neq R$. Then $(f_{R'}) \in P_D$, but $<(f_{R'}), \omega> = \mathrm{Res}_R(f\omega) = \mathrm{Res}_R(\omega') \neq 0$, a contradiction.

Hence $\omega \in \Omega_{-D}$, and we have proved the duality theorem.

In case K is a perfect field we obtain a canonical isomorphism

$$\Omega_{-D} \cong \mathrm{Hom}_K(P/L+P_D, K) = J_D$$

for each divisor D of L/K, and $\Omega^1_{L/K}$ is canonically identified by i with the vector space J of Chevalley differentials of L/K.

<u>Exercises.</u>

1) (Poincaré-residue) Let R be a discrete valuation ring
   with quotient field L and residue field ℓ. Let $t \in R$
   generate the maximal ideal of R, and let Ω be an exterior
   differential algebra of R such that Ω^1 is a free R-
   module of rank $\dim_\ell \Omega^1_\ell + 1$. Choose $x_1, \ldots, x_n \in R$ such that
   $\{dt, dx_1, \ldots, dx_n\}$ is a basis of Ω^1. Any $\omega \in \Omega_L$ can be
   written in the form $\omega = t^\nu \cdot \omega_o$ with $\nu \in \mathbb{Z}$ and $\omega_o \in \Omega$,
   $\omega_o \notin t\Omega$. Put $\nu_R(\omega) := \nu$. Finally, let $Z(\Omega_L)$ denote the
   kernel of the differentiation $d : \Omega_L \to \Omega_L$.
   a) Show that any $\omega \in Z(\Omega_L)$ with $\nu_R(\omega) \geq -1$ can be expressed in the form

$$\omega = \varphi \, \frac{dt}{t} + \psi \quad \text{with } \varphi, \psi \in R[dx_1, \ldots, dx_n].$$

   b) Let $h : \Omega \to \Omega_\ell$ be the functorial epimorphism. Show
   that

$$\text{Res}^*_R \omega := h(\varphi)$$

   is independent of the choice of t and $x_1, \ldots, x_n$.
   $\text{Res}^*_R \omega$ is called the <u>Poincaré-residue</u> of ω.

2) a) In the situation of definition 17.9 show that there
   is an $e \in \mathbb{N}$ such that

$$\nu_{R_e}(\sigma_{L/L_e}(\omega)) \geq -1.$$

   b) If e is chosen as in a), show that

$$\text{Res}_R \omega = \sigma_{\ell_e/K}(\text{Res}^*_{R_e}(\sigma_{L/L_e}(\omega)))$$

   with the Poincaré-residue Res^* defined in exercise 1).

3) (Cartier operator and residues) Let K be a perfect field
   with char.$K =: p > 0$, and let L/K be an algebraic function

field with Trdeg(L/K) = 1. Let X be the non-singular model of L/K. The Cartier operator (§ 5, exercise 6) defines a canonical map

$$C : \Omega^1_{L/K} \to \Omega^1_{L/K} \qquad \text{with ker } C = dL$$

as follows: Let $x \in L$ a separating transcendental element over K (hence $\{x\}$ is a p-basis of L/L^p). Write $\omega \in \Omega^1_{L/K}$ in the form $\omega = \sum_{\nu=0}^{p-1} \lambda_\nu^p x^\nu dx$ $(\lambda_\nu \in L)$. Then $C\omega = \lambda_{p-1} dx$, and this differential is independent of the choice of x. Show that

a) $C(x_1^p \omega_1 + x_2^p \omega_2) = x_1 C\omega_1 + x_2 C\omega_2$ for $x_1, x_2 \in K, \omega_1, \omega_2 \in \Omega^1_{L/K}$.

b) $\nu_R(\omega) \leq p \cdot \nu_R(C\omega) + p - 1$ for each $\omega \in \Omega^1_{L/K}$ and each $R \in X$.

c) $\text{Res}_R \omega = (\text{Res}_R C\omega)^p$ for each $\omega \in \Omega^1_{L/K}$ and each $R \in X$.

4) Under the assumptions of exercise 3) a differential $\omega \in \Omega^1_{L/K}$ is called

a) <u>regular on X</u> (or <u>of the first kind</u>), if $\nu_R(\omega) \geq 0$ for all $R \in X$,

b) <u>of the second kind</u>, if for each $R \in X$ some $y \in L$ exists such that $\nu_R(\omega - dy) \geq 0$,

c) <u>residue-free</u>, if $\text{Res}_R \omega = 0$ for each $R \in X$,

d) <u>pseudo-exact</u>, if $\omega = dx_0 + \sum_{i=1}^{n} \mathbf{x}_i x_i^{p^{\nu_i}-1} dx_i$ with certain $x_0, \ldots, x_n \in L$, $\nu_1, \ldots, \nu_n \in \mathbb{N}$ and $\mathbf{x}_1, \ldots, \mathbf{x}_n \in K$.

Consider the following K-vector spaces in $\Omega^1_{L/K}$:

D_1 := vector space of differentials of the first kind,

D_2 := vector space of differentials of the second kind,

F := vector space of residue-free differentials,

P := vector space of pseudo-exact differentials.

For $\omega \in \Omega^1_{L/K}$ show:

a) ω is of the second kind if and only if $C\omega \in D_1$.

b) ω is residue-free if and only if there is an $n \in \mathbb{N}$ such that $C^n\omega \in D_1$.

c) ω is pseudo-exact if and only if $C^n\omega = 0$ for some $n \in \mathbb{N}$.

Further show:

d) $\dim_K(D_2/dL) = \dim_K(D_1)$.

e) $F = D_1 + P$.

Appendices

Our general reference for results of commutative algebra is Matsumura $[M_1]$ or Bourbaki $[B_2]$. The following appendices collect some material that is frequently needed in the text, and is not immediately available from the above textbooks. The first appendix describes the language and the notations used without further explanation in the whole text.

A. Commutative Algebras

If not stated otherwise a ring R is always a commutative ring with 1,and a ring homomorphism $\rho : R \to S$ maps 1_R to 1_S. Modules M are left modules with $1 \cdot m = m$ for all $m \in M$. Spec(R) denotes the spectrum, Max(R) the maximal spectrum of a ring R,and Min(R) the set of its minimal prime ideals. If the kernel of the natural ring homomorphism $\mathbb{Z} \to R$ is generated by $n \in \mathbb{N}$,then n is called the <u>characteristic</u> of R. Q(R) will always denote the full ring of quotients of R.

An <u>algebra</u> is a triple (S,R,ρ) where S and R are (commutative) rings and $\rho : R \to S$ is a ring homomorphism, called the <u>structure homomorphism</u> of the algebra. The algebra is usually denoted by S/R,or simply by S. If R and S are local rings and $\rho : R \to S$ is a local homomorphism,we call S/R a <u>local algebra</u>. If S/R and S'/R are two algebras with structure homomorphisms ρ and ρ',then an <u>R-homomorphism</u> $\varphi : S \to S'$ (R-algebra homomorphism) is a ring homomorphism with $\varphi \circ \rho = \rho'$. S'/R is a <u>subalgebra</u> of S/R,if S' is a subring of S with $\rho(R) \subset S'$; the structure homomorphism of S'/R being given by ρ. It is clear how the <u>residue class</u>

<u>algebra</u> S/I and the <u>algebra of fractions</u> S_N are defined,if I is an ideal of S and N is a multiplicatively closed subset of S. We always assume that $1 \in N$ for such a set. In the category of R-algebras,limits and colimits of diagrams always exist, in particular, direct and inverse limits, products and coproducts. The products are the direct products, the coproducts are the tensor products.

Let S/R be an algebra. For a family $\{x_\lambda\}_{\lambda \in \Lambda}$ of elements $x_\lambda \in S$ we write $R[\{x_\lambda\}_{\lambda \in \Lambda}]$ for the subalgebra of S/R generated by $\{x_\lambda\}_{\lambda \in \Lambda}$ (i.e. the intersection of all subalgebras containing all x_λ). In case $S = R[\{x_\lambda\}_{\lambda \in \Lambda}]$ we call $\{x_\lambda\}$ a <u>generating system</u> of S/R. If in this case I denotes the kernel of the R-homomorphism

$$\varphi : R[\{X_\lambda\}_{\lambda \in \Lambda}] \to S \qquad (X_\lambda \mapsto x_\lambda)$$

of the polynomial algebra in the indeterminates X_λ onto S, then φ induces an R-isomorphism

A.1. $\qquad\qquad S \cong R[\{X_\lambda\}_{\lambda \in \Lambda}]/I.$

We call A.1 the <u>presentation</u> of S/R given by the generating system $\{x_\lambda\}_{\lambda \in \Lambda}$. If $\{f_\mu\}_{\mu \in M}$ is a system of generators of the ideal I,we say that S/R is given by the <u>generators</u> x_λ and <u>relations</u> f_μ. It is frequently desirable to have a description of an algebra in terms of generators and relations.

S/R is said to be <u>finitely generated</u>,or <u>an algebra of finite type</u>,if S/R has a finite generating system. An algebra of finite type over a field K is also called an <u>affine</u> K-algebra. Observe that for field extensions L/K the notions "generating system" and "finitely generated" have a slightly different meaning. S/R is said to be _essentially of finite type_, if $x_1,\ldots,x_n \in S$ and a multiplicatively closed subset $N \subset R[x_1,\ldots,x_n]$ exist such that $S = R[x_1,\ldots,x_n]_N$. In this case there is a natural R-isomorphism

$$\text{A.2.} \qquad S \cong R[X_1,\ldots,X_n]_{N*}/I$$

with a multiplicatively closed subset $N^* \subset R[X_1,\ldots,X_n]$ (the preimage of N in $R[X_1,\ldots,X_n]$) and an ideal I of $R[X_1,\ldots,X_n]_{N*}$. A.2 is also called a _presentation_ of the given algebra.

S/R is called _finite_ (free, projective, flat, faithfully flat), if S as an R-module has the corresponding property. Here _finite R-modules_ are modules with a finite system of generators (modules of finite type). S/R is said to be _essentially finite_, if $S = T_N$ with a finite algebra T/R and a multiplicatively closed subset $N \subset T$. The notion of a _quasifinite algebra_ will be discussed in Appendix B.

If S/R and R'/R are algebras, then $S' := R' \otimes_R S$ is an R'-algebra with respect to the natural map $R' \to R' \otimes_R S$ ($a \mapsto a \otimes 1$). We say that S'/R' is obtained from S/R by the _base change_ $R \to R'$. If T/S is an S-algebra, then T is an R-algebra via the composition of structure homomorphisms $R \to S$ and $S \to T$. In studying algebras, it is frequently of basic importance to decide whether a property of an algebra is compatible with the fundamental operations on algebras, such as base change, localization, product decomposition etc., and whether it is transitive. In the text such statements are sometimes proved with "differential methods".

If S/R is an algebra and $\mathfrak{p} \in \mathrm{Spec}(S)$, then $\mathfrak{p} \cap R$ will

always denote the preimage of $\mathfrak{p}$ under the structure homo-

morphism $\rho : R \to S$ (even if ρ is not injective). With

$\mathfrak{q} := \mathfrak{p} \cap R$ there is a local homomorphism $\rho_{\mathfrak{p}} : R_{\mathfrak{q}} \to S_{\mathfrak{p}}$ in-

duced by ρ. We call the local algebra $S_{\mathfrak{p}}/R_{\mathfrak{q}}$ the _localization_

of S/R at $\mathfrak{p}$. We write

$$k(\mathfrak{p}) := S_{\mathfrak{p}}/\mathfrak{p}S_{\mathfrak{p}}$$

for the residue field of the local ring $S_{\mathfrak{p}}$. It is an ex-

tension field of $k(\mathfrak{q})$. For any $\mathfrak{q} \in \mathrm{Spec}(R)$

$$S_{\mathfrak{q}}/\mathfrak{q}S_{\mathfrak{q}} = k(\mathfrak{q}) \otimes_R S \quad \text{or} \quad \mathrm{Spec}(S_{\mathfrak{q}}/\mathfrak{q}S_{\mathfrak{q}})$$

is called the _fiber of S/R at $\mathfrak{q}$_. $S_{\mathfrak{q}}/\mathfrak{q}S_{\mathfrak{q}}$ is an algebra over

the field $k(\mathfrak{q})$, and an affine $k(\mathfrak{q})$-algebra, if S/R is of

finite type. As is well-known $\mathrm{Spec}(S_{\mathfrak{q}}/\mathfrak{q}S_{\mathfrak{q}})$ can be canonically

identified with the inverse image of $\mathfrak{q}$ under

$\mathrm{Spec}(\rho) : \mathrm{Spec}(S) \to \mathrm{Spec}(R)$. It is a basic problem how the

fibers $S_{\mathfrak{q}}/\mathfrak{q}S_{\mathfrak{q}}$ of an algebra S/R change as $\mathfrak{q}$ varies. For the

dimension of the fibers of an algebra this is studied in

appendix B.

Numerous notions about algebras are defined "locally".

An algebra S/R is called _flat at $\mathfrak{p} \in \mathrm{Spec}(S)$_, if its locali-

zation $S_{\mathfrak{p}}/R_{\mathfrak{q}}$ at $\mathfrak{p}$ is a flat algebra. It is known that S/R

is flat, if it is flat at all $\mathfrak{p} \in \mathrm{Spec}(S)$. S/R is said to be

regular (smooth) at $\mathfrak{p} \in \mathrm{Spec}(S)$, if it is flat at $\mathfrak{p}$ and

$S_{\mathfrak{p}}/\mathfrak{q}S_{\mathfrak{p}}$ with $\mathfrak{q} := \mathfrak{p} \cap R$ is a regular local ring (resp. S/R

is of finite type and $S_{\mathfrak{p}}/\mathfrak{q}S_{\mathfrak{p}}$ is geometrically regular over

$k(\mathfrak{q})$). S/R is defined to be _regular (smooth)_, if S/R is re-

gular (smooth) at all $\mathfrak{p} \in \mathrm{Spec}(S)$. $\mathfrak{p} \in \mathrm{Spec}(S)$ is called a

Cohen-Macaulay point of S/R (of type r), if S/R is flat at $\mathfrak{p}$

and $S_\mathfrak{p}/\mathfrak{q}S_\mathfrak{p}$ a Cohen-Macaulay local ring (of type r). (See [Ku],Chap.VI,Def.3.18, for the definition of the type of a Cohen-Macaulay ring. Some of its properties are given in [HK]). Cohen-Macaulay points of type 1 are called <u>Gorenstein points</u>. S/R is said to be a <u>Cohen-Macaulay algebra</u> (<u>Gorenstein alge-</u><u>bra</u>), if all $\mathfrak{p} \in \mathrm{Spec}(S)$ are Cohen-Macaulay points (Gorenstein points) of S/R. Algebras that are <u>locally</u> or <u>globally</u> <u>complete intersections</u> will be discussed in appendix C. S/R is called <u>unramified</u> at $\mathfrak{p} \in \mathrm{Spec}(S)$, if $\mathfrak{p}S_\mathfrak{p} = \mathfrak{q}S_\mathfrak{p}$ with $\mathfrak{q} := \mathfrak{p} \cap R$ and $k(\mathfrak{p})$ is a separable algebraic extension field of $k(\mathfrak{q})$. S/R is called <u>étale</u> at $\mathfrak{p}$, if it is unramified and flat at $\mathfrak{p}$. S/R is said to be unramified (étale), if it is so at all $\mathfrak{p} \in \mathrm{Spec}(S)$.

For an S-module M let the minimal number of generators of M be denoted by $\mu(M)$, and for $\mathfrak{p} \in \mathrm{Spec}(S)$ let $\mu_\mathfrak{p}(M)$ be the minimal number of generators of the $S_\mathfrak{p}$-module $M_\mathfrak{p}$.

<u>A.3. Definition.</u> A <u>noetherian normalization</u> of an algebra S/R is a subalgebra $R[X_1,\ldots,X_n]/R$ such that
a) $\{X_1,\ldots,X_n\}$ is algebraically independent over R
(in particular, $R \subset S$).
b) $S/R[X_1,\ldots,X_n]$ is finite.

The notion of <u>quasi-normalization</u> will be introduced in appendix B. An algebra that has a noetherian normalization is of finite type, but the converse is not true in general, if R is not a field. The Noether Normalization Theorem can be expressed in the following form.

<u>A.4. Proposition.</u> Let S/R be an algebra of finite type where R is a domain and $R \subset S$. Then there are elements

$X_1, \ldots, X_n \in S$ that are algebraically independent over R, and there exists an $f \in R \setminus \{0\}$ such that $S_f / R_f[X_1, \ldots, X_n]$ is finite.

For a noetherian normalization $R[X_1, \ldots, X_n]$ of an algebra S/R the following condition frequently plays a role

(1) If $\mathfrak{p} \in \mathrm{Min}(S)$, then $\mathfrak{p} \cap R[X_1, \ldots, X_n] \in \mathrm{Min}(R[X_1, \ldots, X_n])$.

Example 5.C in Matsumura $[M_1]$ shows that this condition is not automatically fulfilled. However, it is so, if S is a domain, or $S/R[X_1, \ldots, X_n]$ satisfies the going-down theorem (e.g. $S/R[X_1, \ldots, X_n]$ is flat).

<u>A.5. Remark.</u>   Let $R[X_1, \ldots, X_n]/R$ be a noetherian normalization of an algebra S/R satisfying (1). Then
a) For any $\mathfrak{p} \in \mathrm{Min}(S)$ we have $\mathfrak{p} \cap R[X_1, \ldots, X_n] = \mathfrak{q} R[X_1, \ldots, X_n]$ with $\mathfrak{q} := \mathfrak{p} \cap R$, and $\mathfrak{q} \in \mathrm{Min}(R)$.
b) If $R[Y_1, \ldots, Y_m]/R$ is another normalization satisfying (1), then $m = n$.

Proof.   a) Certainly $\mathfrak{q} R[X_1, \ldots, X_n]$ is a prime ideal of $R[X_1, \ldots, X_n]$ and $\mathfrak{q} R[X_1, \ldots, X_n] \subset \mathfrak{p} \cap R[X_1, \ldots, X_n]$. By (1) we must have equality and $\mathfrak{q} \in \mathrm{Min}(R)$.
b) For $\mathfrak{p} \in \mathrm{Min}(S)$, $\mathfrak{q} := \mathfrak{p} \cap R$, the $R/\mathfrak{q}$-algebra $S/\mathfrak{p}$ has the normalizations $R/\mathfrak{q}[X_1, \ldots, X_n]$ and $R/\mathfrak{q}[Y_1, \ldots, Y_m]$. Then $\{X_1, \ldots, X_n\}$ and $\{Y_1, \ldots, Y_m\}$ are transcendence bases of $Q(S/\mathfrak{p})$ over $Q(R/\mathfrak{q})$, hence $n = m$.

Let S/R be an algebra, $K := Q(R)$, $L := Q(S)$. If the image of each non-zero-divisor $x \in R$ under the structure homomorphism $\rho : R \to S$ is a non-zero-divisor of S, then ρ

induces a ring-homomorphism $K \to L$ and L may be regarded as a K-algebra.

A.6. <u>Lemma.</u>   Suppose R and S are noetherian.

a) Assume that for any $\mathfrak{p} \in \mathrm{Ass}(S)$ we have $\mathfrak{p} \cap R \in \mathrm{Ass}(R)$. Then $R \to S$ induces a ring-homomorphism $K \to L$.

b) Assume that $\mathrm{Ass}(S) = \mathrm{Min}(S)$, $\mathrm{Ass}(R) = \mathrm{Min}(R)$, and that for each $\mathfrak{p} \in \mathrm{Spec}(S)$ we have $\mathfrak{p} \cap R \in \mathrm{Min}(R)$ if and only if $\mathfrak{p} \in \mathrm{Min}(S)$. Then the canonical homomorphism $K \otimes_R S \to L$ is an isomorphism.

c) Under the assumptions of b) L/K is finite if and only if $k(\mathfrak{p})/k(\mathfrak{q})$ is finite for each $\mathfrak{p} \in \mathrm{Min}(S)$, $\mathfrak{q} := \mathfrak{p} \cap R$.

Proof. a) The conditions of a) imply that any non-zero-divisor of R is mapped to a non-zero-divisor of S. Under the assumptions of b) the conditions of a) are satisfied, hence $K \otimes_R S \to L$ is defined. The only prime ideals of S that survive in $K \otimes_R S$ and L are the elements of $\mathrm{Min}(S)$. Therefore $K \otimes_R S \to L$ is locally an isomorphism, hence an isomorphism.

c) We have $L = \prod\limits_{\mathfrak{p} \in \mathrm{Min}(S)} S_{\mathfrak{p}}$ and $K = \prod\limits_{\mathfrak{q} \in \mathrm{Min}(R)} R_{\mathfrak{q}}$. For $\mathfrak{p} \in \mathrm{Min}(S)$, $\mathfrak{q} := \mathfrak{p} \cap R$, the local rings $S_{\mathfrak{p}}$ and $R_{\mathfrak{q}}$ are complete. Hence $S_{\mathfrak{p}}/R_{\mathfrak{q}}$ is finite if and only if $k(\mathfrak{p})/k(\mathfrak{q})$ is finite. From this the statement of c) follows.

If N is a multiplicatively closed subset of a ring R, then the canonical homomorphism $R \to R_N$ induces a ring-homomorphism $Q(R) \to Q(R_N)$ and a ring-homomorphism

$$\alpha : Q(R) \otimes_R R_N \to Q(R_N).$$

A.7. Lemma. Assume R is noetherian with $\mathrm{Ass}(R) = \mathrm{Min}(R)$. Then α is an isomorphism. Moreover

$$Q(R) = \prod_{\mathfrak{p} \in \mathrm{Min}(R)} R_{\mathfrak{p}} \quad , \quad Q(R_N) = \prod_{\substack{\mathfrak{p} \in \mathrm{Min}(R) \\ \mathfrak{p} \cap N = \phi}} R_{\mathfrak{p}}$$

and $Q(R) \to Q(R_N)$ is identified with the canonical projection

$$\prod_{\mathfrak{p}} R_{\mathfrak{p}} \to \prod_{\mathfrak{p} \cap N = \phi} R_{\mathfrak{p}}.$$

This results from the fact that for $\mathfrak{p} \in \mathrm{Min}(R)$

$$R_{\mathfrak{p}} \otimes_R R_N = \begin{cases} R_{\mathfrak{p}} & \text{if } \mathfrak{p} \cap N = \phi \\[2mm] \{0\} & \text{if } \mathfrak{p} \cap N \neq \phi. \end{cases}$$

For a ring R, an element $f \in R$ and an ideal I of R we use the standard notations

$$D(f) := \{\mathfrak{p} \in \mathrm{Spec}(R) \mid f \notin \mathfrak{p}\}, \quad \mathcal{V}(I) := \{\mathfrak{p} \in \mathrm{Spec}(R) \mid I \subset \mathfrak{p}\}.$$

A.8. Lemma. Let M and N be R-modules, and let $\{f_i\}_{i \in I}$ be a family of elements of R with $\mathrm{Spec}(R) = \bigcup_{i \in I} D(f_i)$.

a) Suppose for each $i \in I$ there is given an $m_i \in M_{f_i}$ such that m_i and m_j have the same image in $M_{f_i f_j}$ for all $i,j \in I$. Then there is precisely one $m \in M$ with the image m_i in M_{f_i} for all $i \in I$.

b) Suppose for each $i \in I$ there is given an R_{f_i}-linear map $\ell_i : M_{f_i} \to N_{f_i}$ such that ℓ_i and ℓ_j induce the same map $\ell_{ij} : M_{f_i f_j} \to N_{f_i f_j}$ for all $i,j \in I$. Then there is exactly one R-linear map $\ell : M \to N$ that induces ℓ_i for each $i \in I$.

It is wellknown and easy to see that M is the limit of the diagram with the arrows $M_{f_i} \to M_{f_i f_j}$ $(i,j \in I)$. From this fact a) and b) follow immediately.

Exercise

Give an example of a noetherian ring R and a multiplicatively closed subset $N \subset R$ such that the canonical map
$\alpha : Q(R) \otimes_R R_N \to Q(R_N)$ is not an isomorphism.

B. <u>Dimension Formulas in Algebras of Finite Type.</u>

<u>Quasifinite and Equidimensional Algebras</u>

For a ring R and $\mathfrak{p} \in \mathrm{Spec}(R)$ <u>the dimension of R at $\mathfrak{p}$</u>
$(\dim_{\mathfrak{p}} R)$ is defined as the supremum of the lengths of all
chains of prime ideals of R that contain $\mathfrak{p}$. It follows from
the definition that

$$(1) \qquad \dim_{\mathfrak{p}} R = \dim R_{\mathfrak{p}} + \dim R/\mathfrak{p}.$$

If R is an affine algebra over a field K, then

$$(2) \qquad \dim_{\mathfrak{p}} R = \dim R_{\mathfrak{p}} + \mathrm{Trdeg}(k(\mathfrak{p})/K)$$

since $\dim R/\mathfrak{p} = \mathrm{Trdeg}(k(\mathfrak{p})/K)$.

A ring R is called <u>equidimensional</u>, if $\dim R/\mathfrak{q}$ for
$\mathfrak{q} \in \mathrm{Min}(R)$ is independent of the $\mathfrak{q}$ chosen. It is known that
for an equidimensional affine algebra R over a field K, and a
field extension K'/K, the ring $R' := K' \otimes_K R$ is equidimensional
too. An affine K-algebra R is equidimensional if and only if
$\dim_{\mathfrak{p}} R = \dim R$ for all $\mathfrak{p} \in \mathrm{Spec}(R)$.

If S/R is a local algebra where R and S are noetherian,
then

$$(3) \qquad \dim S \leq \dim R + \dim S/\mathfrak{m}S$$

where $\mathfrak{m}$ denotes the maximal ideal of R. Here equality holds
e.g., if S/R is flat.

Let R/K be an affine algebra, and let $R' := K' \otimes_K R$ be as
above. For each $\mathfrak{p}' \in \mathrm{Spec}(R')$, $\mathfrak{p} := \mathfrak{p}' \cap R$ we have

$$(4) \qquad \dim_{\mathfrak{p}'} R' = \dim_{\mathfrak{p}} R.$$

In fact, $\dim R'_{\mathfrak{p}'} = \dim R_{\mathfrak{p}} + \dim R'_{\mathfrak{p}'}/\mathfrak{p}R'_{\mathfrak{p}'}$, since $R'_{\mathfrak{p}'}/R_{\mathfrak{p}}$
is flat. Here $\dim R'_{\mathfrak{p}'}/\mathfrak{p}R'_{\mathfrak{p}'} = \dim(K' \otimes_K R/\mathfrak{p})_{\overline{\mathfrak{p}}'}$, where $\overline{\mathfrak{p}}'$ is
the image of $\mathfrak{p}'$ in $K' \otimes_K R/\mathfrak{p}$. This ring is equidimensional,

hence $\dim(K' \otimes_K R/\mathfrak{p})_{\overline{\mathfrak{p}}'} = \dim R/\mathfrak{p} - \dim(K' \otimes_K R/\mathfrak{p})/\overline{\mathfrak{p}}' =$

$\dim R/\mathfrak{p} - \dim R'/\mathfrak{p}' = \mathrm{Trdeg}(k(\mathfrak{p})/K) - \mathrm{Trdeg}(k(\mathfrak{p}')/K')$, and

we obtain $\dim_{\mathfrak{p}'} R' = \dim R'_{\mathfrak{p}'} + \mathrm{Trdeg}(k(\mathfrak{p}')/K') =$

$\dim R_{\mathfrak{p}} + \mathrm{Trdeg}(k(\mathfrak{p})/K) = \dim_{\mathfrak{p}} R$.

Consider now an algebra S/R of finite type where R and S are noetherian domains and $R \subset S$. Let $K := Q(R)$ and $L := Q(S)$. $K \otimes_R S$ is called the <u>generic fiber of S/R</u>. It is, of course, an affine K-algebra. The dimension of the generic fiber will be denoted by d.

<u>B.1. Theorem.</u> For $\mathfrak{p} \in \mathrm{Spec}(S)$, $\mathfrak{q} := \mathfrak{p} \cap R$, we have

$$\dim S_{\mathfrak{p}} + \mathrm{Trdeg}(k(\mathfrak{p})/k(\mathfrak{q})) \leq \dim R_{\mathfrak{q}} + \mathrm{Trdeg}(L/K).$$

If $R_{\mathfrak{q}}$ is universally catenary, or S a polynomial algebra over R, the formula holds with the equality sign.

For the proof, see Matsumura $[M_1]$, 14.C.

Write $\dim_{\mathfrak{p}} S_{\mathfrak{q}}/\mathfrak{q} S_{\mathfrak{q}}$ for $\dim_{\overline{\mathfrak{p}}} S_{\mathfrak{q}}/\mathfrak{q} S_{\mathfrak{q}}$, if $\overline{\mathfrak{p}}$ is the prime ideal of the fiber $S_{\mathfrak{q}}/\mathfrak{q} S_{\mathfrak{q}}$ of $\mathfrak{q}$ corresponding to $\mathfrak{p}$. Since $d = \mathrm{Trdeg}(L/K)$ and $\dim_{\mathfrak{p}} S_{\mathfrak{q}}/\mathfrak{q} S_{\mathfrak{q}} = \dim S_{\mathfrak{p}}/\mathfrak{q} S_{\mathfrak{p}} + \dim S_{\mathfrak{q}}/\mathfrak{p} S_{\mathfrak{q}} =$

$\dim S_{\mathfrak{p}}/\mathfrak{q} S_{\mathfrak{p}} + \mathrm{Trdeg}(k(\mathfrak{p})/k(\mathfrak{q}))$, the formula of B.1 can also be written

$$(5) \quad \dim S_{\mathfrak{p}} \leq \dim R_{\mathfrak{q}} + \dim S_{\mathfrak{p}}/\mathfrak{q} S_{\mathfrak{p}} - (\dim_{\mathfrak{p}}(S_{\mathfrak{q}}/\mathfrak{q} S_{\mathfrak{q}}) - d).$$

The equality sign holds under the same conditions as in the theorem.

<u>B.2. Corollary.</u> If $R_{\mathfrak{q}}$ is universally catenary, or S/R a polynomial algebra, then $\dim_{\mathfrak{p}} S_{\mathfrak{q}}/\mathfrak{q} S_{\mathfrak{q}} \geq d$.

Since (5) holds with the equality sign, this formula is a consequence of (3).

If $\mathfrak{q}S_\mathfrak{q} \neq S_\mathfrak{q}$, then B.2 implies that the dimension of all irreducible components of the fiber $\mathrm{Spec}(S_\mathfrak{q}/\mathfrak{q}S_\mathfrak{q})$ is at least d. (This can also be shown, if R is not universally catenary, see EGA IV, 13.1.1).

__B.3. Corollary.__   a) $\dim S \leq \dim R + \mathrm{Trdeg}(L/K)$.

b) If R is universally catenary and $\dim R = \mathrm{Sup}\{\dim R_\mathfrak{q} \,|\, \mathfrak{q}S_\mathfrak{q} \neq S_\mathfrak{q}\}$, then

$$\dim S = \dim R + \mathrm{Trdeg}(L/K).$$

Proof. We may assume that $\dim R < \infty$. Then $\dim S < \infty$ as well.

a) Choose $\mathfrak{p} \in \mathrm{Spec}(S)$ with $\dim S_\mathfrak{p} = \dim S$. Now apply the formula of B.1.

b) Choose $\mathfrak{q} \in \mathrm{Spec}(R)$ with $\mathfrak{q}S_\mathfrak{q} \neq S_\mathfrak{q}$ and $\dim R_\mathfrak{q} = \dim R$. Choose $\mathfrak{p} \in \mathrm{Spec}(S)$ lying over $\mathfrak{q}$ such that the prime ideal corresponding to $\mathfrak{p}$ in $S_\mathfrak{q}/\mathfrak{q}S_\mathfrak{q}$ is maximal. Then $\mathrm{Trdeg}(k(\mathfrak{p})/k(\mathfrak{q})) = 0$. By B.1 and a) we have $\dim S_\mathfrak{p} = \dim R + \mathrm{Trdeg}(L/K) \geq \dim S$. Since $\dim S_\mathfrak{p} \leq \dim S$ we obtain the formula in b).

__B.4. Definition.__   An algebra S/R has __constant fiber dimension__ $d \in \mathbb{N}$, if $\dim S_\mathfrak{q}/\mathfrak{q}S_\mathfrak{q} = d$ for all $\mathfrak{q} \in \mathrm{Spec}(R)$.

__B.5. Corollary.__   Let R be a noetherian ring which is universally catenary, and let S/R be an algebra of finite type. If S is a domain and S/R has constant fiber dimension $d \in \mathbb{N}$, then

$$\dim S = \dim R + d.$$

Proof. Since $\mathfrak{p}S_\mathfrak{p} \neq S_\mathfrak{p}$ for all $\mathfrak{p} \in \mathrm{Spec}(R)$ the mapping $\mathrm{Spec}(S) \to \mathrm{Spec}(R)$ is surjective, and therefore the kernel of $R \to S$ is nilpotent. Hence we may assume that $R \subset S$. Now B.3b) can be applied.

B.6. **Proposition.** Under the assumptions of B.1 let R be universally catenary. Then there exists an $f \in R\setminus\{0\}$ such that for all $\mathfrak{p} \in \mathrm{Spec}(R)$ with $f \notin \mathfrak{p}$ and all $\mathfrak{P} \in \mathrm{Spec}(S)$ with $\mathfrak{P} \cap R = \mathfrak{p}$ we have $\dim_\mathfrak{P} S_\mathfrak{p}/\mathfrak{p}S_\mathfrak{p} = d$. (This holds also, if R is not universally catenary).

Proof. By B.2 we have $\dim_\mathfrak{P} S_\mathfrak{p}/\mathfrak{p}S_\mathfrak{p} \geq d$ for all $\mathfrak{P} \in \mathrm{Spec}(S)$, $\mathfrak{p} := \mathfrak{P} \cap R$. By the noetherian normalization theorem (A.4) there are elements $Y_1,\ldots,Y_d \in S$ that are algebraically independent over R, and there is an $f \in R\setminus\{0\}$ such that S_f is finite over $R_f[Y_1,\ldots,Y_d]$. For each $\mathfrak{p} \in \mathrm{Spec}(R)$ with $f \notin \mathfrak{p}$ the ring $S_\mathfrak{p}/\mathfrak{p}S_\mathfrak{p}$ is finite over $k(\mathfrak{p})[Y_1,\ldots,Y_d]$, hence $\dim S_\mathfrak{p}/\mathfrak{p}S_\mathfrak{p} \leq d$. Now the claim follows.

B.7. **Theorem** (Semicontinuity theorem of Chevalley).
Let R be noetherian and universally catenary, and let S/R be an algebra of finite type. Then

$$\mathrm{Spec}(S) \to \mathbb{N} \qquad (\mathfrak{P} \mapsto \dim_\mathfrak{P} S_\mathfrak{p}/\mathfrak{p}S_\mathfrak{p})$$

with $\mathfrak{p} := \mathfrak{P} \cap R$ is an upper semicontinuous function. In other words: For each $n \in \mathbb{N}$ the set

$$F_n(S/R) := \{\mathfrak{P} \in \mathrm{Spec}(S) \mid \dim_\mathfrak{P} S_\mathfrak{p}/\mathfrak{p}S_\mathfrak{p} \geq n\}$$

is closed in $\mathrm{Spec}(S)$.
(The theorem holds for arbitrary rings R, see EGA IV, 13.1.3).

Proof. We may assume that R and S are reduced rings. Using noetherian recursion, it is sufficient to prove the theorem under the assumptions that it already holds for all R/I-algebras S/IS where I is an ideal $\neq$(O) of R. Let $\mathfrak{p}_1,\ldots,\mathfrak{p}_s$ be the minimal prime ideals of S, $\mathfrak{q}_i := \mathfrak{p}_i \cap R$, $S_i := S/\mathfrak{p}_i$, and $R_i := R/\mathfrak{q}_i$ (i=1,$\ldots$,s). The formula

$$(6) \qquad \dim_{\mathfrak{p}} S_{\mathfrak{q}}/\mathfrak{q} S_{\mathfrak{q}} = \max_{\mathfrak{p}\supset\mathfrak{p}_i}\{\dim_{\mathfrak{p}}(S_i)_{\mathfrak{q}}/\mathfrak{q}(S_i)_{\mathfrak{q}}\}$$

shows that

$$F_n(S/R) = \bigcup_{i=1}^{s} F_n(S_i/R_i).$$

We may therefore assume that R and S are domains and that R $\subset$ S. Let d be the dimension of the generic fiber of S/R.

For n $\leq$ d we have $F_n(S/R) = \mathrm{Spec}(S)$ by B.2. Assume now that n > d. By B.6 there exists an $f \in R\smallsetminus\{O\}$ such that, for all $\mathfrak{q} \in \mathrm{Spec}(R)$ with $f \notin \mathfrak{q}$, and all $\mathfrak{p} \in \mathrm{Spec}(S)$ with $\mathfrak{p} \cap R = \mathfrak{q}$, we have $\dim_{\mathfrak{p}} S_{\mathfrak{q}}/\mathfrak{q} S_{\mathfrak{q}} = d$. Then $F_n(S/R) \subset \mathrm{Spec}(S/fS) \subset \mathrm{Spec}(S)$. By hypothesis, $F_n(S/R)$ is closed in $\mathrm{Spec}(S/fS)$ and consequently also in $\mathrm{Spec}(S)$.

<u>B.8. Corollary.</u>   If $\varphi : \mathrm{Spec}(S) \to \mathrm{Spec}(R)$ is a closed mapping, then

$$\mathrm{Spec}(R) \to \mathbb{Z} \qquad (\mathfrak{q} \mapsto \dim S_{\mathfrak{q}}/\mathfrak{q} S_{\mathfrak{q}})$$

is an upper semicontinuous function.

For each $n \in \mathbb{N}$ we have $\{\mathfrak{q}\,|\,\dim S_{\mathfrak{q}}/\mathfrak{q} S_{\mathfrak{q}} \geq n\} = \varphi(F_n(S/R))$.

<u>B.9. Definition.</u> An algebra S/R which is essentially of finite type is called <u>quasifinite</u> at $\mathfrak{p} \in \mathrm{Spec}(S)$, if

$$\dim_{\mathfrak{p}} S_{\mathfrak{q}}/\mathfrak{q} S_{\mathfrak{q}} = O \qquad (\mathfrak{q} := \mathfrak{p} \cap R).$$

S/R is called <u>quasifinite</u>, if $\dim_{\mathfrak{p}} S_{\mathfrak{q}}/\mathfrak{q} S_{\mathfrak{q}} = 0$ for all $\mathfrak{p} \in \mathrm{Spec}(S)$.

Of course, if S/R is finite or, more generally, essentially finite, then it is also quasifinite.

<u>B.10. Corollary.</u>   The set of all $\mathfrak{p} \in \mathrm{Spec}(S)$ at which S/R is quasifinite is open in $\mathrm{Spec}(S)$.

It is the complement of $F_1(S/R)$ in $\mathrm{Spec}(S)$.

<u>B.11. Definition.</u>   A module M over a local ring $(R,\mathfrak{m})$ is called <u>quasifinite</u>, if $M/\mathfrak{m}M$ is a finite dimensional vector space over $R/\mathfrak{m}$.

It is known that, if R is complete and M is quasifinite and separated in the $\mathfrak{m}$-topology, then M is even finite.

<u>B.12. Proposition.</u>   Let S/R be an algebra of finite type, $\mathfrak{p} \in \mathrm{Spec}(S)$, $\mathfrak{q} := \mathfrak{p} \cap R$. Then the following statements are equivalent:

a) S/R is quasifinite at $\mathfrak{p}$.

b) $\mathfrak{p}$ is an isolated point of $\mathrm{Spec}(S_{\mathfrak{q}}/\mathfrak{q} S_{\mathfrak{q}})$ (that is, the prime ideal of $S_{\mathfrak{q}}/\mathfrak{q} S_{\mathfrak{q}}$ corresponding to $\mathfrak{p}$ is both maximal and minimal).

c) $S_{\mathfrak{p}}$ is a quasifinite $R_{\mathfrak{q}}$-module.

d) $\mathfrak{q} S_{\mathfrak{p}}$ is a $\mathfrak{p} S_{\mathfrak{p}}$-primary ideal and $k(\mathfrak{p})/k(\mathfrak{q})$ is finite.

Proof.   For an algebra S/R of finite type, $\dim_{\mathfrak{p}} S_{\mathfrak{q}}/\mathfrak{q} S_{\mathfrak{q}}$ is the supremum of the dimensions of the irreducible components of $\mathrm{Spec}(S_{\mathfrak{q}}/\mathfrak{q} S_{\mathfrak{q}})$ that contain $\mathfrak{p}$. From this a) $\leftrightarrow$ b) immediately follows. The formula $\dim_{\mathfrak{p}} S_{\mathfrak{q}}/\mathfrak{q} S_{\mathfrak{q}} = \dim S_{\mathfrak{p}}/\mathfrak{q} S_{\mathfrak{p}} + \mathrm{Trdeg}(k(\mathfrak{p})/k(\mathfrak{q}))$ shows the equivalence of a) with c), and with d).

B.13. <u>Corollary</u>.   a) Let S/R be quasifinite at $\mathfrak{p}$ and $R_\mathfrak{q}$ a noetherian ring. Let $\hat{S}_\mathfrak{p}$ and $\hat{R}_\mathfrak{q}$ denote the completions of $S_\mathfrak{p}$ and $R_\mathfrak{q}$. Then $\hat{S}_\mathfrak{p}/\hat{R}_\mathfrak{q}$ is finite.

b) S/R is quasifinite if and only if $S_\mathfrak{q}/\mathfrak{q}S_\mathfrak{q}$ is a finite dimensional $k(\mathfrak{q})$-vector space for all $\mathfrak{q} \in \mathrm{Spec}(R)$.

Proof.   a) Clearly $\hat{S}_\mathfrak{p}$ is a quasifinite $\hat{R}_\mathfrak{q}$-module. Since $S_\mathfrak{p}$ is noetherian, $\hat{S}_\mathfrak{p}$ is separated in the $\mathfrak{q}\hat{S}_\mathfrak{p}$-topology, and the finiteness of $\hat{S}_\mathfrak{p}/\hat{R}_\mathfrak{q}$ follows.

b) $S_\mathfrak{q}/\mathfrak{q}S_\mathfrak{q}$ is a finite $k(\mathfrak{q})$-algebra if and only if all prime ideals of $S_\mathfrak{q}/\mathfrak{q}S_\mathfrak{q}$ are both maximal and minimal. Now use the equivalence of the statements a) and b) of B.12.

B.14. <u>Example</u>.   Let S/R be a local algebra, and let $\mathfrak{n}$ and $\mathfrak{m}$ be the maximal ideals of S and R respectively. Suppose $S/\mathfrak{m}S$ is noetherian and $S/\mathfrak{n}$ is finite over $R/\mathfrak{m}$. Let $t_1,\ldots,t_d \in \mathfrak{n}$ be elements whose images in $S/\mathfrak{m}S$ form a system of parameters of this ring. Then S is quasifinite over $R[t_1,\ldots,t_d]_\mathfrak{q}$ where $\mathfrak{q} := \mathfrak{n} \cap R[t_1,\ldots,t_d]$. The $\hat{R}$-homomorphism $\hat{R}[\![T_1,\ldots,T_d]\!] \to \hat{S}$ sending T_i to t_i $(i=1,\ldots,d)$ is finite.

We now collect some properties of quasifinite algebras that easily follow from the definition.

B.15. <u>Rules</u>.   Let S/R be a quasifinite algebra of finite type.

a) For each $f \in S$ the algebra S_f/R is quasifinite.

b) If T/S is another quasifinite algebra of finite type, then T/R is quasifinite.

c) For an arbitrary algebra R'/R the algebra $R' \otimes_R S/R'$

is quasifinite.

d) Suppose R is noetherian, $\mathrm{Ass}(S) = \mathrm{Min}(S)$, $\mathrm{Ass}(R) = \mathrm{Min}(R)$, and each $\mathfrak{p} \in \mathrm{Min}(S)$ contracts to an element of $\mathrm{Min}(R)$. Let $L := Q(S)$, $K := Q(R)$. Then $R \to S$ induces a ring homomorphism $K \to L$. The algebra L/K is finite, and $L \cong K \otimes_R S$.

Proof. a) is clear, since quasifiniteness is defined locally.

b) For $\mathfrak{O} \in \mathrm{Spec}(T)$ let $\mathfrak{p} := \mathfrak{O} \cap S$, $\mathfrak{q} := \mathfrak{O} \cap R = \mathfrak{p} \cap R$. Since $\mathfrak{q}S_{\mathfrak{p}}$ is primary for $\mathfrak{p}S_{\mathfrak{p}}$ and $\mathfrak{p}T_{\mathfrak{O}}$ is primary for $\mathfrak{O}T_{\mathfrak{O}}$, we have that $\mathfrak{q}T_{\mathfrak{O}}$ is $\mathfrak{O}T_{\mathfrak{O}}$-primary. Clearly $k(\mathfrak{O})/k(\mathfrak{q})$ is finite, since $k(\mathfrak{O})/k(\mathfrak{p})$ and $k(\mathfrak{p})/k(\mathfrak{q})$ are. Now we can apply B.12.

c) Write $S' := R' \otimes_R S$. For $\mathfrak{q}' \in \mathrm{Spec}(R')$, $\mathfrak{q} := \mathfrak{q}' \cap R$, we have $S'_{\mathfrak{q}'}/\mathfrak{q}'S'_{\mathfrak{q}'} = k(\mathfrak{q}') \otimes_{k(\mathfrak{q})} S_{\mathfrak{q}}/\mathfrak{q}S_{\mathfrak{q}}$. The claim follows from B.13b).

d) For each $\mathfrak{p} \in \mathrm{Spec}(S)$, $\mathfrak{q} := \mathfrak{p} \cap R$, the $R_{\mathfrak{q}}$-module $S_{\mathfrak{p}}$ is quasifinite. If $\mathfrak{q} \in \mathrm{Min}(R)$, then $\dim R_{\mathfrak{q}} = 0$, and $R_{\mathfrak{q}}$ is an artinian ring, so it is complete. But then $S_{\mathfrak{p}}$ is finite over $R_{\mathfrak{q}}$ and $\mathfrak{p} \in \mathrm{Min}(S)$. The claim follows now from A.6b).

One of the most important results about quasifinite algebras is

B.16. Theorem (Zariski's Main Theorem). Let S/R be a quasifinite algebra of finite type. Then there exists a subalgebra T/R of S/R, which is finite, such that $\mathrm{Spec}(S) \to \mathrm{Spec}(T)$ is an open immersion, that is, a homeomorphism onto an open subset of $\mathrm{Spec}(T)$ such that for $\mathfrak{p} \in \mathrm{Spec}(S)$ and $\overline{\mathfrak{p}} := \mathfrak{p} \cap T$ the canonical homomorphism

$$T_{\overline{\mathfrak{p}}} \to S_{\mathfrak{p}}$$

is an isomorphism.

For proofs in the language of algebras we refer to
Peskine [Pe], Raynaud [Ray], Evans [Ev],or Iversen [I].

In the rest of this appendix we shall discuss equidimensional algebras.

B.17. <u>Definition</u>.　An algebra S/R is said to be <u>equidimensional at</u> $\mathscr{p} \in \text{Spec}(S)$ <u>of dimension</u> $d \in \mathbb{N}$,if there is an
$f \in S \setminus \mathscr{p}$ such that the following holds:

a) Each minimal prime ideal of S_f contracts to a minimal prime ideal of R.

b) All irreducible components of the non-empty fibers of S_f/R have dimension d.

S/R is said to be <u>equidimensional of dimension</u> $d \in \mathbb{N}$,if it is so at all $\mathscr{p} \in \text{Spec}(S)$.

Observe that in case S/R is of finite type condition b) of B.17 is equivalent with

$$\dim_{\mathscr{q}}(S_f)_{\mathscr{q}}/\mathscr{q}(S_f)_{\mathscr{q}} = d$$

for all $\mathscr{q} \in \text{Spec}(S_f)$, $\mathscr{q} := \mathscr{q} \cap R$.

B.18. <u>Examples.</u>　a) Suppose S/R is of finite type and quasifinite. Then S/R is equidimensional (of dimension O) if and only if each $\mathscr{p} \in \text{Min}(S)$ contracts to a minimal prime ideal of R.

b) Assume S/R has a noetherian normalization $R[X_1,\ldots,X_d] \to S$ (A.3) such that each $\mathscr{p} \in \text{Min}(S)$ contracts to a minimal prime of $R[X_1,\ldots,X_d]$. Then S/R is equidimensional of dimension d, as will follow from B.20.

B.19. <u>Definition</u>.　Let S/R be an algebra of finite type and $\mathscr{p} \in \text{Spec}(S)$. A <u>quasinormalization of S/R in the neighbourhood of</u> $\mathscr{p}$ is an R-algebra homomorphism

$$R[X_1, \ldots, X_d] \to S_f$$

where $f \in S \setminus \mathfrak{p}$ and the following conditions are satisfied:

a) Each minimal prime of S_f contracts to a minimal prime of $R[X_1, \ldots, X_d]$.

b) $S_f / R[X_1, \ldots, X_d]$ is quasifinite.

<u>B.20. Theorem.</u> Let S/R be of finite type (and R noetherian and universally catenary). For $\mathfrak{p} \in \mathrm{Spec}(S)$, $\mathfrak{p} := \mathfrak{p} \cap R$, the following conditions are equivalent:

a) S/R is equidimensional of dimension d at $\mathfrak{p}$.

b) There exists a quasinormalization $R[X_1, \ldots, X_d] \to S_f$ of S/R in the neighbourhood of $\mathfrak{p}$.

Proof. b) $\to$ a). Let $R[X_1, \ldots, X_d] \to S_f$ be a quasinormalization of S/R in the neighbourhood of $\mathfrak{p}$. For $\mathfrak{q}_o \in \mathrm{Min}(S_f)$, $\mathfrak{q}_o' := \mathfrak{q}_o \cap R[X_1, \ldots, X_d]$, and $\mathfrak{q}_o := \mathfrak{q}_o \cap R$ we have $\mathfrak{q}_o' = \mathfrak{q}_o R[X_1, \ldots, X_d]$, since $\mathfrak{q}_o'$ is a minimal prime ideal of $R[X_1, \ldots, X_d]$. It is therefore clear that $\mathfrak{q}_o \in \mathrm{Min}(R)$. Since $R/\mathfrak{q}_o[X_1, \ldots, X_d] \subseteq S_f/\mathfrak{q}_o$, we have $\mathrm{Trdeg}(k(\mathfrak{q}_o)/k(\mathfrak{q}_o)) \geq d$. By B.2

$$\dim_{\mathfrak{q}} (S_f)_{\mathfrak{q}}/\mathfrak{q}(S_f)_{\mathfrak{q}} \geq d \text{ for all } \mathfrak{q} \in \mathrm{Spec}(S_f),\ \mathfrak{q} := \mathfrak{q} \cap R.$$

On the other hand, let $S' := R[X_1, \ldots, X_d]$, $\mathfrak{q}' := \mathfrak{q} \cap S'$. Then by (3)

$$\dim\ (S_f)_{\mathfrak{q}}/\mathfrak{q}(S_f)_{\mathfrak{q}} \leq \dim\ S'_{\mathfrak{q}'}/\mathfrak{q}S'_{\mathfrak{q}'} + \dim(S_f)_{\mathfrak{q}}/\mathfrak{q}'(S_f)_{\mathfrak{q}}.$$

Here $\dim(S_f)_{\mathfrak{q}}/\mathfrak{q}'(S_f)_{\mathfrak{q}} = 0$ and $\mathrm{Trdeg}(k(\mathfrak{q})/k(\mathfrak{q}')) = 0$, since S_f/S' is quasifinite (B.12d)). We conclude that

$$\dim_{\mathfrak{q}}(S_f)_{\mathfrak{q}}/\mathfrak{q}(S_f)_{\mathfrak{q}} = \dim(S_f)_{\mathfrak{q}}/\mathfrak{q}(S_f)_{\mathfrak{q}} + \mathrm{Trdeg}(k(\mathfrak{q})/k(\mathfrak{q}))$$

$$\leq \dim\ S'_{\mathfrak{q}'}/\mathfrak{q}S'_{\mathfrak{q}'} + \mathrm{Trdeg}(k(\mathfrak{q}')/k(\mathfrak{q})) = d.$$

We therefore have equality, and S/R is equidimensional of dimension d in the neighbourhood of $\mathfrak{p}$.

a) $\to$ b). Let $f \in S \setminus \mathfrak{p}$ be given as in B.17. For simplicity we write S for S_f and $\mathfrak{p}$ for $\mathfrak{p}S_f$. Let $k(\mathfrak{q})[\xi_1,\ldots,\xi_n]$ be a noetherian normalization of the affine $k(\mathfrak{q})$-algebra $S_\mathfrak{q}/\mathfrak{q}S_\mathfrak{q}$. Here $n=d$ by assumption a). We may assume that the ξ_i are images of some $x_i \in S$ in $S_\mathfrak{q}/\mathfrak{q}S_\mathfrak{q}$. Consider the R-homomorphism $R[X_1,\ldots,X_d] \to S$ $(X_i \mapsto x_i)$ which is clearly quasifinite at $\mathfrak{p}$. By B.10 there is an $f_o \in S \smallsetminus \mathfrak{p}$ such that $S_{f_o}/R[X_1,\ldots,X_d]$ is quasifinite. In order to show that $R[X_1,\ldots,X_d] \to S_{f_o}$ is a quasinormalization of S/R in the neighbourhood of $\mathfrak{p}$, it remains to be proven that any $\mathfrak{a} \in \mathrm{Min}(S_{f_o})$ contracts to a minimal prime ideal of $R[X_1,\ldots,X_d]$.

We have $\mathfrak{w} := \mathfrak{a} \cap R \in \mathrm{Min}(R)$ by assumption. $(S_{f_o})_\mathfrak{w}/\mathfrak{w}(S_{f_o})_\mathfrak{w}$ is quasifinite over $k(\mathfrak{w})[X_1,\ldots,X_d]$ (B.15c)). Assume $\mathfrak{w}' := \mathfrak{a} \cap R[X_1,\ldots,X_d]$ is not a minimal prime ideal of $R[X_1,\ldots,X_d]$. Then the prime ideal corresponding to $\mathfrak{w}'$ in $k(\mathfrak{w})[X_1,\ldots,X_d]$ is not zero. Since $k(\mathfrak{a})/k(\mathfrak{w}')$ is finite (B.12d)), this would imply that $\mathrm{Trdeg}(k(\mathfrak{a})/k(\mathfrak{w})) = \mathrm{Trdeg}(k(\mathfrak{w}')/k(\mathfrak{w})) < d$, and hence $\dim_\mathfrak{a}(S_{f_o})_\mathfrak{w}/\mathfrak{w}(S_{f_o})_\mathfrak{w} = \mathrm{Trdeg}(k(\mathfrak{a})/k(\mathfrak{w})) < d$, a contradiction. Thus we have shown that b) holds.

We wish to study now the kernel of the presentation of an equidimensional algebra as a homomorphic image of a polynomial algebra. Let S/R be an algebra of finite type, and let

$$(7) \qquad\qquad S = R[X_1,\ldots,X_n]/I$$

be a presentation of S/R.

B.21. **Proposition.** Assume R is noetherian, and let $\mathfrak{p}$ be a minimal prime ideal of S such that $\mathfrak{q} := \mathfrak{p} \cap R \in \mathrm{Min}(R)$. Let $\mathfrak{a}$ be the preimage of $\mathfrak{p}$ in $R[X_1,\ldots,X_n]$. Then

$$h(I_\mathfrak{a}) = n - \dim_\mathfrak{p} S_\mathfrak{q}/\mathfrak{q}S_\mathfrak{q}.$$

If S/R is equidimensional of dimension d, then all minimal
prime divisors of I have height $n-d$.

Proof. $\mathcal{O}$ is a minimal prime divisor of I with $\mathcal{O} \cap R = \wp$.
Therefore we have
$$h(I_{\mathcal{O}}) = \dim R[X]_{\mathcal{O}} = \dim k(\wp)[X]_{\overline{\mathcal{O}}}$$
where $\overline{\mathcal{O}}$ is the prime ideal corresponding to $\mathcal{O}$ in $k(\wp)[X]$.
To obtain this formula we used that $\dim R_{\wp} = 0$ and that
$R[X]_{\mathcal{O}}$ is flat over $R_{\wp}$. It implies
$$h(I_{\mathcal{O}}) = n-\dim k(\wp)[X]/\overline{\mathcal{O}} = n-\dim S_{\wp}/\wp S_{\wp} =$$
$$n-\dim_{\wp} S_{\wp}/\wp S_{\wp}.$$
The second statement of the proposition now follows at once
from the definition of an equidimensional algebra.

B.22. <u>Proposition.</u> Let S/R be an algebra of finite type for
which a presentation (7) is given. Assume that R is noetherian,
and that each minimal prime ideal of S contracts to a minimal
prime ideal of R. Then the following statements are equivalent:
a) S/R is equidimensional of dimension d.
b) For each $\wp \in \operatorname{Spec}(R)$ with $\wp S_{\wp} \neq S_{\wp}$ all minimal prime
divisors of the image $\overline{I_{\wp}}$ of $I_{\wp}$ in $k(\wp)[X_1,\ldots,X_n]$ have
height $n-d$.

Proof. The presentation (7) induces a presentation
$$S_{\wp}/\wp S_{\wp} = k(\wp)[X_1,\ldots,X_n]/\overline{I_{\wp}}\ .$$
Since the minimal prime divisors of $\overline{I_{\wp}}$ are in one-to-one
correspondence with the irreducible components of $\operatorname{Spec} S_{\wp}/\wp S_{\wp}$
the claim follows immediately.

Let R be a ring, M an R-module. Remember that a sequence
$\{a_1,\ldots,a_m\} \subseteq R$ is called <u>M-quasiregular</u>, if $(a_1,\ldots,a_m)M \neq M$, and
if $\{a_1,\ldots,a_m\}$ is an $M_{\mathfrak{m}}$-regular sequence for each $\mathfrak{m} \in \operatorname{Max}(R)$

with $(a_1,\ldots,a_m) \subset \mathcal{m}$. As a consequence of the local criterion of flatness one has the following (Matsumura $[M_1]$,20.F).

B.23. Lemma.   Let R and P be noetherian local rings and
$R \to P$ a flat local homomorphism. Let $\mathcal{u}$ be an ideal of R,
put $\overline{R} := R/\mathcal{u}$, $\overline{P} := P/\mathcal{u}P$. Given a sequence $\{t_1,\ldots,t_h\}$ of
elements of P, let $\overline{t_i}$ be the image of t_i in $\overline{P}$ (i=1,...,h).
Then the following conditions are equivalent:
a) $\{t_1,\ldots,t_h\}$ is a P-regular sequence, and $P/(t_1,\ldots,t_h)$
is a flat R-module.
b) $\{\overline{t_1},\ldots,\overline{t_h}\}$ is a $\overline{P}$-regular sequence, and $\overline{P}/(\overline{t_1},\ldots,\overline{t_h})$
is a flat $\overline{R}$-module.

B.24. Corollary.   Under the assumptions of B.22 let S/R
be equidimensional of dimension d, let $\mathcal{g}$ be a prime ideal
of R with $\mathcal{g}S_{\mathcal{g}} \neq S_{\mathcal{g}}$, and let h := n-d. Then $I_{\mathcal{g}}$ contains elements $t_1,\ldots,t_h$ having the following property: For each
$\mathcal{p} \in \mathrm{Spec}(R[X])$ lying over $\mathcal{g}$ and containing I, the sequence
$\{t_1,\ldots,t_h\}$ is $R[X]_{\mathcal{p}}$-regular and $R[X]_{\mathcal{p}}/(t_1,\ldots,t_h)$ is flat
over $R_{\mathcal{g}}$.

Proof.   By B.22b) $\overline{I}_{\mathcal{g}}$ is an ideal of height h of the Cohen-
Macaulay ring $k(\mathcal{g})[X_1,\ldots,X_n]$. It contains a regular sequence $\{\overline{t_1},\ldots,\overline{t_h}\}$ of length h. Let $\{t_1,\ldots,t_h\}$ be a sequence of preimages of the $\overline{t_i}$ in $I_{\mathcal{g}}$. We conclude from B.23
that $\{t_1,\ldots,t_h\}$ has the desired property.

B.25. Proposition.   Let S/R and T/S be algebras of finite
type where S/R is equidimensional of dimension d and T/S

is equidimensional of dimension d' (R noetherian and universally catenary). Then T/R is equidimensional of dimension d+d'.

Proof. We shall use the criterion B.20. Consider $\mathfrak{p} \in \text{Spec}(T)$ and $\mathfrak{s} := \mathfrak{p} \cap S$. Let $S[Y_1, \ldots, Y_d'] \to T_f$ and $R[X_1, \ldots, X_d] \to S_g$ be quasinormalizations in the neighbourhood of $\mathfrak{p}$ resp. $\mathfrak{s}$ ($f \in T \smallsetminus \mathfrak{p}$, $g \in S \smallsetminus \mathfrak{s}$). It will turn out that the composed map

$$R[X_1, \ldots, X_d, Y_1, \ldots, Y_d'] \to S_g[Y_1, \ldots, Y_d'] \to T_{fg}$$

is a quasinormalization of T/R in the neighbourhood of $\mathfrak{p}$. Since quasifiniteness is compatible with base change (B.15c)) and is a transitive notion (B.15b)), we see that T_{fg} is quasifinite over R[X,Y]. For $\mathfrak{a} \in \text{Min}(T_{fg})$ we have $\mathfrak{a} \cap S_g[Y] \in \text{Min}(S_g[Y])$, since $T_f/S[Y]$ has the corresponding property. Then $\mathfrak{a} \cap S_g[Y] = \mathfrak{y} S_g[Y]$ with some $\mathfrak{y} \in \text{Min}(S_g)$ and $\mathfrak{y} \cap R[X] \in \text{Min}(R[X])$. We conclude that $\mathfrak{a} \cap R[X,Y] = \mathfrak{y} S_g[Y] \cap R[X,Y] \in \text{Min}(R[X,Y])$, and we have shown that $T_{fg}/R[X,Y]$ is indeed a quasinormalization.

<u>B.26. Proposition.</u>   Let S/R be an algebra of finite type and R'/R an arbitrary algebra. Let $S' := R' \otimes_R S$, and assume that each minimal prime ideal of S' contracts to a minimal prime ideal of S. If S/R is equidimensional of dimension d at $\mathfrak{p} \in \text{Spec}(S)$, then S'/R' is equidimensional of dimension d at all $\mathfrak{p}' \in \text{Spec}(S')$ lying over $\mathfrak{p}$.

Proof. Replacing S by S_f with a suitable $f \in S \smallsetminus \mathfrak{p}$, we may assume that the conditions a) and b) of definition B.17 are already satisfied for S/R. For $\mathfrak{a}' \in \text{Spec}(S')$ let $\mathfrak{a} := \mathfrak{a}' \cap S$,

$\mathfrak{q}' := \mathfrak{Q}' \cap R'$, and $\mathfrak{q} := \mathfrak{Q}' \cap R = \mathfrak{Q} \cap R$. Then $S'_{\mathfrak{Q}'}/\mathfrak{q}'S'_{\mathfrak{Q}'} = k(\mathfrak{q}') \otimes_{k(\mathfrak{q})} S_{\mathfrak{Q}}/\mathfrak{q}S_{\mathfrak{Q}}$, and all irreducible components of $\mathrm{Spec}(S'_{\mathfrak{Q}'}/\mathfrak{q}'S'_{\mathfrak{Q}'})$ have dimension d, since this is the case for $\mathrm{Spec}(S_{\mathfrak{Q}}/\mathfrak{q}S_{\mathfrak{Q}})$ (see (4)). If $\mathfrak{Q}' \in \mathrm{Min}(S')$, then $\mathfrak{Q} \in \mathrm{Min}(S)$ by assumption, and $\mathfrak{q} \in \mathrm{Min}(R)$, since S/R is equidimensional. $S'_{\mathfrak{Q}'}/\mathfrak{q}S'_{\mathfrak{Q}'}$ is a localization of $R'_{\mathfrak{q}'}/\mathfrak{q}R'_{\mathfrak{q}'} \otimes_{k(\mathfrak{q})} S_{\mathfrak{Q}}/\mathfrak{q}S_{\mathfrak{Q}}$ and is therefore flat over $R'_{\mathfrak{q}'}/\mathfrak{q}R'_{\mathfrak{q}'}$. Since $\dim S'_{\mathfrak{Q}'}/\mathfrak{q}S'_{\mathfrak{Q}'} = 0$, we also have $\dim R'_{\mathfrak{q}'} = \dim R'_{\mathfrak{q}'}/\mathfrak{q}R'_{\mathfrak{q}'} = 0$ and $\mathfrak{q}' \in \mathrm{Min}(R')$, q.e.d.

An example (EGA IV. 13.3.9) shows that the condition on the minimal primes of S' is necessary for the proposition to be valid. It is satisfied e.g., if R'/R is flat, in other words: The notion of equidimensionality is compatible with <u>flat</u> base change.

In connection with quasinormalizations the following statement is sometimes of interest.

<u>B.27. Proposition.</u>   Let S/R and P/R be algebras where R,S and P are noetherian, and let $\alpha : P \to S$ be an R-homomorphism. For $\mathfrak{p} \in \mathrm{Spec}(S)$ let $\mathfrak{q} := \mathfrak{p} \cap P$, $\mathfrak{r} := \mathfrak{p} \cap R$. Assume that S/P is quasifinite at $\mathfrak{p}$, P/R is regular at $\mathfrak{q}$, and $\dim S_{\mathfrak{p}}/\mathfrak{r}S_{\mathfrak{p}} = \dim P_{\mathfrak{q}}/\mathfrak{r}P_{\mathfrak{q}}$. If $\mathfrak{p}$ is a Cohen-Macaulay point of S/R, then

a) $\mathfrak{p}$ is a Cohen-Macaulay point of S/P, in particular, $S_{\mathfrak{p}}/P_{\mathfrak{q}}$ is flat.

b) $r(S_{\mathfrak{p}}/\mathfrak{q}S_{\mathfrak{p}}) = r(S_{\mathfrak{p}}/\mathfrak{r}S_{\mathfrak{p}})$ where r denotes the Cohen-Macaulay type.

Proof. In order to show that $S_{\mathfrak{p}}/P_{\mathfrak{q}}$ is flat it suffices to prove that

α) $S_{\mathfrak{p}}/\mathfrak{g}S_{\mathfrak{p}}$ is flat over $P_{\mathfrak{q}}/\mathfrak{g}P_{\mathfrak{q}}$,

β) the canonical homomorphism

$$gr_{\mathfrak{g}P_{\mathfrak{q}}}(P_{\mathfrak{q}}) \otimes_{P_{\mathfrak{q}}} S_{\mathfrak{p}} \to gr_{\mathfrak{g}S_{\mathfrak{p}}}(S_{\mathfrak{p}})$$

is an isomorphism.

(Bourbaki [B_2], Chap.III, § 5, Thm.1).

In our situation the Cohen-Macaulay ring $S_{\mathfrak{p}}/\mathfrak{g}S_{\mathfrak{p}}$ is quasi-finite over the regular local ring $P_{\mathfrak{q}}/\mathfrak{g}P_{\mathfrak{q}}$, and both rings have the same dimension. For their completions we obtain that $\widehat{S_{\mathfrak{p}}/\mathfrak{g}S_{\mathfrak{p}}}$ is finite over $\widehat{P_{\mathfrak{q}}/\mathfrak{g}P_{\mathfrak{q}}}$, and since $\widehat{S_{\mathfrak{p}}/\mathfrak{g}S_{\mathfrak{p}}}$ is Cohen-Macaulay and $\widehat{P_{\mathfrak{q}}/\mathfrak{g}P_{\mathfrak{q}}}$ is regular, $\widehat{S_{\mathfrak{p}}/\mathfrak{g}S_{\mathfrak{p}}}$ is free over $\widehat{P_{\mathfrak{q}}/\mathfrak{g}P_{\mathfrak{q}}}$. Hence α) is satisfied.

The mapping of β) is the composition of natural homomorphisms $gr_{\mathfrak{g}P_{\mathfrak{q}}}(P_{\mathfrak{q}}) \otimes_{P_{\mathfrak{q}}} S_{\mathfrak{p}} \to (gr_{\mathfrak{g}R_{\mathfrak{g}}}(R_{\mathfrak{g}}) \otimes_{R_{\mathfrak{g}}} P_{\mathfrak{q}}) \otimes_{P_{\mathfrak{q}}} S_{\mathfrak{p}} \xrightarrow{\sim} gr_{\mathfrak{g}R_{\mathfrak{g}}}(R_{\mathfrak{g}}) \otimes_{R_{\mathfrak{g}}} S_{\mathfrak{p}} \to gr_{\mathfrak{g}S_{\mathfrak{p}}}(S_{\mathfrak{p}})$ which are all isomorphisms, because $P_{\mathfrak{q}}/R_{\mathfrak{g}}$ and $S_{\mathfrak{p}}/R_{\mathfrak{g}}$ are flat. Hence β) is satisfied.

Since any regular system of parameters of $P_{\mathfrak{q}}/\mathfrak{g}P_{\mathfrak{q}}$ is by α) a regular sequence of $S_{\mathfrak{p}}/\mathfrak{g}S_{\mathfrak{p}}$, it follows that with $S_{\mathfrak{p}}/\mathfrak{g}S_{\mathfrak{p}}$ the ring $S_{\mathfrak{p}}/\mathfrak{q}S_{\mathfrak{p}}$ is Cohen-Macaulay as well, and both rings have the same type.

Exercises

1) Let R be a noetherian ring and S/R an algebra of finite type. Then

   $$\dim S \leq \dim R + \sup\{\dim S_{\mathfrak{g}}/\mathfrak{g}S_{\mathfrak{g}} \mid \mathfrak{g} \in \operatorname{Spec}(R)\}.$$

2) Let S/R be an algebra of finite type where $R \subset S$ are noetherian domains. For $\mathfrak{p} \in \mathrm{Spec}(S)$ let $\mathfrak{q} := \mathfrak{p} \cap R$.

a) If S/R is equidimensional at $\mathfrak{p}$, then
$$\dim S_{\mathfrak{p}} = \dim R_{\mathfrak{q}} + \dim S_{\mathfrak{p}}/\mathfrak{q}S_{\mathfrak{p}}.$$

b) If this formula holds and $R_{\mathfrak{q}}$ is universally catenary, then S/R is equidimensional at $\mathfrak{p}$.

3) Let S/R be an algebra of finite type, $\mathfrak{p} \in \mathrm{Spec}(S)$, $\mathfrak{q} := \mathfrak{p} \cap R$. Let $\mathfrak{p}_1, \dots, \mathfrak{p}_h$ denote the elements of $\mathrm{Min}(S)$ that are contained in $\mathfrak{p}$, let $\mathfrak{q}_i := \mathfrak{p}_i \cap R$, $S_i := S/\mathfrak{p}_i$, and $R_i := R/\mathfrak{q}_i$ $(i=1,\dots,h)$. Write d_i for the dimension of the generic fiber of S_i/R_i $(i=1,\dots,h)$. Then the following statements are equivalent.

a) S/R is equidimensional at $\mathfrak{p}$ of dimension d.

b) $\mathfrak{q}_i \in \mathrm{Min}(R)$, $d_i = d$, and S_i/R_i is equidimensional at $\mathfrak{p}$ (of dimension d) for $i=1,\dots,h$.

C. Complete Intersections

In this appendix various notions of "complete intersections" will be discussed. We first consider a local ring
$(R, \mathcal{m})$ with $\mathcal{m}$-adic completion $\hat{R}$.

C.1. Definition. R is called a complete intersection, if
R is noetherian, and if $\hat{R} \cong S/I$ where S is a regular local
ring and the ideal $I \subset S$ can be generated by an S-regular
sequence.

It is wellknown that this last condition is equivalent
with $\mu(I) = h(I) = \dim S - \dim R$ or, in other words, with I
being a complete intersection ideal. Clearly a noetherian
local ring R is a complete intersection if and only if $\hat{R}$ is.

If R is a complete intersection and $R = T/J$ with a regular local ring T, then it is known that J is a complete
intersection ideal of T. In this case, with R the ring $R_{\mathcal{p}}$ is
likewise a complete intersection for any $\mathcal{p} \in \mathrm{Spec}(R)$. By
Avramov [Av] this is true in general. Ferrand [Fe] and
Vasconcelos [Va$_1$] have shown

C.2. Theorem. Let I be an ideal of a noetherian local ring
S and $R := S/I$. Let $\mathrm{pd}_S(R)$ be the projective dimension of
the S-module R. Then the following statements are equivalent:
a) I can be generated by an S-regular sequence.
a') Each minimal system of generators of I is an S-regular
sequence.
b) $\mathrm{pd}_S(R) < \infty$, and I/I^2 is a free R-module.
c) $\mathrm{pd}_S(R) < \infty$, and $I/I^2 \cong R^t \oplus K$ for some R-module K and
some $t \in \mathbb{N}$ such that I does not contain an S-regular sequence of length t+1.

In case S is regular the condition $pd_S(R) < \infty$ is automatically satisfied. Gulliksen [Gu] proved that the conditions a)-c) of C.2 are also equivalent with

d) $pd_S(R) < \infty$, and $pd_R(I/I^2) \leq 1$.

It is an open problem whether it suffices to assume $pd_R(I/I^2) < \infty$ (Conjecture of Vasconcelos).

An arbitrary ring R is called a <u>complete intersection at</u> $\mathfrak{q} \in \underline{Spec(R)}$, if $R_{\mathfrak{q}}$ is a complete intersection in the sense of C.1. If this is the case for all $\mathfrak{q} \in Spec(R)$, we say that R is <u>locally a complete intersection</u>.

We now introduce a notion of "relative" complete intersection. Let S/R be an algebra.

<u>C.3. Definition.</u> S/R is <u>locally at $\mathfrak{p} \in Spec(S)$ a complete intersection</u>, if $S_{\mathfrak{p}}/R_{\mathfrak{q}}$ (with $\mathfrak{q} := \mathfrak{p} \cap R$) is flat, and $S_{\mathfrak{p}}/\mathfrak{q}S_{\mathfrak{p}}$ is a complete intersection in the sense of C.1. If this is so for all $\mathfrak{p} \in Spec(S)$, we say that S/R is <u>locally a complete intersection</u>.

If S/R is locally a complete intersection, so is S_T/R for each multiplicatively closed subset T of S.

We are mainly interested in the case of S/R being essentially of finite type. Given a presentation

(1) $$S = R[X_1,\ldots,X_n]_N/I$$

we wish to know what it means for I, if S/R is a complete intersection at some $\mathfrak{p} \in Spec(S)$:

<u>C.4. Proposition.</u> Let $(R,\mathfrak{m})$ and P be local rings and $R \to P$ a flat local homomorphism. Suppose P is noetherian, and let I be an ideal of P. Let $S := P/I$, $\bar{S} := S/\mathfrak{m}S$, $\bar{P} := P/\mathfrak{m}P$, and let $\bar{I}$ be the image of I in $\bar{P}$. Then the following conditions are

equivalent:

a) S/R is flat,and $\bar{I}$ is generated by a $\bar{P}$-regular sequence.

b) I is generated by preimages of the elements of some $\bar{P}$-regular sequence in $\bar{I}$ (and the preimages form a P-regular sequence (B.23)).

If $\bar{P}$ is a Cohen-Macaulay ring,these conditions are also equivalent with each of the following:

c) Any minimal system of generators $\{t_1,\ldots,t_h\}$ of I is a P-regular sequence,and the images $\bar{t}_i$ of the t_i in $\bar{P}$ form a $\bar{P}$-regular sequence.

d) $\mu(I) = \dim \bar{P} - \dim \bar{S}$.

Proof. a) $\to$ b) Choose $t_i \in I$ (i=1,...,h) such that $\bar{I}=(\bar{t}_1,..,\bar{t}_h)$ where $\bar{t}_i$ is the image of t_i in $\bar{P}$ and $\{\bar{t}_1,\ldots,\bar{t}_h\}$ is a $\bar{P}$-regular sequence. Tensorize the exact sequence $0 \to I \to P \to S \to 0$ with $R/\mathfrak{m}$ over R. Since S/R is flat,we see that $\bar{I} = I/\mathfrak{m}I$. By Nakayama I = $(t_1,\ldots,t_h)$.

b) $\to$ a) is an immediate consequence of B.23.

Now assume that $\bar{P}$ is a Cohen-Macaulay ring.

b) $\to$ d) If I = $(t_1,\ldots,t_h)$ and the images $\bar{t}_i$ of the t_i in $\bar{P}$ form a $\bar{P}$-regular sequence, then S/R is flat by B.23. From $\bar{S} = \bar{P}/(\bar{t}_1,\ldots,\bar{t}_h)$ we conclude that h=dim $\bar{P}$-dim $\bar{S}$ = dim P-dim S. Moreover, $h \geq \mu(I) \geq \mu(\bar{I})$ = h,and d) follows.

d) $\to$ c) Let $\{t_1,\ldots,t_h\}$ be a minimal system of generators of I. Then h = dim $\bar{P}$-dim $\bar{S}$ by assumption. The images $\bar{t}_i$ of the t_i in $\bar{P}$ form a $\bar{P}$-regular sequence, because $\bar{S} = \bar{P}/(\bar{t}_1,\ldots,\bar{t}_h)$ and $\bar{P}$ is Cohen-Macaulay. By B.23, $\{t_1,..,t_h\}$ is a P-regular sequence.

Since c) $\to$ b) is trivial,the proof is complete.

<u>C.5. Corollary.</u>   Let P/R be an algebra, I an ideal of P,and S := P/I. For some $\mathfrak{p} \in \mathrm{Spec}(S)$ let $\mathfrak{q}$ be the preimage of $\mathfrak{p}$

in P and $\mathfrak{g} := \mathfrak{p} \cap R$. Suppose that $P_{\mathfrak{q}}$ is noetherian, $P_{\mathfrak{q}}/R_{\mathfrak{g}}$ is flat, and $P_{\mathfrak{q}}/\mathfrak{g}P_{\mathfrak{q}}$ is a regular local ring. Then the following conditions are equivalent:

a) S/R is locally at $\mathfrak{p}$ a complete intersection.

b) $S_{\mathfrak{p}}/R_{\mathfrak{g}}$ is flat, and $I_{\mathfrak{q}}$ is generated by a $P_{\mathfrak{q}}$-regular sequence.

c) Each minimal system of generators $\{t_1,\ldots,t_h\}$ of $I_{\mathfrak{q}}$ is a $P_{\mathfrak{q}}$-regular sequence, and the images $\bar{t}_i$ of the t_i in $P_{\mathfrak{q}}/\mathfrak{g}P_{\mathfrak{q}}$ form a regular sequence of that ring.

C.6. <u>Corollary.</u> Let S/R be an algebra for which a presentation (1) is given. Write $P := R[X_1,\ldots,X_n]_N$. For $\mathfrak{p} \in \mathrm{Spec}(S)$ let $\mathfrak{q}$ be the preimage of $\mathfrak{p}$ in P and $\mathfrak{g} := \mathfrak{p} \cap R$. Suppose $R_{\mathfrak{g}}$ is noetherian. Then the conditions a)-c) of C.5 are equivalent.

C.7. <u>Corollary.</u> Under the assumptions of C.6 let R be noetherian. Then the following conditions are equivalent:

a) S/R is locally a complete intersection.

b) S/R is flat, and for each $\mathfrak{n} \in \mathrm{Max}(P) \cap \mathcal{V}(I)$ the ideal $I_{\mathfrak{n}}$ is generated by a $P_{\mathfrak{n}}$-regular sequence.

C.8. <u>Corollary.</u> Under the assumptions of C.6 let R be noetherian, and let S/R be locally a complete intersection. Then I/I^2 is a projective S-module and

$$\mathrm{rank}_{\mathfrak{p}}(I/I^2) = \dim P_{\mathfrak{q}}/\mathfrak{g}P_{\mathfrak{q}} - \dim S_{\mathfrak{p}}/\mathfrak{g}S_{\mathfrak{p}} = n - \dim_{\mathfrak{p}} S_{\mathfrak{p}}/\mathfrak{g}S_{\mathfrak{p}}.$$

Proof. The first assertion about $\mathrm{rank}_{\mathfrak{p}}(I/I^2)$ follows from C.4d) and the second from the equations

$$\dim P_{\mathfrak{P}}/\mathfrak{q}P_{\mathfrak{P}} - \dim S_{\mathfrak{p}}/\mathfrak{q}S_{\mathfrak{p}} =$$
$$\dim k(\mathfrak{q})[X_1,\ldots,X_n] - \dim P_{\mathfrak{q}}/\mathfrak{P}P_{\mathfrak{q}} - \dim S_{\mathfrak{p}}/\mathfrak{q}S_{\mathfrak{p}} =$$
$$n - \dim_{\mathfrak{p}} S_{\mathfrak{q}}/\mathfrak{q}S_{\mathfrak{q}}.$$

We shall now introduce a notion of global complete intersection.

<u>C.9. Definition.</u>   An algebra S/R is called a <u>global complete intersection</u>, if it is flat and has a presentation

$$(2) \qquad S = R[X_1,\ldots,X_n]/(t_1,\ldots,t_h)$$

where $\{t_1,\ldots,t_h\}$ is a quasiregular sequence of the polynomial ring $R[X_1,\ldots,X_n]$. (2) is then called a <u>presentation of S/R as a complete intersection</u>.

If S/R is a global complete intersection, and R is noetherian, then S/R is of finite type and locally a complete intersection by C.7. If we write $P := R[X_1,\ldots,X_n]$, $I := (t_1,\ldots,t_h)$, then the S-epimorphism

$$S[T_1,\ldots,T_h] \to \mathrm{gr}_I(P) \qquad (T_i \mapsto t_i+I^2)$$

is an isomorphism and, in particular, I/I^2 is a free S-module of rank h. If R is noetherian, then I can even be generated by a P-regular sequence ([Ku], Chap.VI,Prop.3.6).

<u>C.10. Examples.</u>   a) Suppose there exists an $x \in S$ such that $S = R \oplus Rx \oplus \ldots \oplus Rx^{n-1}$. Then S/R is a global complete intersection, the kernel of the R-epimorphism $R[X] \to S$ ($X \mapsto x$) being generated by a monic polynomial.

b) Let $S = R[X_1,\ldots,X_n]$ be a polynomial algebra where R is noetherian and let $R[Y_1,\ldots,X_n] \subset S$ be a noetherian normalization of S/R. Then $S/R[Y_1,\ldots,Y_n]$ is a global complete intersection, since $S/R[Y]$ is flat by B.27 and the kernel of the $R[Y_1,\ldots,Y_n]$-epimorphism

$$R[Y_1,\ldots,Y_n,T_1,\ldots,T_n] \to S \qquad (T_i \mapsto X_i)$$

is generated by $\{Y_1-f_1,\ldots,Y_n-f_n\}$ where f_i is the preimage of Y_i in $R[T_1,\ldots,T_n]$ $(i=1,\ldots,n)$.

<u>C.11. Proposition.</u>   Suppose S/R is a global complete intersection where R is noetherian. Then:

a) S/R is equidimensional (B.17).

b) Given an arbitrary presentation $S = R[X_1,\ldots,X_n]/I$ of S/R, the S-module I/I^2 is projective of rank $n-d$ where d is the dimension of the non-empty fibers of S/R.

c) If in this case I is generated by $h := n-d$ elements $t_1,\ldots,t_h$, then they form a quasiregular sequence. In other words: $S = R[X_1,\ldots,X_n]/(t_1,\ldots,t_h)$ is a presentation of S/R as a complete intersection.

Proof.   Consider a presentation $S = R[X_1,\ldots,X_n]/I$ as in b). Since S/R is locally a complete intersection, we have

$$\operatorname{rank}_{\mathfrak{p}}(I/I^2) = n - \dim_{\mathfrak{p}} S_{\mathfrak{q}}/\mathfrak{q}S_{\mathfrak{q}}$$

for each $\mathfrak{p} \in \operatorname{Spec}(S)$, $\mathfrak{q} := \mathfrak{p} \cap R$, by C.8. If I is generated by a quasiregular sequence, then I/I^2 is a free S-module, and hence $\dim_{\mathfrak{p}} S_{\mathfrak{q}}/\mathfrak{q}S_{\mathfrak{q}} =: d$ is independent of $\mathfrak{p}$. Since S/R is flat, any minimal prime ideal of S contracts to a minimal prime ideal of R. It follows that S/R is equidimensional of dimension d.

For arbitrary I we again have

$$\operatorname{rank}_{\mathfrak{p}}(I/I^2) = n-d$$

by C.8, which proves b).

If I is generated by h := n-d elements $t_1,\ldots,t_h$, then they form a minimal system of generators of $I_\mathcal{O}$ for each $\mathcal{O} \in \mathrm{Spec}(R[X_1,\ldots,X_n])$ with $I \subset \mathcal{O}$. By C.6 they form locally at those $\mathcal{O}$ a regular sequence, hence a quasiregular sequence.

<u>C.12. Theorem.</u> For an algebra S/R, where R is a noetherian ring, the following conditions are equivalent:

a) S/R is a global complete intersection.

b) S/R is locally a complete intersection, and there exists a presentation $S = R[X_1,\ldots,X_n]/I$ where I/I^2 is a free S-module.

Proof.   It suffices to show that b) implies a). Let $t_1,\ldots,t_h \in I$ be elements whose images in I/I^2 form a basis of this S-module, and let $I' := (t_1,\ldots,t_h)$. Then $I = I'+I^2$, and in $\overline{S} := R[X_1,\ldots,X_n]/I'$ we have $\overline{I} = \overline{I}^2$ where $\overline{I} := I/I'$. Since $\overline{S}$ is noetherian, $\overline{I}$ is generated by an idempotent element $e \in \overline{S}$. Since $\overline{I}$ is the kernel of the canonical epimorphism $\overline{S} \to S$, we see that $S = \overline{S}_{\tilde{e}}$, the localization of $\overline{S}$ with respect to the element $\tilde{e} := 1-e$.

We extend the canonical epimorphism $R[X_1,\ldots,X_n] \to \overline{S}_{\tilde{e}}$ to an epimorphism $\alpha : R[X_1,\ldots,X_n,Y] \to \overline{S}_{\tilde{e}}$ by sending Y to $\tilde{e}^{-1}$. Let $f \in R[X_1,\ldots,X_n]$ be a preimage of $\tilde{e}$. Then $\ker \alpha = (t_1,\ldots,t_h,1-Yf)$. From C.4 we conclude that $\{t_1,\ldots,t_h,1-Yf\}$ is a regular sequence in every localization of $R[X_1,\ldots,X_n,Y]$ with respect to a prime ideal containing the elements of the sequence. Hence $\{t_1,\ldots,t_h,1-Yf\}$ is a quasiregular sequence, and S/R is a global complete intersection.

C.13. Corollary. Let S/R be of finite type, equidimensional, and locally a complete intersection. If S is semilocal, then S/R is a global complete intersection.

Proof. Let S/R be equidimensional of dimension d. By C.8, for any presentation $S = R[X_1,\ldots,X_n]/I$, the module I/I^2 is projective of rank n-d. But, since S is semilocal, this module is then even free, and C.12 can be applied.

C.14. Example. Let $(R,\mathfrak{m})$ and S be complete noetherian local rings and $R \to S$ a flat local homomorphism where $S/\mathfrak{m}S$ is finite over $R/\mathfrak{m}$ and a complete intersection. Then S/R is a global complete intersection.

The assumptions imply that S/R is finite and locally a complete intersection. Thus C.13 can be applied.

C.15. Corollary. Suppose S/R is finite and locally a complete intersection. Then for any $\wp \in \mathrm{Spec}(R)$ there exists an $f \in R \smallsetminus \wp$ such that S_f/R_f is a global complete intersection.

Proof. Given a presentation $S = R[X_1,\ldots,X_n]/I$, the $S_\wp$-module $I_\wp/I_\wp^2$ is projective of rank n. Since $S_\wp$ is semilocal, $I_\wp/I_\wp^2$ is even free, and there exists an $f \in R \smallsetminus \wp$ such that I_f/I_f^2 is a free S_f-module. We can apply C.12, because with S/R the algebra S_f/R_f is likewise locally a complete intersection.

__C.16. Proposition.__   Let S/R be an algebra which is locally a complete intersection, and let $K := Q(R)$, $L := Q(S)$.

a) If R and S are noetherian and $Ass(R) = Min(R)$, then also $Ass(S) = Min(S)$, and each minimal prime of S contracts to a minimal prime of R.

b) $R \to S$ induces a ring homomorphism $K \to L$, and L/K is locally a complete intersection.

c) If K is noetherian with $Ass(K) = Min(K)$ and L/K is finite, then L/K is a global complete intersection.

Proof.   a) Let $\mathfrak{p} \in Ass(S)$ and $\mathfrak{q} := \mathfrak{p} \cap R$. Since $S_{\mathfrak{p}}/R_{\mathfrak{q}}$ is flat, the ideal $\mathfrak{q}R_{\mathfrak{q}}$ consists only of zerodivisors of $R_{\mathfrak{q}}$. From $Ass(R) = Min(R)$ it follows that $\dim R_{\mathfrak{q}} = 0$. As a complete intersection, $S_{\mathfrak{p}}/\mathfrak{q}S_{\mathfrak{p}}$ is Cohen-Macaulay. Since $R_{\mathfrak{q}}$ and $S_{\mathfrak{p}}/\mathfrak{q}S_{\mathfrak{p}}$ are Cohen-Macaulay, $S_{\mathfrak{p}}$ is Cohen-Macaulay as well. But $\mathfrak{p}S_{\mathfrak{p}}$ consists only of zerodivisors. We conclude that $\dim S_{\mathfrak{p}} = 0$ and $\mathfrak{p} \in Min(S)$.

b) Since S/R is a flat algebra, any non-zero-divisor of R is a non-zero-divisor of S, and there is an induced homomorphism $K \to L$. The localizations of L are also localizations of S, and with S/R the algebra L/K is locally a complete intersection too.

c) If K is noetherian with $Ass(K) = Min(K)$ and L/K is finite, then L is semilocal, and L/K is by C.13 a global complete intersection.

We shall study now the behaviour of local and global complete intersection algebras under various ring theoretic operations.

C.17. __Transitivity.__   Let S/R and T/S be algebras that are essentially of finite type, and assume that R is noetherian.

a) If T/S is locally a complete intersection at $\mathfrak{q} \in \mathrm{Spec}(T)$ and S/R is so at $\mathfrak{p} := \mathfrak{q} \cap S$, then T/R is locally a complete intersection at $\mathfrak{q}$.

b) If T/S and S/R are global complete intersections, so is T/R.

Proof.   Let us show b), (the proof of a) being similar). With T/S and S/R the algebra T/R is flat too. Choose presentations

$S = R[X_1, \ldots, X_n]/(t_1, \ldots, t_h)$ and $T = S[Y_1, \ldots, Y_m]/(u_1, \ldots, u_k)$

of S/R and T/S as complete intersections, and choose representatives U_i of the u_i in $R[X_1, \ldots, X_n, Y_1, \ldots, Y_m]$.

Then $T = R[X_1, \ldots, X_n, Y_1, \ldots, Y_m]/(t_1, \ldots, t_h, U_1, \ldots, U_k)$. We shall show that $\{t_1, \ldots, t_h, U_1, \ldots, U_k\}$ is a quasiregular sequence.

For a maximal ideal $\mathfrak{N}$ containing $(t_1, \ldots, t_h, U_1, \ldots, U_k)$ the sequence $\{t_1, \ldots, t_h\}$ is regular in $R[X_1, \ldots, X_n]_{\mathfrak{m}}$ where $\mathfrak{m} := \mathfrak{N} \cap R[X_1, \ldots, X_n]$. Then this sequence is also regular in $R[X,Y]_{\mathfrak{N}}$, and we have $R[X,Y]_{\mathfrak{N}}/(t_1, \ldots, t_h) = S[Y]_{\overline{\mathfrak{N}}}$ with the image $\overline{\mathfrak{N}}$ of $\mathfrak{N}$ in $S[Y]$. Since $\{u_1, \ldots, u_k\}$ is a regular sequence of $S[Y]_{\overline{\mathfrak{N}}}$, we obtain that $\{t_1, \ldots, t_h, U_1, \ldots, U_k\}$ is a regular sequence of $R[X,Y]_{\mathfrak{N}}$.

C.18. __Base change.__   Let R be a noetherian ring, and let S/R be an algebra that is essentially of finite type. Given an algebra R'/R where R' is noetherian, put

$S' := R' \otimes_R S$.

a) If S/R is locally a complete intersection at $\mathfrak{p} \in \mathrm{Spec}(S)$, then S'/R' is so at all $\mathfrak{p}' \in \mathrm{Spec}(S')$ lying over $\mathfrak{p}$.

b) If S/R is a global complete intersection, so is S'/R'.

Proof. We shall show b); the proof of a) is analogous. With S/R the algebra S'/R' is flat as well. Given a presentation $S = R[X_1,\ldots,X_n]/(t_1,\ldots,t_h)$ of S/R as a complete intersection, we have $S' = R'[X_1,\ldots,X_n]/(t_1,\ldots,t_h)$, and it suffices to show that $\{t_1,\ldots,t_h\}$ is a quasiregular sequence of $R'[X_1,\ldots,X_n]$.

Let $\mathfrak{M}' \in \mathrm{Max}(R'[X])$ with $t_1,\ldots,t_h \in \mathfrak{M}'$ be given, and put $\mathfrak{M} := \mathfrak{M}' \cap R[X]$, $\mathfrak{m}' := \mathfrak{M}' \cap R'$, and $\mathfrak{m} := \mathfrak{M}' \cap R$. Then $R'[X]_{\mathfrak{M}'}$ is a localization of $T := R'_{\mathfrak{m}'} \otimes_{R_{\mathfrak{m}}} R[X]_{\mathfrak{M}}$ at a prime ideal lying over $\mathfrak{m}'R'_{\mathfrak{m}'}$. Let $\overline{\mathfrak{M}}$ be the prime ideal of $k(\mathfrak{m})[X]$ corresponding to $\mathfrak{M}$, and let $\{\bar{t}_1,\ldots,\bar{t}_h\}$ be the set of images of the t_i in $\overline{T} := k(\mathfrak{m}') \otimes_{k(\mathfrak{m})} k(\mathfrak{m})[X]_{\overline{\mathfrak{M}}}$. The $\bar{t}_i$ form, by assumption and by B.23, a regular sequence of $k(\mathfrak{m})[X]_{\overline{\mathfrak{M}}}$, hence also of $\overline{T}$. Again by B.23 the t_i form a regular sequence of $R'[X]_{\mathfrak{M}'}$.

C.19. <u>Direct products.</u>   Assume that the algebras S_i/R $(i=1,\ldots,r)$ are essentially of finite type, and let $S := S_1 \times \ldots \times S_r$ be their direct product.

a) S/R is locally a complete intersection if and only if the S_i/R are so for $i=1,\ldots,r$.

b) If S/R is a global complete intersection, then the S_i/R are so for $i=1,\ldots,r$.

c) If the S_i/R are global complete intersections and finite $(i=1,\ldots,r)$, so is S/R.

C.20. Lemma.   Let $S = S_1 \times \ldots \times S_r$ be a direct product of rings S_i. Then S_i/S, considered as an algebra with respect to the canonical projection, is a global complete intersection.

Let $1 = e_1 + \ldots + e_r$ be the decomposition of 1 into idempotents corresponding to the product decomposition of S. Then $S_i = S_{e_i} = S[X]/(e_iX-1)$. As a localization of S the ring S_i is flat over S. For each $\mathfrak{M} \in \text{Max}(S[X])$ with $e_iX-1 \in \mathfrak{M}$ we have $e_i \notin \mathfrak{M}$, and hence e_iX-1 is a regular element of $S[X]_{\mathfrak{M}}$.

C.21. Lemma.   Given a ring R, consider $R \times R$ as an R-algebra with respect to the mapping $r \mapsto (r,r)$. Then $R \times R/R$ is a global complete intersection.

The mapping $R[X] \to R \times R$, with $r \to (r,r)$ for $r \in R$ and $X \mapsto (1,0)$, is an epimorphism with kernel (X^2-X). $R \times R$ is a free R-module, and for $\mathfrak{M} \in \text{Max}(R[X])$ with $X^2-X \in \mathfrak{M}$ we have $X \notin \mathfrak{M}$ or $X-1 \notin \mathfrak{M}$. Hence X^2-X is a regular element of $R[X]_{\mathfrak{M}}$.

C.22. Lemma.   Assume the algebra S/R is finite and a global complete intersection. Then $S \times T/R \times T$ is a global complete intersection for any ring T.

If $S = R[X_1, \ldots, X_n]/(t_1, \ldots, t_n)$ is a presentation of S/R as a complete intersection and x_i is the image of X_i in S, then the $R \times T$-homomorphism

$$(R \times T)[X_1, \ldots, X_n] \to S \times T \qquad (X_i \mapsto (x_i, 0))$$

is an epimorphism, and its kernel I is generated by the elements $(t_i, 0)$ and $(0, X_i)$ for $i = 1, \ldots, n$. (Here we identify

(R×T)[X] with R[X]×T[X]). But then I is generated by the
elements (t_i, X_i) (i=1,...,n) as well, since with
e_1 := (1,0) and e_2 := (0,1) we have

$$e_1 \cdot (t_i, X_i) = (t_i, 0) \text{ and } e_2 \cdot (t_i, X_i) = (0, X_i).$$

We shall show that the elements (t_i, X_i) form a quasiregular
sequence of (R×T)[X]. Since S×T/R×T is flat, this proves
the lemma.

Let $\mathfrak{M} \in$ Max((R×T)[X]) with $(t_i, X_i) \in \mathfrak{M}$ (i=1,...,n) be
given, and assume $e_1 \notin \mathfrak{M}$, $e_2 \in \mathfrak{M}$. Then $(R×T)[X]_{\mathfrak{M}} \cong R[X]_{\overline{\mathfrak{M}}}$
where $\overline{\mathfrak{M}}$ is the image of $\mathfrak{M}$ in R[X] and (t_i, X_i) is identified
with t_i. Likewise, if $e_1 \in \mathfrak{M}$ and $e_2 \notin \mathfrak{M}$, we have
$(R×T)[X]_{\mathfrak{M}} \cong T[X]_{\overline{\mathfrak{M}}}$ where (t_i, X_i) is identified with X_i. In
either case $\{(t_1, X_1), ..., (t_n, X_n)\}$ is a regular sequence of
$(R×T)[X]_{\mathfrak{M}}$.

<u>Proof of C.19.</u>   It suffices to consider the case r = 2.
S/R is flat if and only if S_i/R is for i=1,2. By C.20, S_i/S
is a global complete intersection. If S/R is locally or
globally a complete intersection, then S_i/R is so by C.17.

Since Spec(S) is the disjoint union of Spec(S_1) and
Spec(S_2), it is clear that S/R is locally a complete intersection if and only if S_i/R is so for i=1,2. Now let S_i/R
(i=1,2) be finite and globally complete intersections. Then
$S_1 × S_2 / S_1 × R$ and $S_1 × R / R × R$ are global complete intersections
by C.22, and R×R/R is so by C.21. By C.17 we obtain that
$S_1 × S_2 / R$ is a global complete intersection, q.e.d.

C.23. **Proposition.** Let S/R and P/R be algebras where R,S and P are noetherian, and let $\alpha : P \to S$ be an R-homomorphism. For $\mathfrak{p} \in \mathrm{Spec}(S)$ let $\mathfrak{q} := \mathfrak{p} \cap P$ and $\mathfrak{r} := \mathfrak{p} \cap R$. Assume that S/P is essentially of finite type and is quasifinite at $\mathfrak{p}$, P/R is regular at $\mathfrak{q}$, and $\dim S_{\mathfrak{p}}/\mathfrak{r}S_{\mathfrak{p}} = \dim P_{\mathfrak{q}}/\mathfrak{r}P_{\mathfrak{q}}$. Then S/R is locally at $\mathfrak{p}$ a complete intersection if and only if S/P is so.

Proof. If S/R is locally at $\mathfrak{p}$ a complete intersection, then $\mathfrak{p}$ is a Cohen-Macaulay point of S/R, and we can apply B.27 to conclude that $S_{\mathfrak{p}}/P_{\mathfrak{q}}$ is flat. Hence $S_{\mathfrak{p}}/\mathfrak{r}S_{\mathfrak{p}}$ is flat over the regular local ring $P_{\mathfrak{q}}/\mathfrak{r}P_{\mathfrak{q}}$, and a regular system of parameters of $P_{\mathfrak{q}}/\mathfrak{r}P_{\mathfrak{q}}$ is also a regular sequence of $S_{\mathfrak{p}}/\mathfrak{r}S_{\mathfrak{p}}$. In that case, with $S_{\mathfrak{p}}/\mathfrak{r}S_{\mathfrak{p}}$ the ring $S_{\mathfrak{p}}/\mathfrak{q}S_{\mathfrak{p}}$ is likewise a complete intersection.

The proof of the converse statement is easy.

For the rest of this appendix let $(R,\mathfrak{m})$ be a noetherian local ring, $\hat{R}$ its completion, and $\mathrm{gr}_{\mathfrak{m}}(R)$ its associated graded ring. For $a \in R$ let La denote the leading form of a in $\mathrm{gr}_{\mathfrak{m}}(R)$.

C.24. **Definition.** A sequence $\{a_1,\ldots,a_n\}$ of elements $a_i \in \mathfrak{m}$ is called **super-regular**, if $\{La_1,\ldots,La_n\}$ is a regular sequence of $\mathrm{gr}_{\mathfrak{m}}(R)$.

It is easy to see that a super-regular sequence is a regular sequence.

C.25. **Definition.** R is called a **strict complete intersection**, if $\hat{R}$ has a presentation

$$\hat{R} = S/(a_1,\ldots,a_n)$$

where $(S,\mathfrak{m})$ is a regular local ring and $\{a_1,\ldots,a_n\}$ is a superregular sequence in $\mathfrak{m}$.

Of course, any strict complete intersection is also a complete intersection. Moreover, in the situation of C.25

$$(3) \qquad gr_{\mathcal{M}}(R) = gr_{\mathcal{M}}(S)/(La_1,\ldots,La_n)$$

as follows i.e. from [Ku], Chap.V, Lemma 5.4, by induction on n.

<u>C.26. Proposition.</u>   If R is a strict complete intersection and R = T/I with a regular local ring T and an ideal I of T, then I can be generated by a super-regular sequence.

The proof is reduced to the corresponding assertion about complete intersection with the help of some lemmas about graded rings. Let $G = \bigoplus_{i \in \mathbb{N}} G_i$ be a positively graded noetherian ring.

<u>C.27. Lemma.</u>   Suppose I is a homogeneous ideal of G. Let M be the set of all $\mathcal{M} \in Max(G)$ with $\bigoplus_{i>0} G_i \subset \mathcal{M}$.

a) A homogeneous element $a \in G$ is contained in I if and only if $a \in I_{\mathcal{M}}$ for all $\mathcal{M} \in M$.

b) Homogeneous elements $a_1,\ldots,a_n \in I$ generate I if and only if $I_{\mathcal{M}} = (a_1,\ldots,a_n)G_{\mathcal{M}}$ for all $\mathcal{M} \in M$.

c) If G_o is a local ring and $I = (a_1,\ldots,a_n)$ with homogeneous elements a_i, then $\{a_1,\ldots,a_n\}$ contains a shortest system of generators of I.

Proof.   The elements of M are the ideals $\mathcal{m} \oplus \bigoplus_{i>0} G_i$ with $\mathcal{m} \in Max(G_o)$.

a) Suppose $a \in I_{\mathcal{M}}$ for some $\mathcal{M} \in M$. Then there exists an $s \in G \setminus \mathcal{M}$ with $sa \in I$. The degree zero component s_o of s does not vanish, and $s_o a \in I$. Since for any $\mathcal{m} \in Max(G_o)$ there is an $s_o \in G_o \setminus \mathcal{m}$ with $s_o a \in I$, we conclude that $a \in I$.

b) follows immediately from a). Under the assumptions of c) M contains only one element, and the claim results from Nakayama's lemma.

$\underline{\text{C.28. Lemma.}}$ Let $a_1,\ldots,a_n \in G$ be homogeneous elements of positive degree, and let M be as in C.27. The following statements are equivalent:

a) $\{a_1,\ldots,a_n\}$ is a G-regular sequence.

b) $\{a_1,\ldots,a_n\}$ is a quasiregular sequence of G.

c) $\{a_1,\ldots,a_n\}$ is a $G_{\mathfrak{m}}$-regular sequence for all $\mathfrak{m} \in M$.

Proof. It is enough to show that a homogeneous element $a \in G$ of positive degree is not a zerodivisor in G, if it is not a zerodivisor of $G_{\mathfrak{m}}$ for all $\mathfrak{m} \in M$. Suppose a has this property and there is a homogeneous element $s \in G$ with $sa = 0$. Then s vanishes in $G_{\mathfrak{m}}$ for all $\mathfrak{m} \in M$. This implies that for each $\mathfrak{m} \in \text{Max}(G_0)$ there is a $\sigma \in G_0 \smallsetminus \mathfrak{m}$ with $\sigma s = 0$. Consequently $s = 0$.

$\underline{\text{Proof of C.26.}}$ We have $\hat{R} = S/(a_1,\ldots,a_n) = T/I$ where $(S,\mathfrak{m})$ and $(T,\mathfrak{m}')$ are regular local rings and $(a_1,\ldots,a_n)$ is a super-regular sequence in $\mathfrak{m}$. Write $G := \text{gr}_{\mathfrak{m}}(S)$, $G' := \text{gr}_{\mathfrak{m}'}(T)$. Then by formula (3)

$$\text{gr}_{\mathfrak{m}}(R) = G/(La_1,\ldots,La_n) = G'/\text{gr}_{\mathfrak{m}'}(I).$$

Let $\mathfrak{n}, \mathfrak{n}$ and $\mathfrak{n}'$ be the homogeneous maximal ideals of $\text{gr}_{\mathfrak{m}}(R)$, G and G' respectively. Then

$$\text{gr}_{\mathfrak{m}}(R)_{\mathfrak{n}} = G_{\mathfrak{n}}/(La_1,\ldots,La_n)G_{\mathfrak{n}} = G'_{\mathfrak{n}'}/\text{gr}_{\mathfrak{m}'}(I)G'_{\mathfrak{n}'}.$$

$\text{gr}_{\mathfrak{m}}(R)_{\mathfrak{n}}$ is a complete intersection in the sense of C.1, since $G_{\mathfrak{n}}$ is a regular local ring and $\{La_1,\ldots,La_n\}$ a $G_{\mathfrak{n}}$-regular sequence.

Let $\{Lb_1, \ldots, Lb_m\}$ be a minimal system of generators of $gr_{\mathfrak{M}'}(I)G'_{\mathfrak{n}}$, consisting of leading forms of elements $b_i \in I$ $(i=1,\ldots,m)$. By C.2, $\{Lb_1, \ldots, Lb_m\}$ is a $G'_{\mathfrak{n}}$-regular sequence, and by C.28 it is also a G'-regular sequence. By C.27 $gr_{\mathfrak{M}'}(I) = (Lb_1, \ldots, Lb_m)$, from which $I = (b_1, \ldots, b_m)$ easily follows, q.e.d.

<u>C.29. Lemma.</u> Let K be a field, and let $a_1, \ldots, a_n \in K[X_1, \ldots, X_n]$ be homogeneous polynomials which form a regular sequence. If $d_i := \deg a_i$ $(i=1,\ldots,n)$, then

$$\dim_K K[X_1, \ldots, X_n]/(a_1, \ldots, a_n) = d_1 \cdot \ldots \cdot d_n.$$

Proof. Put $P_i := K[X_1, \ldots, X_n]/(a_1, \ldots, a_i)$ $(i=0,\ldots,n)$, and let $\chi(P_i, m)$ denote the Hilbert polynomial of P_i:

$$\chi(P_i, m) = \dim_K \bigoplus_{\nu \leq m} P_i^{(\nu)} = e(P_i) \cdot \frac{m^{n-i}}{(n-i)!} + \ldots \quad \text{for large } m,$$

where $P_i^{(\nu)}$ denotes the homogeneous component of P_i of degree ν, and $e(P_i)$ is the multiplicity of P_i. Here $e(P_0) = 1$. From the exact sequence

$$0 \to P_i \xrightarrow{\ a_{i+1}\ } P_i \to P_{i+1} \to 0$$

we have (for large m) the formulas

$$\chi(P_{i+1}, m) = \chi(P_i, m) - \chi(P_i, m-d_{i+1}) =$$

$$\frac{e(P_i)}{(n-i)!}[m^{n-i} - (m-d_{i+1})^{n-i}] + \ldots = \frac{d_{i+1} \cdot e(P_i)}{(n-i-1)!} m^{n-i-1} + \ldots$$

and we find

$$e(P_{i+1}) = d_{i+1} \cdot e(P_i) \qquad (i=0,\ldots,n).$$

For large m

$$\dim_K K[X_1,\ldots,X_n]/(a_1,\ldots,a_n) = \chi(P_n,m)$$

and by induction $\chi(P_n,m) = d_1 \cdot \ldots \cdot d_n$.

Exercises

1) Let R be a ring. Show that $R[X] \times R$ is not a global complete intersection over $R \times R$.

2) For a ring R consider the algebras $S_1 := R[X]/(X^2)$ and $S_2 := R[Y]/(Y^3)$. Let ξ denote the image of X in S_1 and η the image of Y in S_2. Write $S := S_1 \times S_2$. Then there is an R-isomorphism

$$S \cong R[X,Y,Z]/(ZX+(1-Z)X^2,(1-Z)Y+ZY^3,Z-Z^2)$$

where Z is mapped onto $(0,1)$, X is mapped onto ξ, and Y onto η. This is a presentation of S/R as a complete intersection.

D. The Fitting Ideals of a Module

The Fitting ideals (Fitting invariants) of a module are generalizations of the elementary divisors of the theory of modules over principal ideal domains. Many structural properties of a module are reflected in its Fitting invariants.

Let R be a ring, M a finite R-module, and $\{m_1,\ldots,m_n\}$ a system of generators of M. The exact sequence

$$(1) \qquad\qquad 0 \to K \to R^n \overset{\alpha}{\to} M \to 0$$

where α maps the i-th canonical basis element e_i onto m_i $(i=1,\ldots,n)$ and $K := \ker \alpha$ is called the <u>presentation of M</u> defined by $\{m_1,\ldots,m_n\}$. Let $\{v_\lambda\}_{\lambda\in\Lambda}$ be a system of generators of K with $v_\lambda = (x_1^\lambda,\ldots,x_n^\lambda) \in R^n$ $(\lambda \in \Lambda)$. Then

$$(2) \qquad\qquad (x_i^\lambda)_{\substack{i=1,\ldots,n \\ \lambda\in\Lambda}}$$

is called a <u>relation matrix of M</u> with respect to $\{m_1,\ldots,m_n\}$.

Given such a matrix, let $F_i(M)$ denote the ideal of R generated by all $(n-i)$-rowed subdeterminants of the matrix $(i=0,\ldots,n-1)$, and let $F_i(M) := R$ for $i \geq n$. Put $F_i(M):=(0)$ for $i < 0$.

<u>D.1. Lemma.</u> $F_i(M)$ $(i \in \mathbb{N})$ does not depend on the special choice of a relation matrix of M with respect to $\{m_1,\ldots,m_n\}$.

Proof. Let $\{v_\nu'\}_{\nu\in N}$ be another system of generators of K with $v_\nu' = (y_1^\nu,\ldots,y_n^\nu) \in R^n$ $(\nu \in \mathbb{N})$. Consider the subdeterminant $\Delta(\nu_1,\ldots,\nu_{n-i},k_1,\ldots,k_{n-i})$ of (y_k^ν) formed by the elements y_k^ν with $\nu \in \{\nu_1,\ldots,\nu_{n-i}\}, \nu_1<..<\nu_{n-i}, k \in \{k_1,\ldots,k_{n-i}\}, k_1<..<k_{n-i}$.

We have relations

$$v_{\nu_j}' = \sum_{\lambda\in\Lambda} r_{j\lambda}v_\lambda \qquad (j=1,\ldots,n-i, r_{j\lambda} \in R)$$

where,for each j,only finitely many $r_{j\lambda}$ are different from
0. Thus only finitely many v_λ actually occur in these relations. It is clear that $\Delta(v_1,\ldots,v_{n-1},k_1,\ldots,k_{n-i})$ is a
linear combination of (n-i)-rowed subdeterminants of the
matrix (2), hence is contained in $F_i(M)$.

Let $F_i'(M)$ be the ideal generated by the (n-i)-rowed
minors of the matrix (y_k^ν). Then $F_i'(M) \subset F_i(M)$,and by
symmetry, $F_i'(M) = F_i(M)$ for all $i \in \mathbb{N}$.

<u>D.2. Lemma.</u> $F_i(M)$ ($i \in \mathbb{N}$) does not depend on the choice
of the generating system $\{m_1,\ldots,m_n\}$ of M.

Proof. It suffices to show that the $F_i(M)$ do not change,if
we replace $\{m_1,\ldots,m_n\}$ by a system $\{m_1,\ldots,m_n,m\}$ with an
arbitrary (superfluous) element $m \in M$,because we can then
obviously pass from one to any other system of generators
of M. It is enough to consider the case i < n.

Write $m = r_1m_1 +\ldots+ r_nm_n$ ($r_i \in R$),and let

$$0 \to K' \to R^{n+1} \to M \to 0$$

be the presentation of M defined by $\{m_1,\ldots,m_n,m\}$. A system
of generators of K' is given by the elements

$$\bar{v}_\lambda := (x_1^\lambda,\ldots,x_n^\lambda,0) \qquad (\lambda \in \Lambda)$$
$$v := (-r_1,\ldots,-r_n,1).$$

The (n+1-i)-rowed subdeterminants of the matrix given by the
$\bar{v}_\lambda$ and v are either (n+1-i)-rowed subdeterminants of the
original relation matrix (x_i^λ) and hence elements of $F_i(M)$,
or they contain a row coming from the element v. In the
second case they are again linear combinations of elements

of $F_i(M)$, and any $(n-i)$-rowed subdeterminant of (x_i^λ) is also an $(n+1-i)$-rowed subdeterminant of the new matrix. It follows that the new relation matrix defines the same ideals.

<u>D.3. Definition.</u>   (Fitting [Fi]) $F_i(M)$ is called the <u>i-th Fitting ideal (i-th Fitting invariant)</u> of M ($i \in \mathbb{Z}$).

By construction we have

$$F_0(M) \subset F_1(M) \subset \ldots \subset F_i(M) \subset \ldots \text{ and } F_i(M) = R \text{ for } i \geq \mu(M).$$

<u>D.4. Rules.</u>   a) If M is finitely presentable, then the $F_i(M)$ are finitely generated ideals ($i \in \mathbb{N}$).

b) For each algebra S/R we have

$$F_i(S \otimes_R M) = S \cdot F_i(M) \qquad (i \in \mathbb{Z}).$$

c) If $N \subset R$ is a multiplicatively closed subset, then

$$F_i(M_N) = F_i(M)_N \qquad (i \in \mathbb{Z}).$$

d) If I is an ideal of R, then

$$F_i(M/IM) = \overline{F_i(M)} \qquad (i \in \mathbb{Z})$$

where $\overline{F_i(M)}$ denotes the image of $F_i(M)$ in R/I.

a) is trivial. For b) consider the exact sequence

$$S \otimes_R K \to S^n \to S \otimes_R M \to 0$$

derived from (1). It is then clear that the images of the x_i^λ in S define a relation matrix of the S-module $S \otimes_R M$ with respect to $\{1 \otimes m_1, \ldots, 1 \otimes m_n\}$. c) and d) are special cases of b).

<u>D.5. Example.</u>   If R is a principal ideal domain M, then

$$M \cong R/(e_1) \oplus \ldots \oplus R/(e_s) \oplus R^r$$

with non-units $e_i \in R \setminus \{0\}$ where e_i divides e_{i+1} for $i=1,..,s-1$.
We obtain

$$F_i(M) = \begin{cases} (0) & \text{for } i=0,\ldots,r-1 \\ (e_1 \cdot \ldots \cdot e_j) & \text{for } i=r+s-j \ (j=1,\ldots,s) \\ R & \text{for } i \geq r+s = \mu(M). \end{cases}$$

Thus the Fitting ideals give us the opportunity to determine

the rank r and the elementary divisors (e_i) (invariant

factors) of M in this case.

D.6. Definition.   We say that M <u>has a rank</u>, if $Q(R) \otimes_R M$ is a

free Q(R)-module. Its rank is then also called the <u>rank of M</u>.

D.7. Proposition.   If M has rank r, then $F_i(M) = (0)$ for

$i = 0,\ldots,r-1$ and $F_i(M) \neq (0)$ for $i \geq r$.

By D.4c), $Q(R) \cdot F_i(M) = F_i(Q(R) \otimes_R M)$ for $i \in \mathbb{N}$. Since

$Q(R) \otimes_R M$ is free of rank r, we obtain

$F_i(Q(R) \otimes_R M) = (0)$ for $i=0,\ldots,r-1$ and $F_i(Q(R) \otimes_R M) = Q(R)$

for $i \geq r$. Since $R \rightarrow Q(R)$ is injective, the claim follows.

D.8. Proposition.   Let $(R, \mathfrak{m})$ be a local ring. Then

$$\mu(M) = \text{Min}\{n \mid F_n(M) = R\}.$$

Moreover, the following statements are equivalent:

a) M is free of rank r.

b) $F_i(M) = (0)$ for $i=0,\ldots,r-1$, and $F_i(M) = R$ for $i \geq r$.

Proof.   Let $\{m_1,\ldots,m_n\}$ be a minimal system of generators

of M,and let (1) be the presentation defined by this system.
The coefficients of the relation matrix (2) are elements
of $\mathcal{M}$ because the generating system is minimal. Therefore
$F_{n-1}(M) \subset \mathcal{M}$ and $F_n(M) = R$. This proves the formula for $\mu(M)$.

It suffices now to show that b) $\to$ a). But b) implies that
$n = r$ and $F_{n-1}(M) = (0)$. Then the relation matrix (2) must
be the zero matrix, and $M \cong R^r$.

D.9. Corollary. For an arbitrary ring R and $\mathscr{g} \in \mathrm{Spec}(R)$
the following conditions are equivalent:
a) $\mu_{\mathscr{g}}(M) = n$.
b) $F_{n-1}(M) \subset \mathscr{g}$ and $F_n(M) \not\subset \mathscr{g}$.

D.10. Corollary (Semicontinuity of μ) For $n \in \mathbb{N}$ the set
of all $\mathscr{g} \in \mathrm{Spec}(R)$ with $\mu_{\mathscr{g}}(M) \leq n$ is open.

In fact, this is the set of all $\mathscr{g}$ with $F_n(M) \not\subset \mathscr{g}$. For, if
$\mu_{\mathscr{g}}(M) = m \leq n$, then $F_m(M) \not\subset \mathscr{g}$, and hence $F_n(M) \not\subset \mathscr{g}$ as well.

D.11. Corollary. $\mathrm{Supp}(M) = \mho(F_0(M)) = \{ \mathscr{g} \in \mathrm{Spec}(R) \mid F_0(M) \subset \mathscr{g} \}$.

D.12. Corollary. If M is finitely presentable, then the
set of all $\mathscr{g} \in \mathrm{Spec}(R)$ for which $M_{\mathscr{g}}$ is a free $R_{\mathscr{g}}$-module
(of rank r) is open in $\mathrm{Spec}(R)$.

By D.8, $M_{\mathscr{g}}$ is free of rank r if and only if
$\mathscr{g} \not\in \mathrm{Supp}(F_i(M))$ for $i=0,\ldots,r-1$ and $F_r(M) \not\subset \mathscr{g}$. Since $F_i(M)$
is finitely generated (D.4a), $\mathrm{Supp}(F_i(M))$ is closed in
$\mathrm{Spec}(R)$, and so is $\mho(F_r(M))$. This proves the corollary,
since a union of open sets is open.

D.13. Corollary.   For a ring R and a finite R-module M the following statements are equivalent:

a) M is locally free of rank r.

b) $F_i(M) = (0)$ for $i=0,\ldots,r-1$, and $F_i(M) = R$ for $i \geq r$.

   This is a consequence of D.8 and the local-global principle.

   D.11 also results from the following proposition.

D.14. Proposition.   If $\mu(M) = n$, then

$$(\text{Ann } M)^n \subset F_0(M) \subset \text{Ann } M.$$

Proof.   Let $\{m_1,\ldots,m_n\}$ be a minimal system of generators of M and $a_1,\ldots,a_n \in \text{Ann } M$. From the relations $a_i m_i = 0$ $(i=1,\ldots,n)$ we obtain $a_1 \cdot \ldots \cdot a_n \in F_0(M)$, and hence $(\text{Ann } M)^n \subset F_0(M)$.

   Let $(a_{ij})_{i,j=1,\ldots,n}$ be an n-rowed submatrix of a relation matrix (2) and $\Delta := \det(a_{ij})$. Let Δ_{ij} be the subdeterminant of (a_{ij}) adjoint to a_{ij}. By Cramer's rule

$$(\Delta_{ij}) \cdot (a_{ij}) = \begin{pmatrix} \Delta & & 0 \\ & \ddots & \\ 0 & & \Delta \end{pmatrix}.$$

From $\sum_{j=1}^{n} a_{ij} m_j = 0$ $(i=1,\ldots,n)$ it follows that $\Delta m_j = 0$ $(j=1,\ldots,n)$, hence $F_0(M) \subset \text{Ann } M$.

   Let us now look at the multiplicative properties of the Fitting ideals.

D.15. Proposition.   Let M_1 and M_2 be finite R-modules. Then

$$F_i(M_1 \oplus M_2) = \sum_{\rho+\sigma=i} F_\rho(M_1) \cdot F_\sigma(M_2) \qquad (i \in \mathbb{N})$$

and in particular

$$F_o(M_1 \oplus M_2) = F_o(M_1) \cdot F_o(M_2).$$

Proof.   There is a relation matrix of $M_1 \oplus M_2$ of the form

$$\left(\begin{array}{c|c} A & O \\ \hline O & B \end{array} \right)$$

where A is a relation matrix of M_1 and B a relation matrix of M_2. The claim follows immediately.

<u>D.16. Corollary.</u>   Let M be a finitely generated R-module and $m \in \mathbb{N}$. Then

$$F_i(R^m \oplus M) = \begin{cases} (O) & i=O,\ldots,m-1 \\[2ex] F_{i-m}(M) & \text{for } i \geq m. \end{cases}$$

<u>D.17. Proposition.</u>   For an exact sequence

$$O \to M_1 \to M_2 \to M_3 \to O$$

of finite R-modules we have

$$F_o(M_1) \cdot F_o(M_3) \subset F_o(M_2).$$

If M_3 has a system of n generators such that the kernel of the corresponding presentation is also generated by n elements, then $F_o(M_3)$ is a prinicpal ideal and

$$F_o(M_1) \cdot F_o(M_3) = F_o(M_2).$$

Proof.   Start with presentations

$$O \to K_i \to R^{n_i} \to M_i \to O$$

for M_i (i=1 or 3). As usual, one constructs a commutative

diagram with exact rows and columns

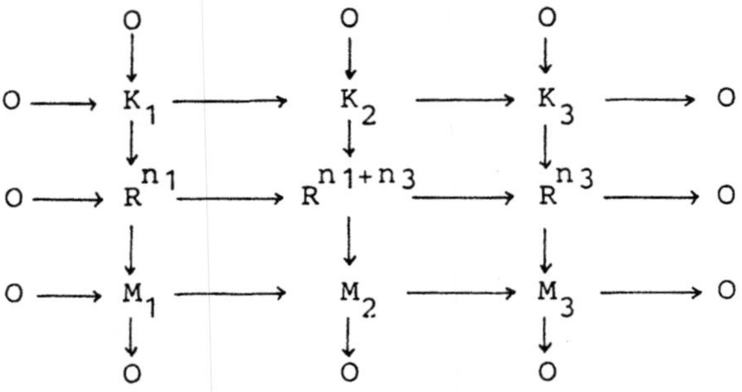

We obtain a system of generators of K_2 by adding to a system of generators of K_1 a system of representatives in K_2 for a generating system of K_3. The relation matrix corresponding to this system is of the form

$$\overbrace{}^{n_1}\quad\overbrace{}^{n_3}$$
$$\begin{pmatrix} A & O \\ * & B \end{pmatrix}$$

where A is a relation matrix of M_1 and B a relation matrix of M_3. We now see that $F_o(M_1)\cdot F_o(M_3)\subset F_o(M_2)$, and it is clear that equality holds, if B is a square matrix.

Let $T(M)$ be the torsion of M, i.e. the set of all $m\in M$ which are annihilated by a non-zero divisor of R (the kernel of the canonical map $M\to Q(R)\otimes_R M$).

D.18. Theorem (Lipman [Lip$_2$]). For a local ring R and $r\in\mathbb{N}$ the following conditions are equivalent:

a) $F_i(M) = O$ for $i=0,\ldots,r-1$, and $F_r(M)$ is a principal ideal generated by a non-zero divisor of R.

b) $M/T(M)\cong R^r$, and there exists an exact sequence

$$O\to P_1\to P_o\to M\to O$$

with free R-modules P_o, P_1 of finite rank.

Proof. b) → a). Since $M = R^r \oplus T(M)$, we have $F_i(M) = O$ for

$i=0,\ldots,r-1$ and $F_r(M) = F_o(T(M))$ by D.16. Moreover, there is

an exact sequence

$$O \to P_1' \to P_o \to T(M) \to O$$

with a free R-module P_1' of finite rank (snake lemma). In fact,

rank P_o = rank P_1', since $T(M)$ is a torsion module (tensorize

with $Q(R)$). Necessarily $F_o(T(M))$ is a principal ideal, and

since $Q(R) \cdot F_o(T(M)) = F_o(Q(R) \otimes_R T(M)) = Q(R)$, any generator

of $F_o(T(M))$ is a non-zero divisor of R.

a) → b). Let $\{m_1,\ldots,m_n\}$ be a system of generators and A a

relation matrix of M with respect to $\{m_1,\ldots,m_n\}$. By Nakayama's

lemma the principal ideal $F_r(M)$ is already generated by an

(n-r)-rowed minor Δ of A which is not a zero divisor of R.

Then there are relations

$$\sum_{j=1}^{n} a_{ij}m_j = O \qquad (i=1,\ldots,n-r, a_{ij} \in R)$$

and we can assume $\Delta = \det(a_{ij})_{i,j=1,\ldots,n-r}$.

By Cramer's rule there are equations

$$\Delta m_h = \sum_{j=n-r+1}^{n} \Delta_{hj}m_j \qquad (h=1,\ldots,n-r)$$

where Δ_{hj} are certain determinants contained in $F_r(M)$. We

have $\Delta_{hj} = r_{hj} \cdot \Delta$ with certain $r_{hj} \in R$ and, since

$$\Delta \cdot (m_h - \sum_{j=n-r+1}^{n} r_{hj}m_j) = O \quad (h=1,\ldots,n-r)$$

the elements $m_h - \sum_{j=n-r+1}^{n} r_{hj}m_j$ belong to $T(M)$.

Hence $M/T(M)$ is generated by the images $\overline{m}_{n-r+1},\ldots,\overline{m}_n$ of

$m_{n-r+1},\ldots,m_n$ in $M/T(M)$.

From $Q(R) \otimes_R M = Q(R) \otimes_R (M/T(M))$,

$F_i(Q(R) \otimes_R M) = Q(R) \cdot F_i(M) = O \ (i=0,\ldots,r-1)$, and

$F_r(Q(R) \otimes_R M) = Q(R) \cdot F_r(M) = Q(R)$ it follows, as in D.8, that $Q(R) \otimes_R M$ is free of rank r. Since $M/T(M)$ is generated by r elements, we necessarily have $M/T(M) \cong R^r$.

Consider now the presentation

$$0 \to K \to R^n \to M \to 0$$

belonging to $\{m_1, \ldots, m_n\}$. The elements $a_i := (a_{i1}, \ldots, a_{in}) \in R^n$ ($i=1, \ldots, n-r$) are in K, and they are linearly independent over R, since $\Delta := \det(a_{ij})_{i,j=1,\ldots,n-r}$ is not a zero divisor of R. We shall show that they span K, which will give us the last statement of b).

For any $(b_1, \ldots, b_n) \in K$ and $h=1, \ldots, n$ we have

$$\det \left(\begin{array}{ccc|c} a_{11}, & \cdots\cdots, & a_{1n-r} & a_{1h} \\ \vdots & & \vdots & \vdots \\ a_{n-r,1}, & \cdots, & a_{n-r,n-r} & a_{n-r,h} \\ \hline b_1, & \cdots\cdots, & b_{n-r} & b_h \end{array} \right) = 0$$

because $F_{r-1}(M) = 0$. The expansion of the determinant with respect to the last column yields an equation

$$\Delta b_h = \sum_{i=1}^{n-r} \Delta_i' \cdot a_{ih}$$

with certain $\Delta_i' \in F_r(M)$ which do not depend on h. Since Δ divides all Δ_i', we see that $(b_1, \ldots, b_n)$ is a linear combination of $a_1, \ldots, a_{n-r}$, q.e.d.

D.19. __Corollary.__ For a noetherian ring R with $\mathrm{Ass}(R) = \mathrm{Min}(R)$ and a finite R-module M the following statements are equivalent:

a) $F_i(M) = 0$ for $i=0, \ldots, r-1$, and $F_r(M)$ is an invertible ideal.

b) $M/T(M)$ is a projective R-module of rank r, and $pd_R(M) \leq 1$.

Use A.7 to show that $T(M)_{\mathscr{G}} = T(M_{\mathscr{G}})$ for each $\mathscr{G} \in \mathrm{Spec}(R)$. Since an ideal in a noetherian ring is invertible if and only if it is locally generated by a non-zero divisor, D.19 follows from D.18 by the local-global principle.

In particular, $F_0(M)$ is an invertible ideal if and only if M is a torsion module with $pd_R(M) \leq 1$.

Exercise

Let R be a ring. Assume there is an exact sequence of R-modules
$$0 \to F_1 \to F_0 \to M \to 0$$
with $F_0 \cong R^n$, $F_1 \cong R$. Let $T(M)$ be the torsion of M and $R : F_{n-1}(M) := \{x \in Q(R) \mid x \cdot F_{n-1}(M) \subset R\}$. Show that
$$T(M) \cong R : F_{n-1}(M)/R.$$

E. The Dual of a Module over a Noetherian Ring

In this section we collect some data about the dual of a finite module M over a noetherian ring R. If S/R is a finite algebra, we are particularly interested in the properties of the S-module $\mathrm{Hom}_R(S,R)$, the <u>canonical module</u> of the algebra S/R.

M* denotes the dual $\mathrm{Hom}_R(M,R)$ of M and M** := $\mathrm{Hom}_R(M^*,R)$ its bidual. It is wellknown that

$$\mathrm{Ass}(\mathrm{Hom}_R(M,R)) = \mathrm{Supp}(M) \cap \mathrm{Ass}(R).$$

We therefore have

<u>E.1. Remark.</u>   M* = O if and only if $\mathrm{Supp}(M) \cap \mathrm{Ass}(R) = \emptyset$.

Let

$$\alpha : M \to M^{**} \qquad (m \mapsto (\ell \mapsto \ell(m)) \text{ for } m \in M, \ell \in M^*)$$

be the canonical mapping of M into its bidual. M is <u>reflexive</u> (i.e. α bijective) if and only if $M_{\wp}$ is reflexive for all $\wp \in \mathrm{Spec}(R)$. In particular, if M is projective, then M is reflexive.

If K is the full ring of quotients of R, we write M_K for $K \otimes_R M$ in this section. M is <u>torsion free</u> if and only if the canonical mapping $M \to M_K$ is injective. It is easy to see that $M^* \to M_K^*$ is injective, hence

<u>E.2. Remark.</u>   M* is always torsion free, in particular, if M is reflexive, then M is torsion free.

<u>E.3. Proposition.</u>   If M_K is a reflexive K-module, then M* is reflexive: $M^{***} \cong M^*$.

Proof. Let $\beta : M^* \to M^{***}$ and $\alpha_K : M_K \to M_K^{**}$ be the canonical mappings, and let $\alpha^* : M^{***} \to M^*$ the transposed (dual) mapping of α.

We first show that $\alpha^* \circ \beta = \mathrm{id}_{M^*}$. Indeed, for $\ell \in M^*$ and $m \in M$ we have $[\alpha^*(\beta(\ell))](m) = \beta(\ell)(\alpha(m))$. Here $\alpha(m)(\ell') = \ell'(m)$ for all $\ell' \in M^*$, and $\beta(\ell)(\lambda) = \lambda(\ell)$ for all $\lambda \in M^{**}$. Thus $\beta(\ell)(\alpha(m)) = \alpha(m)(\ell) = \ell(m)$, and we obtain $[\alpha^*(\beta(\ell))](m) = \ell(m)$, hence $(\alpha^* \circ \beta)(\ell) = \ell$. In particular, α^* is surjective.

In the commutative diagram of natural mappings

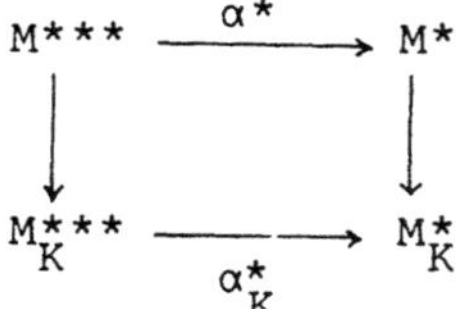

the vertical mappings are injective by E.2, and α_K^* is bijective, since M_K is reflexive. It follows that α^* is injective and hence an isomorphism. Then β is an isomorphism too.

If S/R is a finite algebra, then $\mathrm{Hom}_R(S,R)$ is a torsion free R-module by E.2, and if $\mathrm{Hom}_K(K \otimes_R S, K)$ is a reflexive K-module, then $\mathrm{Hom}_R(S,R)$ is a reflexive R-module by E.3.

Next we assume R is <u>normal</u>, that is, reduced and integrally closed in K. If $\mathscr{p}_1, \ldots, \mathscr{p}_n$ are the minimal prime ideals of R, then $R = R_1 \times \ldots \times R_n$ with $R_i := R/\mathscr{p}_i$. The R_i are integrally closed in their quotient fields K_i $(i=1, \ldots, n)$, and $K = K_1 \times \ldots \times K_n$. M has a canonical decomposition

$$M = M_1 \times \ldots \times M_n$$

where the M_i are R_i-modules $(i=1, \ldots, n)$, and M is reflexive (torsion free) if and only if M_i is reflexive (torsion free) as an R_i-module ($i=1,\ldots,n$).

Consider first the case that R is a normal domain and M is a torsion free R-module. Then M and $M_\wp$ for $\wp \in \operatorname{Spec}(R)$ can be identified with their images in M_K. Similarly M* and $M_\wp^*$ will be identified with their images in M_K^*, and M** and $M_\wp^{**}$ with their images in M_K^{**}.

<u>E.4. Proposition.</u>   If R is a normal domain and M is torsion free, then

$$M^* = \bigcap_{h(\wp)=1} M^*$$

where the intersection is taken for all $\wp \in \operatorname{Spec}(R)$ of heigth 1.

Proof. $\ell \in \operatorname{Hom}_K(M_K,K)$ belongs to $M_\wp^*$ (to M*) if and only if $\ell(x) \in R_\wp$ ($\ell(x) \in R$) for all $x \in M_\wp$ (all $x \in M$). Clearly $M^* \subset \bigcap_{h(\wp)=1} M_\wp^*$. Conversely, for $\ell \in \bigcap_{h(\wp)=1} M_\wp^*$ we have $\ell(x) \in \bigcap_{h(\wp)=1} R_\wp$ for all $x \in M$. For a normal domain R it is wellknown that $\bigcap_{h(\wp)=1} R_\wp = R$, hence $\ell(x) \in R$ and $\ell \in M^*$.

<u>E.5. Corollary.</u> $M^{**} = \bigcap_{h(\wp)=1} M_\wp$.

If we apply E.4 to M*, we obtain $M^{**} = \bigcap_{h(\wp)=1} M_\wp^{**}$. But $R_\wp$ is a discrete valuation ring and $M_\wp$ a free $R_\wp$-module for all $\wp$ with $h(\wp) = 1$. Hence $M_\wp^{**} = M_\wp$, and the formula of E.5 follows.

<u>E.6. Corollary.</u>   M is reflexive if and only if $M = \bigcap_{h(\wp)=1} M_\wp$.

Now let R = $R_1 \times \ldots \times R_n$ be again an arbitrary normal ring and M = $M_1 \times \ldots \times M_n$ a torsion free R-module. Each $\wp \in$ Spec(R) is in a canonical manner an element of Spec(R_i) for exactly one i $\in \{1,\ldots,n\}$ and $M_\wp = (M_i)_\wp$ for this i. M can be embedded into $(M_i)_{K_i}$ ($K_i := Q(R_i)$).

<u>E.7. Proposition.</u>   Let R be a normal ring, K := Q(R), and let V be a finite K-module. Let M,N be finite reflexive R-modules contained in V with $M_K = N_K = V$. Then M = N if and only if $M_\wp = N_\wp$ for all $\wp \in$ Spec(R) with h($\wp$) = 1.

Proof.   Let N = $N_1 \times \ldots \times N_n$ be the canonical decomposition of N. Clearly M = N if and only if $M_i = N_i$ for i=1,$\ldots$,n. By E.6, $M_i = \cap (M_i)_\wp$, and $N_i = \cap (N_i)_\wp$ where $\wp$ ranges over the elements of Spec(R_i) with h($\wp$) = 1. The claim follows.

We now apply the previous results to the canonical module of a finite algebra.

<u>E.8. Theorem.</u>   Let S/R be a finite algebra where R $\subset$ S and both rings are normal. Then Hom_R(S,R) is a reflexive S-module.

Proof.   R is a direct product of normal domains. We may assume without loss of generality that R is itself a domain. Since S is a direct product of normal domains which are R-algebras, we may also assume that S is a domain. With K := Q(R), L := Q(S), we have L = K $\otimes_R$ S, and Hom_K(K $\otimes_R$ S,K) = Hom_K(L,K) is certainly a reflexive L-module. Hom_R(S,R) is reflexive as an R-module by E.3, and by E.6 we have in

$Hom_K(L,K)$ the equation

$$Hom_R(S,R) = \bigcap_{\substack{\mathcal{g} \in Spec(R), h(\mathcal{g})=1}} Hom_R(S,R)_{\mathcal{g}}.$$

For a given $\mathcal{g} \in Spec(R)$ with $h(\mathcal{g}) = 1$ let $\mathcal{p}_1,\ldots,\mathcal{p}_h$ be the prime ideals of S lying over $\mathcal{g}$. $S_{\mathcal{g}}$ is a semilocal ring with the maximal ideals $\mathcal{p}_1 S_{\mathcal{g}},\ldots,\mathcal{p}_h S_{\mathcal{g}}$. Therefore

$$Hom_R(S,R)_{\mathcal{g}} = \bigcap_{i=1}^{h} Hom_R(S,R)_{\mathcal{p}_i}$$

and since each $\mathcal{p} \in Spec(S)$ with $h(\mathcal{p}) = 1$ is lying over some $\mathcal{g} \in Spec(R)$ with $h(\mathcal{g}) = 1$, we obtain

$$Hom_R(S,R) = \bigcap_{h(\mathcal{p})=1} Hom_R(S,R)_{\mathcal{p}}.$$

By E.6, $Hom_R(S,R)$ is a reflexive S-module.

In the following let R again be an arbitrary noetherian ring. We shall give some criteria in terms of the depth for a module to be reflexive. We first prove two lemmas.

E.9. Lemma.   Let R be a semilocal ring with radical $\mathcal{m}$, and let M and N be R-modules. If $\{f,g\}$ is an N-regular sequence in $\mathcal{m}$, then $\{f,g\}$ is $Hom_R(M,N)$-regular as well.

Proof.   In the commutative diagram with exact rows

$$0 \to Hom_R(M,N) \xrightarrow{\mu_f} Hom_R(M,N) \xrightarrow{\alpha} Hom_R(M,N/fN)$$
$$\downarrow{\mu_g} \qquad\qquad \downarrow{\mu_g} \qquad\qquad \downarrow{\mu_g}$$
$$0 \to Hom_R(M,N) \xrightarrow{\mu_f} Hom_R(M,N) \xrightarrow{\alpha} Hom_R(M,N/fN)$$

the mappings μ_g are injective, because g is not a zerodivisor of N and of N/fN. The mapping α has a factorization

$$Hom_R(M,N) \twoheadrightarrow Hom_R(M/N)/f Hom_R(M,N) \hookrightarrow Hom_R(M,N/fN)$$

and therefore

$$\mu_g \,:\, \mathrm{Hom}_R(M,N)/f\,\mathrm{Hom}_R(M,N) \;\to\; \mathrm{Hom}_R(M,N)/f\,\mathrm{Hom}_R(M,N)$$

is injective as well.

E.10. Lemma. Let R be a noetherian ring, M a finite R-module, and $N \subset M$ a submodule. Suppose that for all $\mathfrak{p} \in \mathrm{Spec}(R)$ with $\mathrm{depth}(R_\mathfrak{p}) \leq 1$ we have $N_\mathfrak{p}=M_\mathfrak{p}$, and for all $\mathfrak{p}$ with $\mathrm{depth}(R_\mathfrak{p}) \geq 2$ we have $\mathrm{depth}(M_\mathfrak{p}) \geq 1$ and $\mathrm{depth}(N_\mathfrak{p}) \geq 2$. Then $N = M$.

Proof. For $D := M/N$ we shall show that $\mathrm{Ass}(D) = \emptyset$, hence $D = 0$. For $\mathfrak{p} \in \mathrm{Spec}(R)$ with $\mathrm{depth}(R_\mathfrak{p}) \leq 1$ we have $D_\mathfrak{p} = 0$, thus $\mathfrak{p} \notin \mathrm{Ass}(D)$. For $\mathfrak{p}$ with $\mathrm{depth}(R_\mathfrak{p}) \geq 2$ we shall see that $\mathrm{depth}(D_\mathfrak{p}) \geq 1$, and hence again $\mathfrak{p} \notin \mathrm{Ass}(D)$. Assume $\mathfrak{p}R_\mathfrak{p} \in \mathrm{Ass}(D_\mathfrak{p})$. Then there is an element $x \neq 0$ in $D_\mathfrak{p}$ annihilated by $\mathfrak{p}R_\mathfrak{p}$. Let $\{f,g\}$ be an $N_\mathfrak{p}$-regular sequence in $\mathfrak{p}R_\mathfrak{p}$ where f is also $M_\mathfrak{p}$-regular. From the commutative diagram with exact rows

$$
\begin{array}{ccccccccc}
0 & \longrightarrow & N_\mathfrak{p} & \longrightarrow & M_\mathfrak{p} & \longrightarrow & D_\mathfrak{p} & \longrightarrow & 0 \\
 & & \downarrow{\mu_f} & & \downarrow{\mu_f} & & \downarrow{\mu_f} & & \\
0 & \longrightarrow & N_\mathfrak{p} & \longrightarrow & M_\mathfrak{p} & \longrightarrow & D_\mathfrak{p} & \longrightarrow & 0
\end{array}
$$

using the snake lemma we get an exact sequence

$$0 \to D' \to N_\mathfrak{p}/fN_\mathfrak{p} \to M_\mathfrak{p}/fM_\mathfrak{p}$$

where D' is the kernel of $\mu_f : D_\mathfrak{p} \to D_\mathfrak{p}$. Since $x \in D'$, we see that there is a nonzero element of $N_\mathfrak{p}/fN_\mathfrak{p}$ annihilated by $\mathfrak{p}R_\mathfrak{p}$, a contradiction, since g is not a zerodivisor of $N_\mathfrak{p}/fN_\mathfrak{p}$.

E.11. Proposition. Let M and N be two finite modules over a noetherian ring R, and let $\ell : N \to M$ be a linear map.

a) If $\ell_{\mathfrak{p}} : N_{\mathfrak{p}} \to M_{\mathfrak{p}}$ is bijective for all $\mathfrak{p} \in \mathrm{Spec}(R)$ with $\mathrm{depth}(R_{\mathfrak{p}}) = 0$, then $\ell^* : M^* \to N^*$ is injective.

b) If $\ell_{\mathfrak{p}} : N_{\mathfrak{p}} \to M_{\mathfrak{p}}$ is bijective for all $\mathfrak{p} \in \mathrm{Spec}(R)$ with $\mathrm{depth}(R_{\mathfrak{p}}) \leq 1$, then ℓ^* is bijective.

Proof. a) From the exact sequence $0 \to K \to N \to \ell(N) \to 0$ with $K := \ker \ell$ we obtain an exact sequence

$$0 \to \ell(N)^* \to N^* \to K^*.$$

Here $\mathrm{Ass}(K^*) = \mathrm{Ass}(\mathrm{Hom}_R(K,R)) = \mathrm{Supp}(K) \cap \mathrm{Ass}(R) = \emptyset$, since $K_{\mathfrak{p}} = 0$ for all $\mathfrak{p} \in \mathrm{Ass}(R)$. Hence $\ell(N)^* \to N^*$ is bijective.

From the exact sequence $0 \to \ell(N) \to M \to C \to 0$ with $C := \mathrm{coker}\, \ell$ we obtain an exact sequence

$$0 \to C^* \to M^* \to \ell(N)^*$$

and, as above, we have $C^* = 0$. Hence $M^* \to N^*$ is injective.

b) By a) we have an injection $\ell^* : M^* \to N^*$, and by assumption $\ell^*_{\mathfrak{p}} : M^*_{\mathfrak{p}} \to N^*_{\mathfrak{p}}$ is bijective for all $\mathfrak{p} \in \mathrm{Spec}(R)$ with $\mathrm{depth}(R_{\mathfrak{p}}) \leq 1$. By E.9, $\mathrm{depth}(M^*_{\mathfrak{p}}) \geq 2$ and $\mathrm{depth}(N^*_{\mathfrak{p}}) \geq 2$ for $\mathfrak{p} \in \mathrm{Spec}(R)$ with $\mathrm{depth}(R_{\mathfrak{p}}) \geq 2$. From E.10 we conclude that ℓ^* is bijective.

E.12. Corollary. Suppose N is reflexive, M is torsion free, M_K is a reflexive K-module $(K := Q(R))$, and $\ell_{\mathfrak{p}}$ is bijective for all $\mathfrak{p} \in \mathrm{Spec}(R)$ with $\mathrm{depth}(R_{\mathfrak{p}}) \leq 1$. Then ℓ is bijective.

Proof. By E.11, ℓ^* and hence also $\ell^{**} : N^{**} \to M^{**}$ is bijective. Consider the commutative diagram of natural mappings

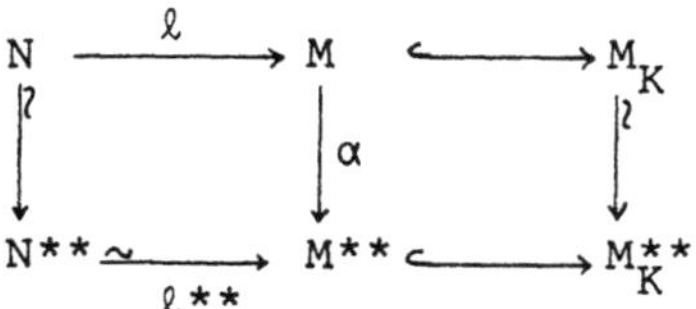

Here ℓ and α are injective and $\alpha \circ \ell$ is an isomorphism. Then ℓ is an isomorphism.

E.13. Theorem.   Let R be a noetherian ring, S/R a finite flat algebra, and M a finite S-module. M is reflexive if and only if the following two conditions are satisfied:

a) $M_{\wp}$ is a reflexive $S_{\wp}$-module for all $\wp \in \mathrm{Spec}(R)$ with depth$(R_{\wp}) \leq 1$.

b) depth$_{R_{\wp}}(M_{\wp}) \geq 2$ for all $\wp \in \mathrm{Spec}(R)$ with depth$(R_{\wp}) \geq 2$.

Proof. We may assume without loss of generality that R is local with maximal ideal $\mathfrak{m}$.

If M is a reflexive S-module, then a) is, of course, satisfied. Moreover, any R-regular sequence $\{f,g\}$ in $\mathfrak{m}$ is S-regular and, by E.9, also a regular sequence for $\mathrm{Hom}_S(M^*,S) = M^{**} = M$.

Conversely, assume that a) and b) are satisfied, and consider the exact sequence $O \to K \to M \xrightarrow{\alpha} M^{**} \to C \to O$ with K := ker α, C := coker α. Using induction on the dimension of R, we may assume that $\alpha_{\wp}$ is bijective for all $\wp \in \mathrm{Spec}(R) \smallsetminus \{\mathfrak{m}\}$. Then Ass(K) and Ass(C) are contained in $\{\mathfrak{m}\}$. By a) we may assume that depth$(R) \geq 2$, hence depth$(M) \geq 2$ by b). But then Ass(K) = $\emptyset$, from which K = O follows. By E.9 depth$(M^{**}) \geq 2$, and by E.10 we conclude that α is bijective, q.e.d.

For $M = \mathrm{Hom}_R(S,R)$ condition b) of the theorem is satisfied by E.9. Condition a) is certainly fulfilled, if $\mathrm{Hom}_{R_{\mathscr{g}}}(S_{\mathscr{g}}, R_{\mathscr{g}})$ is a projective $S_{\mathscr{g}}$-module for the $\mathscr{g} \in \mathrm{Spec}(R)$ with $\mathrm{depth}(R_{\mathscr{g}}) \leq 1$. We shall investigate now when this can happen.

E.14. Theorem.   Let R be noetherian and S/R a finite flat algebra. For $\mathscr{p} \in \mathrm{Spec}(S)$ let $\mathscr{g} := \mathscr{p} \cap R$. Then

$$\mu_{\mathscr{p}}(\mathrm{Hom}_R(S,R)) = r(S_{\mathscr{p}}/\mathscr{g}S_{\mathscr{p}})$$

where r denotes the Cohen-Macaulay type of the artinian ring $S_{\mathscr{p}}/\mathscr{g}S_{\mathscr{p}}$.

Proof.   Let $\widehat{S_{\mathscr{p}}}$ and $\widehat{R_{\mathscr{g}}}$ denote the completions of $S_{\mathscr{p}}$ and $R_{\mathscr{g}}$. Since $S_{\mathscr{p}}/R_{\mathscr{g}}$ is flat, $\widehat{S_{\mathscr{p}}}/\widehat{R_{\mathscr{g}}}$ is flat too. But this extension is finite, hence $\widehat{S_{\mathscr{p}}}$ is a free $\widehat{R_{\mathscr{g}}}$-module. It is easily seen, using the Chinese remainder theorem, that

$$\widehat{S_{\mathscr{p}}} \otimes_S \mathrm{Hom}_R(S,R) \cong \mathrm{Hom}_{\widehat{R_{\mathscr{g}}}}(\widehat{S_{\mathscr{p}}}, \widehat{R_{\mathscr{g}}})$$

hence

$$\mu_{\mathscr{p}}(\mathrm{Hom}_R(S,R)) = \mu(\mathrm{Hom}_{\widehat{R_{\mathscr{g}}}}(\widehat{S_{\mathscr{p}}}, \widehat{R_{\mathscr{g}}})).$$

We can assume that R and S are (complete) local rings and that S is free over R. Let $\mathscr{m}$ be the maximal ideal of R and let $\mathscr{n}$ be the maximal ideal of S. Since S/R is free, we have

$$\mathrm{Hom}_R(S,R)/\mathscr{m}\mathrm{Hom}_R(S,R) \cong \mathrm{Hom}_R(S,R/\mathscr{m}) = \mathrm{Hom}_{R/\mathscr{m}}(S/\mathscr{m}S, R/\mathscr{m})$$

and

$$\mu_{\mathscr{n}}(\mathrm{Hom}_R(S,R)) = \mu_{\mathscr{n}}(\mathrm{Hom}_{R/\mathscr{m}}(S/\mathscr{m}S, R/\mathscr{m})) =$$

$$\frac{1}{[S/\mathscr{n}:R/\mathscr{m}]} \cdot \dim_{R/\mathscr{m}}(\mathrm{Hom}_{R/\mathscr{m}}(S/\mathscr{m}S, R/\mathscr{m})/\mathscr{n}\cdot\mathrm{Hom}_{R/\mathscr{m}}(S/\mathscr{m}S, R/\mathscr{m})).$$

Since $r(S/\mathscr{m}S)$ is the dimension of the socle $0 : (\mathscr{n}/\mathscr{m}S)$ of

$S/\mathfrak{m}S$ as a vector space over $S/\mathfrak{m}$, the formula of the theorem follows, if we apply the following lemma with $k := R/\mathfrak{m}$, $A := S/\mathfrak{m}S$, and $I := \mathfrak{n}/\mathfrak{m}S$.

E.15. Lemma. Let A be a finite dimensional algebra over a field k. For each ideal I of A the module $I \cdot \mathrm{Hom}_k(A,k)$ consists of exactly those linear forms $\ell : A \to k$ which vanish on $0 : I = \{a \in A \mid Ia = 0\}$. In particular

$$\dim_k(I \cdot \mathrm{Hom}_k(A,k)) = \dim_k(A/0:I).$$

Proof. Since the elements of $I \cdot \mathrm{Hom}_k(A,k)$ vanish on $0:I$, it suffices to prove the dimension formula. This will be done by induction on $\mu(I)$.

For $I = (0)$ the formula certainly holds. Assume that it holds for an ideal I of A, and let $J = (I,a)$ with some $a \in A \setminus I$. Write $A = V \oplus (0:I)$ with a k-subspace $V \subset A$ and put $r := \dim_k V = \dim_k A/0:I$. Further let

$$0:I = k\omega_1 \oplus \ldots \oplus k\omega_s \oplus 0:J \qquad (\omega_j \in 0:I).$$

By assumption there are linear forms

$$a_1\ell_1,\ldots,a_r\ell_r \in I \cdot \mathrm{Hom}_k(A,k) \qquad (a_i \in I,\ \ell_i \in \mathrm{Hom}_k(A,k))$$

that are linearly independent over k. The elements $a\omega_1,\ldots,a\omega_s$ are also linearly independent over k, because from a relation $\sum_{j=1}^{s} \xi_j a\omega_j = 0$ $(\xi_j \in k)$ we obtain $\sum_{j=1}^{s} \xi_j \omega_j \in 0:J$, and hence $\xi_j = 0$ for $j=1,\ldots,s$.

Choose linear forms $\ell_1',\ldots,\ell_s' \in \mathrm{Hom}_k(A,k)$ with $\ell_i'(a\omega_j) = \delta_{ij}$ $(i,j=1,\ldots,s)$. Then

$$a_1\ell_1,\ldots,a_r\ell_r,a\ell_1',\ldots,a\ell_s'$$

are linearly independent over k, because from a relation

$$\ell := \sum_{i=1}^{r} \xi_i a_i \ell_i + \sum_{j=1}^{s} \lambda_j a\ell_j' = 0 \qquad (\xi_i, \lambda_i \in k)$$

we first conclude that $\lambda_j = \ell(\omega_j) = 0$ (j=1,...,s), and then, since the $a_i \ell_i$ are linearly independent, $\xi_i = 0$ for i=1,...,r.

We therefore have $\dim_k (J \cdot Hom_k(A,k)) \geq r+s = \dim_k A/O:J$. On the other hand for $\ell \in J \cdot Hom_k(A,k)$ we have $\ell(O:J) = O$, and hence

$$\dim_k (J \circ Hom_k(A,k)) \leq \dim_k A/O:J, \quad q.e.d.$$

E.16. <u>Corollary.</u> For a finite flat algebra S/R the following assertions are equivalent:

a) S/R is a Gorenstein algebra.

b) $Hom_R(S,R)$ is a projective S-module of rank 1.

Proof. It suffices to prove this when R is local.

b) → a) follows from E.14. Conversely, if S/R is a Gorenstein algebra, then $Hom_R(S,R) = S\sigma$ by E.14 for some linear form $\sigma : S \to R$. We shall show that σ is a basis element of $Hom_R(S,R)$.

Suppose $s \cdot \sigma = O$ for some $s \in S$. Then $s' \cdot \sigma(s) = \sigma(s's) = (s\sigma)(s') = O$ for all $s' \in S$, that is, $\tau(s) = O$ for all $\tau \in Hom_R(S,R)$. Since S has a basis over R, this can only happen for s = O, hence $Hom_R(S,R) \cong S$.

E.17. <u>Corollary.</u>   If S/R is a finite Gorenstein algebra and R is local, then $Hom_R(S,R) = S \cdot \sigma$ with a basis element σ (a "trace").

E.18. <u>Corollary.</u>   Let R be a noetherian ring, and let S/R be a finite flat algebra. Suppose that each $\mathcal{P} \in$ Spec(S) with depth($S_{\mathcal{P}}$) $\leq$ 1 is a Gorenstein point of S/R. Then $\text{Hom}_R(S,R)$ is a reflexive S-module.

Proof.   Given $\mathcal{q} \in$ Spec(R) with depth($R_{\mathcal{q}}$) $\leq$ 1 for any $\mathcal{P} \in$ Spec(S) lying over $\mathcal{q}$, we have depth($S_{\mathcal{P}}$) $\leq$ 1, since S/R is flat and finite. By E.17, $\text{Hom}_{R_{\mathcal{q}}}(S_{\mathcal{q}},R_{\mathcal{q}})$ is a projective $S_{\mathcal{q}}$-module. As remarked at the end of the proof of E.13 this shows that $\text{Hom}_R(S,R)$ is a reflexive S-module.

The following considerations will lead to a duality (E.21) that is useful on many occasions. Let M be a finite module over a noetherian ring R, and let $a := \{a_1,\ldots,a_n\}$ and $b := \{b_1,\ldots,b_n\}$ be two M-quasiregular sequences in R with $(b_1,\ldots,b_n) \subset (a_1,\ldots,a_n)$. Write

$$(1) \qquad b_i = \sum_{k=1}^{n} c_{ik} a_k \qquad (i=1,\ldots,n; c_{ik} \in R)$$

and put $\Delta := \det(c_{ik})$. By Cramer's rule the multiplication by Δ induces a well-defined R-linear map

$$\mu_\Delta : M/(a)M \to M/(b)M \qquad (m+(a)M \mapsto \Delta m + (b)M).$$

E.19. <u>Lemma.</u> μ_Δ does not depend on the special choice of the coefficients c_{ik} in the relations (1).

Proof.   We may assume $(R,\mathcal{m})$ is a local ring with $(a) \subset \mathcal{m}$. It suffices to show that the same mapping μ_Δ is obtained, if one of the equations in (1) is replaced by a new equation (for some $i \in \{1,\ldots,n\}$)

$$b_i = \sum_{k=1}^{n} d_{ik} a_k \qquad (d_{ik} \in R).$$

Let Δ' be the determinant of the new system of relations. Using the equation

$$\sum_{k=1}^{n} (c_{ik}-d_{ik})a_k = 0$$

and applying Cramer's rule we obtain relations

$$a_k \cdot (\Delta-\Delta') \in (b_1,\ldots,\hat{b}_i,\ldots,b_n) \qquad (k=1,\ldots,n).$$

Since $b_i \in (a_1,\ldots,a_n)$ and b_i is not a zero divisor of $M/(b_1,\ldots,\hat{b}_i,\ldots,b_n)M$, we have

$(\Delta-\Delta')m \in (b_1,\ldots,\hat{b}_i,\ldots,b_n)M \subset (b_1,\ldots,b_n)M$ for all $m \in M$

hence $\mu_\Delta = \mu_{\Delta'}$.

We write μ_b^a for the mapping μ_Δ. If $c := \{c_1,\ldots,c_n\}$ is another M-quasiregular sequence with $(c) \subset (b)$, then clearly

$$(2) \qquad\qquad \mu_c^b \circ \mu_b^a = \mu_c^a .$$

If $(a) = (b)$, then μ_b^a is an automorphism of $M/(a)M$.

<u>E.20. Theorem.</u>   Put $\overline{M} := M/(b)M$. μ_b^a is injective and

$$\mathrm{im}(\mu_b^a) = \{\overline{m} \in \overline{M} \mid (a_1,\ldots,a_n)\overline{m} = 0\}.$$

Proof. We may suppose $(R,\mathcal{W})$ is a local ring with $(a) \subset \mathcal{W}$. For n=0 the statement of the theorem is trivial, and for n=1 the proof is easy. Let n>1, and assume the theorem is true for quasiregular sequences of length n-1.

The ideal $(a_1,\ldots,a_n)$ contains a non-zerodivisor of $\tilde{M} := M/(b_2,\ldots,b_n)M$, namely b_1. Therefore there is an M-regular sequence $a' := \{a_1',\ldots,a_n'\}$ with $(a') = (a)$ such that a_1' is not a zerodivisor of $\tilde{M}$. Since $\mu_b^a = \mu_b^{a'} \circ \mu_{a'}^a$, and since $\mu_a^{a'}$ is an automorphism of $M/(a)M$, it suffices to prove the theorem for $\mu_b^{a'}$. We may therefore assume that already

a_1 is not a zerodivisor of $\widetilde{M}$.

Let Δ_1 be the determinant of the system

$$a_1 = a_1$$
$$b_i = \sum_{k=1}^{n} c_{ik} a_k \qquad (i=2,\ldots,n).$$

By Cramer's rule, applied to (1), we obtain

$$b_1 \Delta_1 - a_1 \Delta \in (b_2,\ldots,b_n).$$

Hence the following diagram commutes

(3)

$$
\begin{array}{ccc}
M/(a_1,\ldots,a_n)M & \xrightarrow{\ \mu_{\Delta_1}\ } & M/(a_1,b_2,\ldots,b_n)M \\
\Big\downarrow{\scriptstyle \mu_\Delta} & & \Big\downarrow{\scriptstyle \mu_{b_1}} \\
M/(b_1,\ldots,b_n)M & \xrightarrow{\ \mu_{a_1}\ } & M/(a_1 b_1,b_2,\ldots,b_n)M
\end{array}
$$

Here μ_{a_1} and μ_{b_1} denote the multiplication by a_1 resp. b_1.

By applying the induction hypothesis to the $R/(a_1)$-module $M/a_1 M$ and the images of the a_i and b_i in $R/(a_1)$, we see that μ_{Δ_1} is injective and

$$\mathrm{im}\,\mu_{\Delta_1} = \{\overline{m} \in M/(a_1,b_2,\ldots,b_n)M \mid (a)\overline{m} = 0\} =$$

$$\{m \in M \mid (a_1,\ldots,a_n)m \subset (a_1,b_2,\ldots,b_n)M\}/(a_1,b_2,\ldots,b_n)M.$$

Since a_1 and b_1 are not zerodivisors of $\widetilde{M}$, the mappings μ_{a_1} and μ_{b_1} are injective. Diagram (3) now shows that $\mu_{b}^{a} = \mu_\Delta$ is injective.

Moreover

$$\{\overline{m} \in \overline{M} \mid (a)\overline{m} = 0\} = \{m' \in M \mid (a)m' \subset (b)M\}/(b)M$$

and, since μ_{a_1} is injective, the image of this module under μ_{a_1} is

$$\{a_1 m' \mid m' \in M, (a)a_1 m' \subset (a_1 b_1,b_2,\ldots,b_n)M\}/(a_1 b_1,b_2,\ldots,b_n)M.$$

Since μ_{b_1} is injective, we obtain

$$\mu_{b_1}(\mathrm{im}\mu_{\Delta_1}) = \{b_1 m \mid m \in M, (a)b_1 m \subset (a_1 b_1, b_2, \ldots, b_n)M\} / (a_1 b_1, b_2, \ldots, b_n)M.$$

Given $m' \in M$ with $(a)a_1 m' \subset (a_1 b_1, b_2, \ldots, b_n)M$ we have $a_1^2 m' \in (a_1 b_1, b_2, \ldots, b_n)M$, and hence

$$a_1 m' \equiv b_1 m \bmod (b_2, \ldots, b_n)M$$

with some $m \in M$ such that $(a)b_1 m \subset (a_1 b_1, b_2, \ldots, b_n)M$. Conversely, if such an $m \in M$ is given, then in particular $b_1^2 m \in (a_1 b_1, b_2, \ldots, b_n)M$, and hence $b_1 m \equiv a_1 m' \bmod (b_2, \ldots, b_n)M$ with some $m' \in M$ such that $(a)a_1 m' \subset (a_1 b_1, b_2, \ldots, b_n)M$.

This proves that the image of $\{\overline{m} \in \overline{M} \mid (a)\overline{m} = 0\}$ under μ_{a_1} is

$$\mu_{b_1}(\mathrm{im}\mu_{\Delta_1}) = \mathrm{im}(\mu_{b_1} \circ \mu_{\Delta_1}) = \mu_{a_1}(\mathrm{im}\mu_{\Delta})$$

and, since μ_{a_1} is injective,

$$\mathrm{im}\mu_{\Delta} = \{\overline{m} \in \overline{M} \mid (a)\overline{m} = 0\}, \quad \text{q.e.d.}$$

<u>E.21. Corollary</u> (Wiebe [Wi]). Let $\{a_1, \ldots, a_n\}$ and $\{b_1, \ldots, b_n\}$ be R-quasiregular sequences with $(b_1, \ldots, b_n) \subset (a_1, \ldots, a_n)$. Let $\overline{\Delta}$ denote the image of Δ in $\overline{R} := R/(b_1, \ldots, b_n)$, and let $\overline{I} := (a_1, \ldots, a_n)/(b_1, \ldots, b_n)$. Then

$$\mathrm{Ann}_{\overline{R}}(\overline{\Delta}) = I \quad \text{and} \quad \mathrm{Ann}_{\overline{R}}(\overline{I}) = (\overline{\Delta}).$$

<u>Exercises</u>

1) Let (R, m) be a noetherian local ring with $\mathrm{depth}(R) = 1$, and let $K := Q(R)$. For an ideal $I \subset \mathit{m}$ of R show that

$$R \underset{K}{:} I := \{x \in K \mid xI \subset R\} \neq R.$$

2) Let $(R,\mathfrak{m})$ be a noetherian local ring with depth$(R) \leq 1$, and let M be a finite R-module. If M* has a basis $\{\ell_1,\ldots,\ell_n\}$, then $\ell_i(M) = R$ for i=1,...,n. Conclude that there exist elements $m_1,\ldots,m_n \in M$ and a submodule $N \subset M$ such that

$$M = \langle m_1,\ldots,m_n\rangle \oplus N \text{ and } N^* = \langle 0\rangle.$$

3) Let R be a noetherian ring, and let S/R be a finite flat algebra. Let M be a finite S-module which, as an R-module, is projective. Then, if $\text{Hom}_S(M,S)$ is a projective (free) S-module, M is a projective (free) S-module. In particular, the following assertions are equivalent:

a) S/R is a Gorenstein algebra.

b) $\text{Hom}_S(\text{Hom}_R(S,R),S)$ is a projective S-module.

F. Traces

a) The canonical trace and norm of a finite projective algebra

If R is a ring and F a free R-module with a basis $\{\omega_1,\ldots,\omega_n\}$, then for $\varphi \in \mathrm{End}(F)$ the <u>trace</u> $\sigma(\varphi)$ and the <u>norm</u> $n(\varphi)$ of φ are defined as follows: Write $\varphi(\omega_i) = \sum_{k=1}^{n} r_{ik}\omega_k$ $(i=1,\ldots,n; r_{ik} \in R)$. Then (independent of the choice of the basis)

$$\sigma(\varphi) := \sum_{i=1}^{n} r_{ii} \text{ and } n(\varphi) := \det(r_{ik}).$$

If $N \subset R$ is a multiplicatively closed subset and $\varphi_N \in \mathrm{End}(F_N)$ the endomorphism induced by φ, then clearly $\sigma(\varphi_N)$ and $n(\varphi_N)$ are the images of $\sigma(\varphi)$ and $n(\varphi)$ in R_N.

Now let P be a finite locally free R-module, and let $\{f_i\}_{i \in I}$ be a family of elements $f_i \in R$ with $(\{f_i\}) = R$ such that P_{f_i} is a free R_{f_i}-module for all $i \in I$. For any $\varphi \in \mathrm{End}(P)$ the traces $\sigma(\varphi_{f_i})$ are defined, and the images of $\sigma(\varphi_{f_i})$ and $\sigma(\varphi_{f_j})$ in $R_{f_if_j}$ coincide for all $i,j \in I$. There is a unique element $\sigma(\varphi) \in R$ whose image in R_{f_i} is $\sigma(\varphi_{f_i})$ for all $i \in I$ (A.8). It is called the <u>trace</u> of φ. The <u>norm</u> $n(\varphi)$ of φ is defined likewise.

The following rules easily follow from the definition of the trace and the norm.

F.1. Rules.

a) $\sigma : \mathrm{End}(P) \to R$ $(\varphi \mapsto \sigma(\varphi))$ is R-linear, and

$$\sigma(\varphi_1 \circ \varphi_2) = \sigma(\varphi_2 \circ \varphi_1)$$

for $\varphi_1,\varphi_2 \in \mathrm{End}(P)$.

b) If P has rank r, then $\sigma(id_P) = r \cdot 1_R$. For general
P we have $n(id_P) = 1_R$.

c) $\varphi \in End(P)$ is an automorphism if and only if $n(\varphi)$ is a
unit of R.

d) (Base change). Let S/R be an algebra. For $\varphi \in End(P)$ we
have $id_S \otimes \varphi \in End(S \otimes_R P)$ and

$$\sigma(id_S \otimes \varphi) = \sigma(\varphi) \cdot 1_S \ , \quad n(id_S \otimes \varphi) = n(\varphi) \cdot 1_S.$$

e) If P' is another finite locally free R-module and
$\varphi \in End(P)$, $\varphi' \in End(P')$, then

$$\sigma(\varphi \oplus \varphi') = \sigma(\varphi) + \sigma(\varphi') \text{ and } n(\varphi \oplus \varphi') = n(\varphi) \cdot n(\varphi').$$

<u>F.2. Proposition.</u>   If $\varphi \in End(P)$ is nilpotent, then $\sigma(\varphi)$
is a nilpotent element of R.

Proof.   For any $\mathfrak{z} \in Spec(R)$ clearly $id_{k(\mathfrak{z})} \otimes \varphi$ is a nilpotent endomorphism of the $k(\mathfrak{z})$-vector space $k(\mathfrak{z}) \otimes_R P$.
It is well-known that its trace is zero, therefore by F.1d)

$$\sigma(\varphi) \cdot 1_{k(\mathfrak{z})} = \sigma(id_{k(\mathfrak{z})} \otimes P) = 0$$

and hence $\sigma(\varphi)$ is contained in any prime ideal of R.

   Assume now that S/R is a finite algebra that is locally
free as an R-module. As usual $\mu_x : S \to S$ denotes the
multiplication by x.

<u>F.3. Definition.</u>   The map

$$\sigma_{S/R} : S \to R \qquad (x \mapsto \sigma(\mu_x))$$

is called the <u>canonical trace</u>, and

$$n_{S/R} : S \to R \qquad (x \mapsto n(\mu_x))$$

is called the <u>canonical norm</u> of S/R.

By the rules F.1 we have $\sigma_{S/R} \in \mathrm{Hom}_R(S,R)$ and $\sigma_{S/R}(1_S) = r \cdot 1_R$, if S has rank r over R. Moreover, $n_{S/R}(x_1 x_2) = n_{S/R}(x_1) \cdot n_{S/R}(x_2)$ for $x_1, x_2 \in S$, and $n(x)$ is a unit of R if and only if x is a unit of S. By F.2 the trace of a nilpotent element of S is a nilpotent element of R. If S/R is a (finite) field extension, then it is well-known that $\sigma_{S/R} \neq 0$ if and only if S/R is separable.

<u>F.4. Proposition.</u> Suppose S/R has a basis $\{\omega_1, \ldots, \omega_n\}$ and $\{\omega_1^*, \ldots, \omega_n^*\}$ is the dual basis of the R-module $\mathrm{Hom}_R(S,R)$. Then

$$\sigma_{S/R} = \sum_{i=1}^{n} \omega_i \omega_i^*.$$

Proof. If $\omega_j \omega_k = \sum_{\ell=1}^{n} \rho_{jk}^{\ell} \omega_\ell$ $(j,k=1,\ldots,n; \rho_{jk}^{\ell} \in R)$, then $\sigma_{S/R}(\omega_k) = \sum_{i=1}^{n} \rho_{ik}^{i}$ for $k=1,\ldots,n$. On the other hand,

$$\left(\sum_{i=1}^{n} \omega_i \omega_i^* \right)(\omega_k) = \sum_{i=1}^{n} \omega_i^*(\omega_i \omega_k) = \sum_{i=1}^{n} \omega_i^* \left(\sum_{\ell=1}^{n} \rho_{ik}^{\ell} \omega_\ell \right) = \sum_{i=1}^{n} \rho_{ik}^{i}.$$

From the definition of the trace, and of the norm, we obtain

<u>F.5. Base change.</u> If T/R is an arbitrary algebra, then

$$\sigma_{T \otimes_R S/T} = \mathrm{id}_T \otimes_R \sigma_{S/R}$$

and for each $s \in S$

$$n_{T \otimes_R S/T}(1 \otimes s) = n_{S/R}(s).$$

<u>F.6. Direct products.</u> Let $S = S_1 \times \ldots \times S_t$ be a direct product of finite locally free R-algebras S_i/R $(i=1,\ldots,t)$. Then for $x = (x_1,\ldots,x_t) \in S_1 \times \ldots \times S_t$ we have

$$\sigma_{S/R}(x) = \sum_{i=1}^{t} \sigma_{S_i/R}(x_i) \quad \text{and} \quad n_{S/R}(x) = \prod_{i=1}^{t} n_{S_i/R}(x_i).$$

This is obvious, since μ_x induces μ_{x_i} on each factor S_i.

<u>F.7. Transitive law.</u>   Suppose R is noetherian, and let T/S be another finite locally free algebra. Then $\sigma_{T/R}$ and $n_{T/R}$ are defined, and

$$\sigma_{T/R} = \sigma_{S/R} \circ \sigma_{T/S} \quad , \quad n_{T/R} = n_{S/R} \circ n_{T/S}.$$

Proof.   Since R is noetherian, "locally free" means the same as "projective". Clearly T is a projective R-module, hence $\sigma_{T/R}$ and $n_{T/R}$ are defined. After localization in R the R-module S has a basis. Passing to the completion of R we may assume that $S = S_1 \times \ldots \times S_n$ is a direct product of (complete) local R-algebras S_i ($i=1,\ldots,n$), which are free as R-modules, being direct summands of S. Let $T = T_1 \times \ldots \times T_n$ be the corresponding decomposition of T. Then T_i/S_i is a finite algebra with a basis ($i=1,\ldots,n$), and for $x = (x_1,\ldots,x_n) \in T_1 \times \ldots \times T_n$ we have

$$\sigma_{T/S}(x) = (\sigma_{T_1/S_1}(x_1),\ldots,\sigma_{T_n/S_n}(x_n))$$

$$n_{T/S}(x) = (n_{T_1/S_1}(x_1),\ldots,n_{T_n/S_n}(x_n)).$$

Thus we may assume that T/S and S/R are free. For the trace the proof of the transitive law is now simple. For the norm we refer to Bourbaki $[B_1]$, Chap.III,§9, Cor. of Prop.6.

We now answer the question under which conditions $\sigma_{S/R}$ is a basis element of the S-module $\mathrm{Hom}_R(S,R)$.

F.8. Proposition. Suppose S is a projective R-module. $\mathrm{Hom}_R(S,R)$ is a free S-module of rank 1 with $\{\sigma_{S/R}\}$ as a basis if and only if S/R is étale.

Proof. We may assume that R is a local ring with maximal ideal $\mathcal{m}$. Let $K := R/\mathcal{m}$ and $\overline{S} := S/\mathcal{m}S$.

a) Suppose $\mathrm{Hom}_R(S,R) = S\cdot\sigma_{S/R}$. Since S/R is free and the trace is compatible with base change, we obtain $\mathrm{Hom}_K(\overline{S},K) = \overline{S}\cdot\sigma_{\overline{S}/K}$. We have $\overline{S} = \overline{S}_1 \times\ldots\times \overline{S}_n$ with local K-algebras $\overline{S}_i$, and $\mathrm{Hom}_K(\overline{S}_i,K) = \overline{S}_i\cdot\sigma_{\overline{S}_i/K}$ $(i=1,\ldots,n)$. We shall show that $\overline{S}_i/K$ is a separable field extension $(i=1,\ldots,n)$, from which it follows that S/R is étale.

The maximal ideal $\mathcal{m}_i$ of $\overline{S}_i$ is nilpotent. Hence for $x \in \mathcal{m}_i$ and $y \in \overline{S}_i$ we have by F.2

$$x\cdot\sigma_{\overline{S}_i/K}(y) = \sigma_{\overline{S}_i/K}(xy) = 0$$

that is, $x\cdot\sigma_{\overline{S}_i/K} = 0$. Since $\mathrm{Hom}_K(\overline{S}_i,K) \cong \overline{S}_i$, we conclude that $\mathcal{m}_i = 0$ and $\overline{S}_i$ is a field. Since $\sigma_{\overline{S}_i/K} \neq 0$, the field extension $\overline{S}_i/K$ is separable.

b) Suppose S/R is étale. Then $\overline{S}$ is a direct product of finite separable extension fields of K. Therefore $\mathrm{Hom}_K(\overline{S},K) = \overline{S}\cdot\sigma_{\overline{S}/K}$, and by Nakayama $\mathrm{Hom}_R(S,R) = S\cdot\sigma_{S/R}$. By E.16 the S-module $\mathrm{Hom}_R(S,R)$ is free with basis $\{\sigma_{S/R}\}$, since S/R is clearly a Gorenstein algebra.

b) Traces in Gorenstein algebras

In this section let S/R be a finite flat (projective) algebra where R is a noetherian ring. By E.16, S/R is a Gorenstein algebra if and only if $\mathrm{Hom}_R(S,R)$ is a projective S-module (of rank 1). In case $\mathrm{Hom}_R(S,R) \cong S$ we call any

$\sigma \in \text{Hom}_R(S,R)$ with $\text{Hom}_R(S,R) = S \cdot \sigma$ a <u>trace of S/R</u>. Observe that in general the canonical trace is not a trace in this sense (F.8). Two traces of S/R differ by a factor that is a unit of S.

Consider at first an arbitrary finite flat (projective) algebra S/R where R is noetherian and let I be the kernel of

$$\mu : S \otimes_R S \to S \qquad (a \otimes b \mapsto a \cdot b).$$

It is generated by $\{s \otimes 1 - 1 \otimes s \mid s \in S\}$. $S \otimes_R S$ has two S-module structures (multiplication in the first and the second factor). On $\text{Ann}_{S \otimes_R S}(I)$ the two structures agree, since

$$\text{Ann}_{S \otimes_R S}(I) \cdot (s \otimes 1 - 1 \otimes s) = O \qquad (s \in S).$$

Since S is a projective R-module, there is a canonical isomorphism of R-modules

$$(1) \qquad \phi : S \otimes_R S \to \text{Hom}_R(\text{Hom}_R(S,R),S)$$

satisfying the formula

$$(2) \qquad \phi(x)(\ell) = \Sigma \ell(a_k) b_k$$

for $x = \Sigma a_k \otimes b_k \in S \otimes_R S$ and $\ell \in \text{Hom}_R(S,R)$.

<u>F.9. Proposition.</u> ϕ induces an isomorphism of S-modules

$$\phi : \text{Ann}_{S \otimes_R S}(I) \xrightarrow{\sim} \text{Hom}_S(\text{Hom}_R(S,R),S).$$

Proof. For $x = \Sigma a_k \otimes b_k \in \text{Ann}(I)$, $s \in S$, and $\ell \in \text{Hom}_R(S,R)$, we have $\Sigma s a_k \otimes b_k = \Sigma a_k \otimes s b_k$ and

$$\phi(x)(s\ell) = \Sigma \ell(s a_k) b_k = \Sigma \ell(a_k) s b_k = s \Sigma \ell(a_k) b_k = s \phi(x)(\ell).$$

Therefore $\phi(x) : \text{Hom}_R(S,R) \to S$ is an S-linear map. Moreover,

$\phi(sx) = s\phi(x)$, hence $\phi|_{Ann(I)}$ is S-linear.

Conversely, if for $x = \Sigma a_k \otimes b_k \in S \otimes_R S$ the map $\phi(x)$ is S-linear, then with $x_1 := \Sigma sa_k \otimes b_k$ and $x_2 := \Sigma a_k \otimes sb_k$ we obtain for any $\ell \in Hom_R(S,R)$

$$\phi(x_1)(\ell) = \phi(x)(s\ell) = s\phi(x)(\ell) = \phi(x_2)(\ell)$$

and hence $x_1 = x_2$, which means that $x \in Ann(I)$.

<u>F.10. Corollary.</u>   For a Gorensteinalgebra S/R the following statements are equivalent:

a) S/R has a trace (i.e. $Hom_R(S,R) \cong S$).

b) $Ann_{S\otimes_R S}(I) \cong S$.

c) $Ann_{S\otimes_R S}(I)$ is generated by one element.

Moreover, for a trace σ of S/R, let $\Delta_\sigma \in Ann(I)$ be the element with $\phi(\Delta_\sigma)(\sigma) = 1$. Then $\sigma \mapsto \Delta_\sigma$ defines a one-to-one correspondence between the traces of S/R and the basis elements of the S-module $Ann(I)$. If $\phi(\Delta_\sigma)(\ell) = s$ for some $\ell \in Hom_R(S,R)$, then $\ell = s \cdot \sigma$. If $\Delta_\sigma = \Sigma s_k' \otimes s_k$ $(s_k, s_k' \in S)$, then $s = \Sigma \ell(s_k') \cdot s_k$.

<u>F.11. Proposition.</u>   Assume S/R has a trace σ and $\Delta_\sigma \in Ann(I)$ is the element with $\phi(\Delta_\sigma)(\sigma) = 1$. Let $\{s_1, \ldots, s_n\}$ be a basis of S as an R-module and write

$$\Delta_\sigma = \sum_{k=1}^{n} s_k' \otimes s_k \qquad (s_k' \in S).$$

Then

$$\sigma(s_i' s_k) = \delta_{ik}$$

that is, $\{s_1', \ldots, s_n'\}$ is the (unique) basis of S/R that is dual to $\{s_1, \ldots, s_n\}$ with respect to σ.

Proof.   From $\phi(\Delta_\sigma)(\sigma) = \sum_{i=1}^{n} \sigma(s_i') s_i = 1$ it follows that

$$s_k = s_k \cdot \phi(\Delta_\sigma)(\sigma) = \phi(\Delta_\sigma)(s_k \sigma) = \sum_{i=1}^{n} \sigma(s_i' s_k) s_i, \text{ and hence}$$

$$\sigma(s_i' s_k) = \delta_{ik}.$$

Between a trace σ and the canonical trace $\sigma_{S/R}$ there is the following connection.

F.12. Corollary. $\quad \sigma_{S/R} = \mu(\Delta_\sigma) \cdot \sigma.$

Proof. We may assume R is local, and hence S has a basis $\{s_1, \ldots, s_n\}$ over R. Write

$$\Delta_\sigma = \sum_{k=1}^{n} s_k' \otimes s_k \qquad\qquad (s_k' \in S)$$

and

$$s_k s_i' = \Sigma r_{ki}^j s_j' \qquad\qquad (r_{ki}^j \in R)$$

which is possible, since $\{s_1', \ldots, s_n'\}$ is a basis of S/R by F.11. Moreover,

$$\sigma(\mu(\Delta_\sigma) s_k) = \sigma\left(\left(\sum_{i=1}^{n} s_i' s_i\right) s_k\right) = \sigma\left(\sum_{i=1}^{n} (s_i' s_k) s_i\right) =$$

$$\sigma\left(\sum_{i,j=1}^{n} r_{ki}^j s_i s_j'\right) = \sum_{i=1}^{n} r_{ki}^i = \sigma_{S/R}(s_k).$$

F.13. Corollary. The following conditions are equivalent:

a) $\sigma_{S/R}$ is a trace.

b) $\mu(\Delta_\sigma)$ is a unit of S for any trace $\sigma : S \to R$.

c) S/R is étale.

If one of the conditions is satisfied, then

$$\mu(\Delta_{\sigma_{S/R}}) = 1.$$

Proof. a) $\leftrightarrow$ c) was shown in F.8, and a) $\leftrightarrow$ b) follows from F.12. If $\sigma_{S/R}$ is a trace, then by F.12

$$\sigma_{S/R} = \mu(\Delta_{\sigma_{S/R}}) \cdot \sigma_{S/R}$$

and hence $\mu(\Delta_{\sigma_{S/R}}) = 1.$

If there is an R-algebra homomorphism $\pi : S \to R$ (a retraction), then π is a multiple of any trace σ of S/R. A more precise statement is given in

F.14. Proposition.   If S/R has a trace σ and a retraction π, then

$$\pi = s \cdot \sigma \quad \text{with} \quad s := \mu((\pi \otimes id_S)(\Delta_\sigma)).$$

Proof.   We may assume that R is local. Then there is a basis $\{s_1, \ldots, s_n\}$ of S/R with $\pi(s_1) = 1$ and $\pi(s_i) = 0$ for $i = 2, \ldots, n$. Write $\Delta_\sigma = \sum_{k=1}^{n} s_k \otimes s_k'$ $(s_k' \in S)$. We have $\mu((\pi \otimes id_S)(\Delta_\sigma)) = s_1'$, that is, $s = s_1'$. Moreover,

$$(s_1' \sigma)(s_k) = \sigma(s_1' s_k) = \delta_{1k} = \pi(s_k) \quad (k = 1, \ldots, n)$$

by F.11, which implies that $s\sigma = \pi$.

F.15. Corollary.   Suppose R is reduced, S is a free R-module of rank [S:R], and the retraction $\pi : S \to R$ has nilpotent kernel. Then for any trace σ of S/R

$$\Delta_\sigma \equiv [S:R] \cdot (\pi \otimes id_S)(\Delta_\sigma) \bmod I$$

Proof.   By F.2 we have $\sigma_{S/R}(x) = 0$ for $x \in \ker \pi$, and hence $\sigma_{S/R} = [S:R] \cdot \pi$. By F.12 and F.14 we conclude that

$$\mu(\Delta_\sigma) \cdot \sigma = [S:R] \cdot \mu(\pi \otimes id_S)(\Delta_\sigma) \cdot \sigma$$

and therefore

$$\mu(\Delta_\sigma - [S:R](\pi \otimes id_S)(\Delta_\sigma)) = 0.$$

F.16. Proposition.   a) If S/R has a trace σ and R'/R is an arbitrary algebra, then

$$id_{R'} \otimes \sigma : R' \otimes_R S \to R' \otimes_R R = R'$$

is a trace of $R' \otimes_R S/R'$.

b) If $S = S_1 \times \ldots \times S_t$ is a direct product of finite projective R-algebras S_i $(i=1,\ldots,t)$, and if σ is a trace of S/R, then $\sigma_i := \sigma|_{S_i}$ is a trace of S_i/R and

$$\sigma(x) = \sum_{i=1}^{t} \sigma_i(x_i)$$

for any $x = (x_1,\ldots,x_t) \in S_1 \times \ldots \times S_t$.

Proof. a) Since S is a projective R-module, the canonical homomorphism

$$R' \otimes_R Hom_R(S,R) \to Hom_{R'}(R' \otimes_R S,R')$$

is an isomorphism of $R' \otimes_R S$-modules. The claim follows.

b) is an immediate consequence of the fact that there is a canonical isomorphism of S-modules

$$Hom_R(S,R) \cong Hom_R(S_1,R) \times \ldots \times Hom_R(S_t,R).$$

<u>F.17. Proposition.</u> Let T/S be another finite projective algebra, let $I := ker(S \otimes_R S \to S)$, $K := ker(T \otimes_R T \to T)$, and $J := ker(T \otimes_S T \to T)$. Assume S/R has a trace σ and T/S has a trace τ. Let Δ_σ be the element associated with σ in $Ann_{S \otimes_R S}(I)$, Δ_τ the element associated with τ in $Ann_{T \otimes_S T}(J)$, and let $\tilde{\Delta}_\tau \in T \otimes_R T$ be a representative of Δ_τ. Then

a) $\sigma \circ \tau$ is a trace of T/R.

b) $\Delta_\sigma \cdot \tilde{\Delta}_\tau$ is contained in $Ann_{T \otimes_R T}(K)$, and is the element associated with $\sigma \circ \tau$.

Proof. We may assume that R is local, then S has a basis over R. Passing to the completion of R we may assume that S

is a direct product of local R-algebras, and using F.16b)
we may also assume that T has a basis over S.

a) There is a canonical isomorphism of T-modules

$$(3) \qquad \psi \ : \ \mathrm{Hom}_R(T,R) \ \xrightarrow{\sim} \ \mathrm{Hom}_S(T,\mathrm{Hom}_R(S,R))$$

with $\psi(\ell)(t)(s) = \ell(st)$ for $\ell \in \mathrm{Hom}_R(T,R)$, $t \in T$, and $s \in S$.
The S-linear map $\varphi \ : \ T \to \mathrm{Hom}_R(S,R)$ with $\varphi(t) = \tau(t)\sigma$ for
all $t \in T$ is a basis element of $\mathrm{Hom}_S(T,\mathrm{Hom}_R(S,R))$ as a T-
module. Since $\psi(\sigma \circ \tau) = \varphi$, we see that $\sigma \circ \tau$ is a basis element of the T-module $\mathrm{Hom}_R(T,R)$.

b) Given a basis $\{s_1,\ldots,s_n\}$ of S/R and a basis $\{t_1,\ldots,t_m\}$
of T/S, write

$$\Delta_\sigma \ = \ \sum_{i=1}^{n} s_i' \otimes_R s_i \qquad\qquad (s_i' \in S)$$

$$\Delta_\tau \ = \ \sum_{k=1}^{m} t_k' \otimes_S t_k \qquad\qquad (t_k' \in T)$$

and choose for $\tilde{\Delta}_\tau$ the element

$$\tilde{\Delta}_\tau \ := \ \sum_{k=1}^{m} t_k' \otimes_R t_k \ \in \ T \otimes_R T.$$

Since the kernel of the canonical map $T \otimes_R T \to T \otimes_S T$ is
generated by the elements $s \otimes_R 1 - 1 \otimes_R s$ $(s \in S)$, and Δ_σ
annihilates that kernel, $\Delta_\sigma \tilde{\Delta}_\tau$ is independent of the special
choice of a representative $\tilde{\Delta}_\tau$ of Δ_τ. Moreover,
$\Delta_\sigma \tilde{\Delta}_\tau \in \mathrm{Ann}_{T \otimes_R T}(K)$ and $(\sigma \circ \tau)(s_i' t_k' s_j t_\ell) = \sigma(s_i' s_j \cdot \tau(t_k' t_\ell)) = $
$\sigma(s_i' s_j \delta_{k\ell}) = \delta_{k\ell}\delta_{ij}$, which implies that $\{s_i' t_k'\}$ is a basis
of T/R, dual to $\{s_i t_k\}$ with respect to $\sigma \circ \tau$. By F.11
$\Delta_\sigma \tilde{\Delta}_\tau \ = \ \Delta_{\sigma \circ \tau}.$

c) Traces in complete intersections

If S/R is a finite complete intersection, there are traces $\sigma : S \to R$ corresponding to the presentations of S/R as a complete intersection (C.9).

F.18. Assumptions.

a) R is noetherian and S/R is a finite Gorenstein algebra.

b) There is a presentation

$$S = P/(t_1, \ldots, t_n)$$

where P/R is a flat algebra, P is a noetherian ring, and $t := \{t_1, \ldots, t_n\}$ is a quasiregular sequence of P.

c) The kernel J of the composed map

$$S \otimes_R P \xrightarrow{\mathrm{id} \otimes \rho} S \otimes_R S \xrightarrow{\mu} S$$

where ρ is the canonical epimorphism, is generated by a quasiregular sequence $\xi := \{\xi_1, \ldots, \xi_m\}$ of $S \otimes_R P$.

Clearly the kernel of $\mathrm{id} \otimes \rho$ is generated by $\{1 \otimes t_1, \ldots, 1 \otimes t_n\}$, and this is a quasiregular sequence of $S \otimes_R P$. Since S is finite and flat over R, it is easily seen that both sequences have the same length $m = n$.

F.19. Example.
Under the assumptions a) and b) of F.18 let P be of the form

$$P = R[\![Y_1, \ldots, Y_\ell]\!][X_1, \ldots, X_m]_N$$

with indeterminates Y_i and X_j and a multiplicatively closed subset $N \subset R[\![Y]\!][X]$. Let y_i, x_j denote the images of Y_i, X_j in S. Then, since S/R is finite,

$$S \otimes_R P = S[\![Y_1, \ldots, Y_\ell]\!][X_1, \ldots, X_m]_N$$

and the ideal J of F.18c) is generated by

$\{Y_1-y_1,\ldots,Y_\ell-y_\ell,X_1-x_1,\ldots,X_m-x_m\}$, which is a regular sequence of $S \otimes_R P$. Moreover, $\ell+m = n$.

Under the assumptions F.18 there are equations

$$(4) \qquad 1\otimes t_i = \sum_{j=1}^{n} a_{ij}\xi_j \qquad (i=1,\ldots,n;a_{ij} \in S \otimes_R P)$$

By E.19

$$(5) \qquad \Delta_\xi^t := (\mathrm{id}\otimes\rho)(\det(a_{ij})) \in S \otimes_R S$$

is independent of the choice of the a_{ij}. If $I := \ker\mu$, then

$$(6) \qquad \mathrm{Ann}_{S\otimes_R S}(I) = (\Delta_\xi^t) \text{ and } \mathrm{Ann}_{S\otimes_R S}(\Delta_\xi^t) = I$$

by E.21. Moreover, $\mathrm{Hom}_R(S,R)$ is a free S-module by F.10. Let ϕ be the isomorphism of S-modules described in F.9.

F.20. Definition. Write $\tau_t^\xi : S \to R$ for the trace with

$$\phi(\Delta_\xi^t)(\tau_t^\xi) = 1.$$

We call it <u>the trace associated with the presentation</u> <u>$S = P/(t)$</u> (and the system ξ).

From F.11 we obtain

F.21. Rule. If S/R has a basis $\{s_1,\ldots,s_m\}$ and

$$\Delta_\xi^t = \sum_{i=1}^{m} s_i' \otimes s_i \qquad (s_i' \in S, i=1,\ldots,m)$$

then

$$\tau_t^\xi(s_i's_k) = \delta_{ik} \qquad (i,k=1,\ldots,m).$$

The most important case for our present considerations is the following: S/R has a presentation

$$(7) \qquad S = R[X_1,\ldots,X_n]/(t_1,\ldots,t_n) = R[x_1,\ldots,x_n]$$

as a finite complete intersection. Then the assumptions F.18 are satisfied with $P := R[X_1,\ldots,X_n]$ and $\xi_i := X_i-x_i$ $(i=1,\ldots,n)$. We write Δ_x^t for Δ_ξ^t and τ_t^x for τ_t^ξ in this case.

F.22. Examples.

a) The Tate trace ($[T_1]$).

Assume $S = R[X]/(t)$ with a monic polynomial

$$t = X^n + r_{n-1}X^{n-1} + \ldots + r_0 \in R[X]$$

and let x be the image of X in S. In S[X] we have

$$t = t - t(x) = (X^n - x^n) + r_{n-1}(X^{n-1} - x^{n-1}) + \ldots + r_1(X-x)$$

and we obtain

$$\Delta_x^t = \sum_{\alpha=0}^{n-1} x^\alpha \otimes x^{n-1-\alpha} + r_{n-1} \cdot \sum_{\alpha=0}^{n-2} x^\alpha \otimes x^{n-2-\alpha} + \ldots + r_1.$$

From

$$1 = \phi(\Delta_x^t)(\tau_t^x) = \sum_{\alpha=0}^{n-1} \tau_t^x(x^\alpha) x^{n-1-\alpha} + r_{n-1} \sum_{\alpha=0}^{n-2} \tau_t^x(x^\alpha) x^{n-2-\alpha} + \ldots + r_1 \tau_t^x(1)$$

we see, by comparing coefficients with respect to the basis

$\{1, \ldots, x^{n-1}\}$ of S/R, that

$$\tau_t^x(x^\alpha) = \begin{cases} 1 & \text{for } \alpha = n-1 \\ 0 & \text{for } \alpha = 0, \ldots, n-2. \end{cases}$$

Hence, for an arbitrary element $y = \rho_0 + \rho_1 x + \ldots + \rho_{n-1} x^{n-1} \in S$ ($\rho_i \in R$)

$$\tau_t^x(y) = \rho_{n-1}.$$

b) Assume S/R has a presentation

$$S = R[X_1, \ldots, X_n]/(X_1^{\nu_1} - r_1, \ldots, X_n^{\nu_n} - r_n)$$

with $r_i \in R$, $\nu_i > 0$ ($i = 1, \ldots, n$). Let $t_i := X_i^{\nu_i} - r_i$ and let

x_i denote the image of X_i in S ($i = 1, \ldots, n$). Then

$\{x_1^{\alpha_1} \ldots x_n^{\alpha_n} \mid 0 \le \alpha_i \le \nu_i - 1\}$ is a basis of S/R. A calculation

similar to the one in a) yields

$$\tau_t^x(x_1^{\alpha_1} \ldots x_n^{\alpha_n}) = \begin{cases} 1 & \text{for } (\alpha_1, \ldots, \alpha_n) = (\nu_1 - 1, \ldots, \nu_n - 1) \\ 0 & \text{otherwise.} \end{cases}$$

(This formula results also from a) and the transitive law

F.28).

c) Let $R[Y_1,\ldots,Y_n] \subset R[X_1,\ldots,X_n]$ be polynomial rings such that $R[X]/R[Y]$ is finite. Then $R[X]/R[Y]$ has a presentation

$$R[X] = R[Y_1,\ldots,Y_n,T_1,\ldots,T_n]/(Y_1-f_1,\ldots,Y_n-f_n)$$

as a complete intersection where $f_i \in R[T_1,\ldots,T_n]$ $(i=1,\ldots,n)$, see C.10b). With $t_i := Y_i-f_i$ $(i=1,\ldots,n)$ there is a trace $\tau_t^x : R[X] \to R[Y]$ and

$$\mathrm{Hom}_{R[Y]}(R[X],R[Y]) = R[X] \cdot \tau_t^x.$$

The same facts hold for power series algebras in place of polynomial algebras.

In the situation of example F.19 we define for each $t \in S[\![Y_1,\ldots,Y_\ell]\!][X_1,\ldots,X_m]_N$ the partial derivatives $\dfrac{\partial t}{\partial Y_j}$ and $\dfrac{\partial t}{\partial X_k}$ as follows: If $t = \dfrac{p}{q}$ with $p \in S[\![Y]\!][X]$ and $q \in N$,

then $\dfrac{\partial t}{\partial Y_j} = \dfrac{q\dfrac{\partial p}{\partial Y_j} - p\dfrac{\partial q}{\partial Y_j}}{q^2}$ where $\dfrac{\partial p}{\partial Y_j}$ denotes the formal partial

derivative of the power series p. Let $\dfrac{\partial t}{\partial y_j}$ denote the image

of $\dfrac{\partial t}{\partial Y_j}$ in S, $\dfrac{\partial t}{\partial x_k}$ the image of $\dfrac{\partial t}{\partial X_k}$, and

$\dfrac{\partial(t_1,\ldots,t_n)}{\partial(Y_1,\ldots,Y_\ell,X_1,\ldots,X_m)}$ the image of the Jacobian deter-

minant $\dfrac{\partial(t_1,\ldots,t_n)}{\partial(Y_1,\ldots,Y_\ell,X_1,\ldots,X_m)}$. Moreover, write $t(y,x)$ for

the image of t in S.

Here the relations (4) are of the form

$$(8) \qquad 1 \otimes t_i = \sum_{j=1}^{\ell} a_{ij}(Y_j-y_j) + \sum_{k=1}^{m} b_{ik}(X_k-x_k)$$

with $a_{ij}, b_{ik} \in S[\![Y]\!][X]_N$, and by differentiating these relations we see that

$$a_{ij}(y,x) = \frac{\partial t_i}{\partial y_j} \quad , \quad b_{ik}(y,x) = \frac{\partial t_i}{\partial x_k}.$$

With $\xi := \{Y_1 - y_1, \ldots, Y_\ell - y_\ell, X_1 - x_1, \ldots, X_m - x_m\}$ we therefore have

$$(9) \qquad \mu(\Delta_\xi^t) = \frac{\partial(t_1, \ldots, t_n)}{\partial(y_1, \ldots, y_\ell, x_1, \ldots, x_m)} \ .$$

From F.12 we obtain

F.23. <u>Relation to the canonical trace</u>: In the situation of example F.19

$$\sigma_{S/R} = \frac{\partial(t_1, \ldots, t_n)}{\partial(y_1, \ldots, y_\ell, x_1, \ldots, x_m)} \cdot \tau_t^\xi .$$

The formula can be applied, if S/R has a presentation as a complete intersection, as for example in F.22,a)-c). In particular, we have

F.24. <u>Corollary.</u>   For $S = R[X]/(t)$, as in F.22a), assume $t'(x)$ is a unit of S. Then

$$\sigma_{S/R}\left(\frac{x^\alpha}{t'(x)}\right) = \begin{cases} 1 & \text{for } \alpha = n-1 \\ \\ 0 & \text{for } \alpha = 0, \ldots, n-2. \end{cases}$$

This formula is closely related to the Lagrange interpolation formula. Since $\tau_t^x = \frac{1}{t'(x)} \sigma_{S/R}$, it results from F.22a).

Assume that we are still in the situation of example F.19. Write $P := R[\![Y]\!][X]_N$, and suppose there is a retraction $\pi: S \to R$ with $\pi(y_j) = 0$ $(j=1, \ldots, \ell)$ and $\pi(x_k) = 0$ $(k=1, \ldots, m)$. Write

$$t_i = \sum_{j=1}^\ell c_{ij} Y_j + \sum_{k=1}^m d_{ik} X_k \qquad (c_{ij}, d_{ik} \in P)$$

and let $d_{y,x}^t$ be the image of $\det(c_{ij}, d_{ik})$ in S, which is independent of the special choice of the coefficients c_{ij} and d_{ik} by E.19. Comparing with (8) we see that in $S \otimes_R S$

$$\Delta_\xi^t = 1 \otimes d_{y,x}^t + q \qquad (q \in \ker(\pi \otimes id_S))$$

By F.14

$$\pi = s \cdot \tau_t^\xi \text{ with } s := \mu(\pi \otimes id_S)(\Delta_\xi^t) = \mu(1 \otimes d_{y,x}^t) = d_{y,x}^t$$

Thus we have the formula

$$(10) \qquad \pi = d_{y,x}^t \cdot \tau_t^\xi, \text{ in particular } \tau_t^\xi(d_{y,x}^t) = 1.$$

<u>F.25. Corollary.</u>   Assume that in the above situation R is reduced, S/R is free, and ker π is nilpotent. Then

$$\frac{\partial(t_1,\ldots,t_n)}{\partial(y_1,\ldots,y_\varrho,x_1,\ldots,x_m)} = [S:R] \cdot d_{y,x}^t.$$

Apply F.15 and use (9).

For the remainder of this appendix let the assumptions F.18 be satisfied. Let S'/R be another finite Gorenstein algebra with a presentation

$$S' = P/(t_1',\ldots,t_n')$$

where $t' := \{t_1',\ldots,t_n'\}$ is a quasiregular sequence of P with $(t_1,\ldots,t_n) \subset (t_1',\ldots,t_n')$. Thus S' is a homomorphic image of S.

Let $\xi' := \{\xi_1',\ldots,\xi_n'\}$ be the sequence of images of the ξ_i in $S' \otimes_R P$. This is a quasiregular sequence of $S' \otimes_R P$ by B.23, and it generates the kernel of the composed homomorphism $S' \otimes_R P \to S' \otimes_R S' \to S'$, as is easily seen. Thus the trace $\tau_{\xi'}^{t'} : S' \to R$ is defined. What is its relation to τ_ξ^t ?

Write

$$t_i = \sum_{j=1}^{n} a_{ij} t_j' \qquad (i=1,\ldots,n; a_{ij} \in P)$$

and let $\Delta_{t'}^t$ denote the image of $\det(a_{ij})$ in S. Since $\Delta_{t'}^t$ annihilates the kernel I of the canonical epimorphism $S \to S'$, the multiplication by $\Delta_{t'}^t$ defines an S-linear map

$$\Delta^{t}_{t'} \; : \; S' \rightarrow S.$$

<u>Ŧ.26. Proposition.</u>   The following diagram commutes

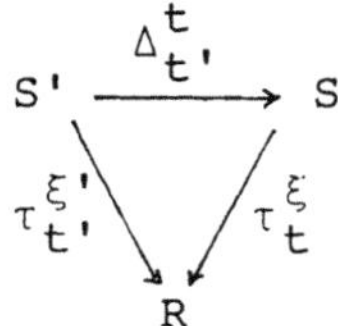

Proof. In $S' \otimes_R P$ there are equations

$$1 \otimes t'_j \;=\; \sum_{k=1}^{n} u_{jk}\xi'_k \quad (j=1,\ldots,n; u_{jk} \in S' \otimes_R P)$$

$$1 \otimes t_i \;=\; \sum_{j,k=1}^{n} (1\otimes a_{ij}) u_{jk}\xi'_k \quad (i=1,\ldots,n).$$

Observe that $\{1\otimes t_1,\ldots,1\otimes t_n\}$ is a quasiregular sequence in $S' \otimes_R P$, therefore by E.19

$$\overline{\Delta^{t}_{\xi}} \;=\; (1\otimes\Delta^{t}_{t'})\cdot\overline{\det(u_{ij})}$$

where the bar denotes the image in $S' \otimes_R S$. By definition, $\Delta^{t'}_{\xi'}$ is the image of $\overline{\det(u_{ij})}$ in $S' \otimes_R S'$. Write $\overline{\det(u_{ij})} = \Sigma a'_i\otimes b_i$ with $a'_i \in S'$, $b_i \in S$. Then $\Delta^{t'}_{\xi'} = \Sigma a'_i\otimes b'_i$ with the images b'_i of b_i in S'.

Consider $\mathrm{Hom}_R(S',R)$ as a submodule of $\mathrm{Hom}_R(S,R)$ in the canonical manner. There is a canonical commutative diagram

$$\begin{array}{ccc}
S \otimes_R S & \xrightarrow{\;\;\phi\;\;} & \mathrm{Hom}_R(\mathrm{Hom}_R(S,R),S) \\
\downarrow & & \downarrow \\
S' \otimes_R S & \xrightarrow{\;\;\phi'\;\;} & \mathrm{Hom}_R(\mathrm{Hom}_R(S',R),S)
\end{array}$$

and $\phi'(\overline{\Delta^{t}_{\xi}})(\tau^{\xi'}_{t'}) = \Delta^{t}_{t'}\cdot\Sigma\tau^{\xi'}_{t'}(a'_i)\cdot b_i$. On the other hand, by the definition of $\tau^{\xi'}_{t'}$ we have $\Sigma\tau^{\xi'}_{t'}(a'_i)b'_i = 1$, hence $\Sigma\tau^{\xi'}_{t'}(a'_i)b_i \equiv 1 \bmod I$. Since $\Delta^{t}_{t'}\cdot I = 0$, we obtain $\phi(\Delta^{t}_{\xi})(\tau^{\xi'}_{t'}) = \Delta^{t}_{t'}$ which leads to

$$(11) \qquad\qquad \tau^{\xi'}_{t'} \;=\; \Delta^{t}_{t'}\cdot\tau^{\xi}_{t}.$$

This formula is particularly important in the special case $S' = S$, $\xi' = \xi$. It then tells us how τ^{ξ}_{t} changes, if we pass from the generating system t of the kernel of P → S to another such system.

The trace τ_t^ξ _is compatible with base change_. Let R'/R be an arbitrary algebra where R' is noetherian. Put S' := R' $\otimes_R$ S, P' := R' $\otimes_R$ P, and assume that P' is noetherian. Then

$$S' = P'/(t_1, \ldots, t_n)$$

and $\{t_1, \ldots, t_n\}$ is a quasiregular sequence of P' (see the argument in the proof of C.18). Clearly S'/R' is a finite Gorenstein algebra. Since S' $\otimes_{R'}$ P' = R' $\otimes_R$ (S $\otimes_R$ P), the kernel J' of S' $\otimes_{R'}$ P' → S' is generated by $\{\xi_1, \ldots, \xi_n\}$, which is a quasiregular sequence of S' $\otimes_{R'}$ P'. Let τ'^ξ_t: S' → R' be the trace corresponding to this situation.

Since Δ_ξ^t generates the annihilator of ker(S' $\otimes_{R'}$ S' → S') and the mapping ϕ is compatible with base change, we obtain

__F.27. Proposition.__ $\quad \tau'^\xi_t = \mathrm{id}_{R'} \otimes_R \tau_t^\xi$.

Finally, we wish to study the _transitive properties of the traces_ τ_t^ξ. Assume T/S is another finite Gorenstein algebra having a presentation

$$T = Q/(u_1, \ldots, u_m)$$

where Q/S is a flat algebra, Q a noetherian ring, and u := $\{u_1, \ldots, u_m\}$ a quasiregular sequence of Q. Also assume that the kernel of the T-homomorphism T $\otimes_S$ Q → T is generated by a quasiregular sequence η := $\{\eta_1, \ldots, \eta_m\}$. Then the trace τ_u^η : T → S is defined. The composition $\tau_t^\xi \circ \tau_u^\eta$: T → R is a trace of T/R by F.17a).

Suppose there is a flat algebra $\tilde{Q}/P$ where $\tilde{Q}$ is a noetherian ring and there is an epimorphism $\tilde{Q} \to Q$ with kernel $(t_1,\ldots,t_n)\tilde{Q}$. Let $U := \{U_1,\ldots,U_m\}$ be a system of pre-images of the u_i in $\tilde{Q}$. Then $t,U := \{t_1,\ldots,t_n,U_1,\ldots,U_m\}$ is a quasiregular sequence of $\tilde{Q}$ and

$$T = \tilde{Q}/(t_1,\ldots,t_n,U_1,\ldots,U_m).$$

Consider the canonical homomorphism $S \otimes_R P \to T \otimes_R \tilde{Q}$. Since T/S and $\tilde{Q}/P$ are flat, so is $T \otimes_R \tilde{Q} = T \otimes_S (S \otimes_R P) \otimes_P \tilde{Q}$ over $S \otimes_R P$ and ξ is therefore a quasiregular sequence of $T \otimes_R \tilde{Q}$. We have

$$T \otimes_R \tilde{Q}/(\xi_1,\ldots,\xi_n) = T \otimes_R Q/(\overline{\xi_1},\ldots,\overline{\xi_n})$$

with the images $\overline{\xi_i}$ of the ξ_i in $T \otimes_R Q$. Since the kernel of $\mu : S \otimes_R S \to S$ is generated by the images of the ξ_i in $S \otimes_R S$, we obtain a canonical isomorphism

$$T \otimes_R Q/(\overline{\xi_1},\ldots,\overline{\xi_n}) \cong T \otimes_S Q$$

Let $\{E_1,\ldots,E_m\}$ be a system of preimages of the η_i in $T \otimes_R \tilde{Q}$. Then $\xi,E := \{\xi_1,\ldots,\xi_n,E_1,\ldots,E_m\}$ is a quasiregular sequence of $T \otimes_R \tilde{Q}$ which generates the kernel of $T \otimes_R \tilde{Q} \to T$. Thus $\tau_{t,U}^{\xi,E} : T \to R$ is defined.

<u>F.28. Proposition.</u> $\tau_{t,U}^{\xi,E} = \tau_t^{\xi} \circ \tau_u^{\eta}.$

Proof. Consider the relations (4) and corresponding relations

$$1 \otimes u_j = \sum_{\ell=1}^{m} b_{jk}\eta_k \qquad (j=1,\ldots,m; b_{jk} \in T \otimes_S Q)$$

in $T \otimes_S Q$. Since ξ generates the kernel of $T \otimes_R \tilde{Q} \to T \otimes_S Q$, one obtains in $T \otimes_R Q$ a system of relations

$$1 \otimes t_i = \sum_{k=1}^{n} a_{ik}\xi_k \quad (i=1,\ldots,n; a_{ik} \in S \otimes_R P \subset T \otimes_R \tilde{Q})$$

$$1 \otimes U_j = \sum_{k=1}^{n} c_{jk} \xi_k + \sum_{\ell=1}^{m} A_{j\ell} E_\ell \quad (j=1,\ldots,m; c_{jk}, A_{j\ell} \in T \otimes_R \widetilde{Q}).$$

From this we can see that in $T \otimes_R T$ there is an equation

$$\Delta_{\xi,E}^{t,U} = \Delta_\xi^t \circ \widetilde{\Delta}_\eta^u$$

where $\widetilde{\Delta}_\eta^u$ has the image Δ_η^u in $T \otimes_S T$. The formula of the proposition now results from F.17b).

The transitive law F.28 can be applied, in particular, if S/R and T/S have presentations (7) as complete intersections; see the proof of C.17. Moreover, F.28 enables us <u>to compare traces of an algebra S/R associated with different presentations</u>, as we shall see now.

Suppose

$$S = Q/(u_1, \ldots, u_m)$$

is another presentation of S/R with a flat algebra Q/R where Q is noetherian and $u := \{u_1, \ldots, u_m\}$ a quasiregular sequence of Q. Assume that the kernel of $S \otimes_R Q \to S$ is generated by a quasiregular sequence $\eta := \{\eta_1, \ldots, \eta_m\}$ of $S \otimes_R Q$. Then $\tau_u^\eta : S \to R$ is defined.

$P \otimes_R Q$ is flat over R. Suppose $P \otimes_R Q$ is a noetherian ring. Let $\widetilde{\eta} := \{\widetilde{\eta}_1, \ldots, \widetilde{\eta}_m\}$ be a system of preimages of the η_i under the canonical epimorphism $P \otimes_R Q \to S \otimes_R Q$, and let $\widetilde{\xi} := \{\widetilde{\xi}_1, \ldots, \widetilde{\xi}_n\}$ be a system of preimages of the ξ_i under $P \otimes_R Q \to P \otimes_R S = S \otimes_R P$. Then

$$S = P \otimes_R Q / (t_1 \otimes 1, \ldots, t_n \otimes 1, \widetilde{\eta}_1, \ldots, \widetilde{\eta}_m) = P \otimes_R Q / (1 \otimes u_1, \ldots, 1 \otimes u_m, \widetilde{\xi}_1, \ldots, \widetilde{\xi}_n)$$

where both

$$t \otimes 1, \widetilde{\eta} := \{t_1 \otimes 1, \ldots, t_n \otimes 1, \widetilde{\eta}_1, \ldots, \widetilde{\eta}_m\}$$

and

$$\widetilde{\xi}, 1 \otimes u := \{\widetilde{\xi}_1, \ldots, \widetilde{\xi}_n, 1 \otimes u_1, \ldots, 1 \otimes u_m\}$$

are quasiregular sequences of $P \otimes_R Q$.

The three composed mappings

$$P \otimes_R S \otimes_R Q \xrightarrow{\mathrm{mod}(\xi \otimes 1_Q)} S \otimes_R Q \xrightarrow{\mathrm{mod}(\eta)} S$$

$$P \otimes_R S \otimes_R Q \xrightarrow{\mathrm{mod}(1_P \otimes \eta)} P \otimes_R S \xrightarrow{\mathrm{mod}(\xi)} S$$

and

$$P \otimes_R S \otimes_R Q \to S \otimes_R P \otimes_R Q \xrightarrow{\mathrm{id}_S \otimes \varepsilon} S \otimes_R S \xrightarrow{\mu} S$$

where $\varepsilon : P \otimes_R Q \to S$ is the epimorphism with kernel
$(t \otimes 1, \tilde{\eta}) = (\tilde{\xi}, 1 \otimes u)$ agree, and their kernel is generated
by

$$\xi \otimes 1, 1 \otimes \eta := \{\xi_1 \otimes 1_Q, \ldots, \xi_n \otimes 1_Q, 1_P \otimes \eta_1, \ldots, 1_P \otimes \eta_m\}$$

which is a quasiregular sequence of $P \otimes_R S \otimes_R Q$. Hence the
traces

$$\tau^{\xi \otimes 1, 1 \otimes \eta}_{t \otimes 1, \tilde{\eta}} : S \to R \quad \text{and} \quad \tau^{\xi \otimes 1, 1 \otimes \eta}_{\tilde{\xi}, 1 \otimes u} : S \to R$$

are defined and by F.26

$$\tau^{\xi \otimes 1, 1 \otimes \eta}_{t \otimes 1, \tilde{\eta}} = \Delta^{\tilde{\xi}, 1 \otimes u}_{t \otimes 1, \tilde{\eta}} \cdot \tau^{\xi \otimes 1, 1 \otimes \eta}_{\tilde{\xi}, 1 \otimes u}.$$

What is the relation between τ^{ξ}_{t} and $\tau^{\xi \otimes 1, 1 \otimes \eta}_{t \otimes 1, \tilde{\eta}}$ (between
τ^{η}_{u} and $\tau^{\xi \otimes 1, 1 \otimes \eta}_{\tilde{\xi}, 1 \otimes u}$)? We may regard

$$S = S \otimes_R Q / (\eta_1, \ldots, \eta_m)$$

as a presentation of the trivial algebra S/S, leading to
the trace $\tau^{\eta}_{\eta} : S \to S$, which is the identity. By F.28

$$\tau^{\xi \otimes 1, 1 \otimes \eta}_{t \otimes 1, \tilde{\eta}} = \tau^{\xi}_{t} \circ \tau^{\eta}_{\eta} = \tau^{\xi}_{t}$$

and similarly

$$\tau^{\xi \otimes 1, 1 \otimes \eta}_{\tilde{\xi}, 1 \otimes u} = \tau^{\eta}_{u}.$$

Thus we have established

F.29. Proposition. $\quad \tau_t^\xi = \Delta_{t \otimes 1, \tilde{\eta}}^{\tilde{\xi}, 1 \otimes u} \cdot \tau_u^\eta.$

We can use the transitive formula F.28 to exhibit the the behaviour of the traces τ_t^ξ under product decomposition of algebras. Let $S = S_1 \times \ldots \times S_n$ be a decomposition of S into algebras S_i/R $(i=1,\ldots,n)$, and let

$$1 = e_1 + \ldots + e_n$$

be the corresponding decomposition of 1 into idempotents e_i. Then $S_i = Se_i = S_{e_i} = S[X]/(e_iX-1)$ is the localization of S with respect to $\{1,e_i\}$, and this relation may also be regarded as a presentation of S_i/S as a complete intersection (C.20).

Put $u_i := e_iX-1$ $(i=1,\ldots,n)$. The kernel of the canonical epimorphism

$$S_i[X] = S_i \otimes_S S[X] \to S_i \otimes_S S_i = S_i$$

is generated by $\eta_i := X-1$. Thus we have a trace $\tau_{u_i}^{\eta_i} : S_i \to S$. But the image of u_i in $S_i[X]$ is η_i, hence $\Delta_{\eta_i}^{u_i} = 1$, and it is easily seen from the definition that $\tau_{u_i}^{\eta_i} : S_i \to S$ is the inclusion mapping of S_i into S.

Let the assumptions F.18 be satisfied. Then the trace $\tau_t^\xi : S \to R$ is defined. Moreover, there is a presentation $S_i = P[X]/(t_1,\ldots,t_n,U_i)$ with $U_i := E_iX-1$ where $E_i \in P$ is a preimage of e_i, and $t,U_i := \{t_1,\ldots,t_n,U_i\}$ is a quasiregular sequence of $P[X]$. The kernel of the canonical map

$$S_i \otimes_S (S \otimes_R P[X]) = (S_i \otimes_R P)[X] \to S_i \otimes_R S_i \to S_i$$

is generated by $\{1 \otimes \xi_1,\ldots,1 \otimes \xi_n, H_i\}$ with a preimage H_i of $\eta_i = X-1$ in $(S_i \otimes_R P)[X]$. In fact, we can choose $H_i = X-1 \otimes 1$.

By F.28 we then have

$$\tau^{1\otimes\xi,H_i}_{t,U_i} = \tau^{\xi}_t \circ \tau^{\eta_i}_{u_i} = \tau^{\xi}_t|_{S_i}$$

and we obtain

F.30. **Proposition.** For $s = (s_1,\ldots,s_n) \in S_1\times\ldots\times S_n = S$

$$\tau^{\xi}_t(s) = \sum_{i=1}^{n} \tau^{1\otimes\xi,H_i}_{t,U_i}(s_i)$$

Indeed, $\tau^{\xi}_t(s) = \sum_{i=1}^{n} \tau^{\xi}_t(se_i) = \sum_{i=1}^{n} \tau^{1\otimes\xi,H_i}_{t,U_i}(s_i)$.

Up to some modifications and generalizations, the results of the sections b) and c) are due to Scheja and Storch ([SS$_3$],[SS$_4$]).

Exercises

1) Let R be a ring and M a finite R-module that has a finite projective resolution P. where the P_i are finitely generated. For $\varphi \in \mathrm{End}(M)$ let $\varphi_\bullet : P_\bullet \to P_\bullet$ be a complex-homomorphism that makes

$$\begin{array}{ccc} P_\bullet & \longrightarrow & M \\ \varphi_\bullet \downarrow & & \downarrow \varphi \\ P_\bullet & \longrightarrow & M \end{array}$$

commutative. Let $\sigma(\varphi_i)$ be the trace of φ_i ($i \in \mathbb{N}$).

a) $\sigma(\varphi) := \sum(-1)^i\sigma(\varphi_i)$ is independent of the choice of the lifting $\varphi_\bullet$ of φ.

b) $\sigma(\varphi)$ does not depend on the choice of the finite projective resolution P. of M.

$\sigma(\varphi)$ is called the trace of φ.

2) Under the assumptions of F.8, what is the support of the S-module $\mathrm{Hom}_R(S,R)/S\cdot\sigma_{S/R}$?

3) Let $P = K[\![X_1,\ldots,X_d]\!]$ be a power series algebra over a field K. For an ideal $I \subset P$ let $S := P/I$ be a finite Gorenstein algebra over K. The socle $\gamma(S)$ of S is the K-vector space of all elements of S annihilated by the maximal ideal of S.

a) For any trace $\sigma : S \to K$ we have $\sigma(\gamma(S)) \neq 0$.

b) Suppose $I = (t_1,\ldots,t_d)$ where $t_1,\ldots,t_d$ is a system of parameters of P. Let $\dfrac{\partial(t_1,\ldots,t_d)}{\partial(x_1,\ldots,x_d)}$ denote the image of $\dfrac{\partial(t_1,\ldots,t_d)}{\partial(x_1,\ldots,x_d)}$ in S and assume the characteristic of K does not divide $\dim_K S$. Then

$$\gamma(S) = K\cdot\frac{\partial(t_1,\ldots,t_d)}{\partial(x_1,\ldots,x_d)} \ .$$

G. Differents

At first we introduce the $\underline{\text{Noether different}}$ which is defined for an arbitrary algebra S/R. Let I denote the kernel of

$$\mu : S \otimes_R S \to S \qquad (a \otimes b \mapsto ab).$$

$\underline{\text{G.1. Definition.}}$ $\vartheta_N(S/R) := \mu(\text{Ann}_{S\otimes_R S}(I))$ is called the $\underline{\text{Noether different}}$ of S/R.

Obviously $\vartheta_N(S/R)$ is an ideal of S. Given a presentation

$$S = R[\{X_\lambda\}_{\lambda\in\Lambda}]/\mathfrak{a}$$

it can also be described as follows.

$\underline{\text{G.2. Remark.}}$ Let x_λ denote the image of X_λ in S. Then

$$\vartheta_N(S/R) = \{f(\{x_\lambda\}) \mid f\in S[\{X_\lambda\}], f\cdot(X_\lambda-x_\lambda)\in\mathfrak{a}S[\{X_\lambda\}] \text{ for all } \lambda\in\Lambda\}.$$

In fact, if

$$\rho : S[\{X_\lambda\}] \to S \otimes_R S$$

is the S-epimorphism corresponding to the canonical epimorphism

$$S \otimes_R R[\{X_\lambda\}] \to S \otimes_R S$$

then

$$\text{Ann}_{S\otimes_R S}(I) = \rho(\mathfrak{a}S[\{X_\lambda\}] : (\{X_\lambda-x_\lambda\}_{\lambda\in\Lambda})).$$

$\underline{\text{G.3. Example.}}$ Assume S/R has a presentation

$$S = R[X_1,\ldots,X_n]/(t_1,\ldots,t_n) = R[x_1,\ldots,x_n]$$

as a finite complete intersection. Then with the notation of appendix F, section c), formula (6) we have

$$\text{Ann}_{S\otimes_R S}(I) = (\Delta_x^t)$$

and formula (9) of the same appendix shows that

$$\vartheta_N(S/R) = \left(\frac{\partial(t_1,\ldots,t_n)}{\partial(x_1,\ldots,x_n)}\right) .$$

The following rule results immediately from the definition G.1:

G.4. Rule. If the ideal I is finitely generated (which is for instance the case, if S/R is essentially of finite type), then

$$\vartheta_N(S_T/R) = \vartheta_N(S/R)_T$$

for each multiplicatively closed subset $T \subset S$.

Next we define the **Dedekind different** whose construction requires some restrictive hypothesis.

G.5. Assumptions.

a) R is noetherian and any non-zerodivisor of R is also one of S.

b) There is a subalgebra T/R of S/R such that T/R is finite and the following conditions hold:

α) For each $\mathfrak{p} \in \mathrm{Spec}(S)$, $\bar{\mathfrak{p}} := \mathfrak{p} \cap T$, the canonical homomorphism $T_{\bar{\mathfrak{p}}} \to S_{\mathfrak{p}}$ is bijective.

β) The canonical homomorphisms $Q(R) \otimes_R T \to Q(T)$ and $Q(T) \to Q(S)$ are bijective.

Observe that S/T is flat by α), hence any non-zerodivisor of T is also one of S and $Q(T) \to Q(S)$ is defined.

The assumptions G.5 are satisfied, if S/R is an extension of domains and S is finite over R. More generally, condition α) of G.5b) holds by Zariski's Main Theorem, if S/R is of finite type and quasifinite (B.16).

We first draw some consequences from G.5.

Putting $K := Q(R)$ and $L := Q(T) = Q(S)$, we have

$$L = K \otimes_R T \subset K \otimes_R S \subset L$$

hence

(1) $$L = K \otimes_R T = K \otimes_R S.$$

By G.5b) the canonical map $\mathrm{Spec}(S) \to \mathrm{Spec}(T)$ is injective. If $\bar{\mathfrak{p}} := \mathfrak{p} \cap T$ for some $\mathfrak{p} \in \mathrm{Spec}(S)$ then $\mathrm{Spec}(S_{\bar{\mathfrak{p}}}) \to \mathrm{Spec}(T_{\bar{\mathfrak{p}}})$ is injective too. Since the composed map $T_{\bar{\mathfrak{p}}} \to S_{\bar{\mathfrak{p}}} \to S_{\mathfrak{p}}$ is bijective, we see that $\mathrm{Spec}(S_{\mathfrak{p}}) \to \mathrm{Spec}(S_{\bar{\mathfrak{p}}})$ is bijective, hence $S_{\bar{\mathfrak{p}}}$ is local and $S_{\bar{\mathfrak{p}}} = S_{\mathfrak{p}}$. Therefore

(2) $$T_{\bar{\mathfrak{p}}} = S_{\bar{\mathfrak{p}}}, \text{ if } \bar{\mathfrak{p}} = \mathfrak{p} \cap T \text{ for some } \mathfrak{p} \in \mathrm{Spec}(S).$$

By (1) there are canonical isomorphisms

$$\mathrm{Hom}_K(L,K) \cong K \otimes_R \mathrm{Hom}_R(T,R) \cong L \otimes_T \mathrm{Hom}_R(T,R).$$

The canonical map $\mathrm{Hom}_R(T,R) \to \mathrm{Hom}_K(L,K)$ is injective. We identify $\mathrm{Hom}_R(T,R)$ with its image in $\mathrm{Hom}_K(L,K)$ and denote by $S \cdot \mathrm{Hom}_R(T,R)$ the S-submodule of $\mathrm{Hom}_K(L,K)$ generated by $\mathrm{Hom}_R(T,R)$:

$$S \cdot \mathrm{Hom}_R(T,R) \cong S \otimes_T \mathrm{Hom}_R(T,R).$$

<u>G.6. Lemma.</u>   Let T'/R be another subalgebra of S/R such that G.5b) is satisfied for T'/R. Then G.5b) is satisfied for the subalgebra $T \cdot T'$ of S/R generated by T and T'. If $T \subset T'$, and if $f_{T'/T}$ denotes the conductor from T' to T, then

$$S \cdot f_{T'/T} = S.$$

Proof.   The assertion about $T \cdot T'$ is trivial. Assume now that $T \subset T'$. For $\mathfrak{p} \in \mathrm{Spec}(S)$ and $\bar{\mathfrak{p}} := \mathfrak{p} \cap T$ we have

$T_{\overline{\mathfrak{p}}} \subset T'_{\overline{\mathfrak{p}}} \subset S_{\overline{\mathfrak{p}}}$, hence

$$T_{\overline{\mathfrak{p}}} = T'_{\overline{\mathfrak{p}}} = S_{\mathfrak{p}}$$

by (2). Thus

$$S_{\mathfrak{p}} \cdot f_{T'/T} = T_{\overline{\mathfrak{p}}} \cdot f_{T'/T} = f_{T'_{\overline{\mathfrak{p}}}/T_{\overline{\mathfrak{p}}}} = T'_{\overline{\mathfrak{p}}} = S_{\mathfrak{p}}$$

for all $\mathfrak{p} \in \mathrm{Spec}(S)$, and $S \cdot f_{T'/T} = S$ follows.

G.7. <u>Corollary.</u> $S \cdot \mathrm{Hom}_R(T,R) = S \cdot \mathrm{Hom}_R(T',R)$.

Proof.   We may assume $T \subset T'$. By G.6 we have an equation

$$1 = \sum_{i=1}^{m} s_i f_i \qquad (s_i \in S, \ f_i \in f_{T'/T}).$$

Clearly $S \cdot \mathrm{Hom}_R(T',R) \subset S \cdot \mathrm{Hom}_R(T,R)$. On the other hand, each $\ell \in \mathrm{Hom}_R(T,R)$ may be written

$$\ell = \sum_{i=1}^{m} s_i \cdot (f_i \ell).$$

Since $f_i \ell \in \mathrm{Hom}_R(T',R)$ we obtain the desired equality.

G.8. <u>Definition.</u>   Under the assumptions G.5 suppose L/K is flat and a trace $\sigma : L \to K$ exists (appendix F, section b)). Write

$$S \cdot \mathrm{Hom}_R(T,R) = \mathcal{L}^{\sigma}_{S/R} \circ \sigma$$

with a fractional S-ideal $\mathcal{L}^{\sigma}_{S/R}$ (independent of T by G.7). $\mathcal{L}^{\sigma}_{S/R}$ is called the <u>Dedekind complementary module of S/R</u> with respect to σ and

$$\vartheta^{\sigma}(S/R) := \{x \in L \mid x \cdot \mathcal{L}^{\sigma}_{S/R} \subset S\}$$

is called the <u>Dedekind different of S/R with respect to σ.</u> If L/K is étale and $\sigma = \sigma_{L/K}$, we write $\mathcal{L}_{S/R}$ and $\vartheta_D(S/R)$ instead of $\mathcal{L}^{\sigma}_{S/R}$ and $\vartheta^{\sigma}(S/R)$.

G.9. <u>Examples.</u>   a) If S/R is finite and L/K is étale, then

$$\mathcal{L}_{S/R} = \{x \in L \mid \sigma_{L/K}(sx) \in R \text{ for all } s \in S\}.$$

This is the classical definition of the Dedekind complementary module and

$$\vartheta_D(S/R) = \{x \in L \mid x\mathcal{L}_{S/R} \subset S\}$$

is the classical Dedekind different.

If $L = K$, then $\sigma_{L/K} = \mathrm{id}_K$ and

$$\mathcal{L}_{S/R} = f_{S/R}$$

the conductor from S to R.

b) Assume S/R is finite and has a presentation

$$S = R[X_1,\ldots,X_n]/(t_1,\ldots,t_n) = R[x_1,\ldots,x_n]$$

as a complete intersection. Let $\tau_t^x : S \to R$ be the trace associated with this presentation (F.20), and let $\tau_t^x : L \to K$ also denote the extension of τ_t^x to L/K. Then $\mathrm{Hom}_R(S,R)$ is the submodule $S \cdot \tau_t^x$ of $\mathrm{Hom}_K(L,K) = L \cdot \tau_t^x$, hence

$$\mathcal{L}_{S/R}^{\tau_t^x} = S \text{ and } \vartheta^{\tau_t^x}(S/R) = S.$$

Under the assumptions of definition G.8 we have

$$\mathrm{Hom}_K(L,K) = L \cdot \sigma = L \cdot \mathrm{Hom}_R(T,R) = L \cdot \mathcal{L}_{S/R}^{\sigma} \circ \sigma$$

hence

(3) $$L \cdot \mathcal{L}_{S/R}^{\sigma} = L \text{ and } L \circ \vartheta^{\sigma}(S/R) = L.$$

If σ and σ' are two traces of L/K, then $\sigma' = x\sigma$ with a unit $x \in L$, and we see that

(4) $$\mathcal{L}_{S/R}^{\sigma'} = x \cdot \mathcal{L}_{S/R}^{\sigma} \text{ and } \vartheta^{\sigma}(S/R) = x \cdot \vartheta^{\sigma'}(S/R).$$

The complementary modules form a class of fractional S-ideals in L which we call the <u>canonical class of the</u>

algebra S/R. Similarly the $\vartheta^{\sigma}(S/R)$ form the <u>class of differents</u> of S/R.

<u>G.10. Proposition.</u>   Under the assumptions of definition G.8 suppose S/R is finite and flat. Then the following assertions are equivalent:

a) S/R is a Gorenstein algebra.

b) $\mathcal{L}^{\sigma}_{S/R}$ is an invertible ideal.

(see also exercise 1)).

Proof.   Since $\mathcal{L}^{\sigma}_{S/R} \cong \mathrm{Hom}_R(S,R)$, the proposition follows from E.16, because a fractional S-ideal I with $I \cdot L = L$ (see (3)) is invertible if and only if I is a projective S-module of rank 1.

In particular, the canonical class of a finite Gorenstein algebra S/R is a class of invertible ideals (the class of principal ideals in case R is local).

Let us now compare the classical Dedekind different with the Noether different.

<u>G.11. Theorem</u> (E.Noether [No]).   Let S/R be finite, assume R is noetherian, and each non-zerodivisor of R is a non-zero-divisor of S. If L/K is étale, then $\vartheta_D(S/R)$ is defined (G.9a) and we have

a) $\vartheta_N(S/R) \subset \vartheta_D(S/R)$.

b) If S/R is projective, then

$$\vartheta_N(S/R) = \vartheta_D(S/R).$$

Proof.   Let J be the kernel of $\mu : L \otimes_K L \to L$. By F.9 there is an isomorphism of L-modules

$$\phi \;:\; \mathrm{Ann}_{L\otimes_K L}(J) \;\xrightarrow{\;\sim\;}\; \mathrm{Hom}_L(\mathrm{Hom}_K(L,K),L)$$

Let

$$\psi \;:\; \mathrm{Hom}_L(\mathrm{Hom}_K(L,K),L) \to L$$

be the isomorphism that associates with each L-linear form

$\varphi : \mathrm{Hom}_K(L,K) \to L$ the element $\varphi(\sigma_{L/K})$. Then with the notation

of F.10 we have by F.13

$$\psi(\phi(\Delta_{\sigma_{L/K}})) \;=\; \phi(\Delta_{\sigma_{L/K}})(\sigma_{L/K}) \;=\; 1 \;=\; \mu(\Delta_{\sigma_{L/K}}).$$

This shows that

$$\psi \circ \phi \;=\; \mu\big|_{\mathrm{Ann}(J)}\,.$$

With $I := \ker(S \otimes_R S \to S)$ there is a natural map

$\mathrm{Ann}_{S\otimes_R S}(I) \to \mathrm{Ann}_{L\otimes_K L}(J)$, and by the definition of $\vartheta_N(S/R)$ we

have

$$(\psi \circ \phi)(\mathrm{Ann}(I)) \;=\; \vartheta_N(S/R).$$

On the other hand, ψ maps the submodule

$\mathrm{Hom}_S(\mathrm{Hom}_R(S,R),S) = \mathrm{Hom}_S(\mathcal{L}_{S/R}\cdot\sigma_{L/K},S)$ of $\mathrm{Hom}_L(\mathrm{Hom}_K(L,K),L)$

onto $\vartheta_D(S/R)$, since $\vartheta_D(S/R) = S \underset{L}{:} \mathcal{L}_{S/R} \cong \mathrm{Hom}_S(\mathcal{L}_{S/R},S)$.

a) For each $x \in \mathrm{Ann}(I)$ the L-linear form $\phi(x) : \mathrm{Hom}_K(L,K) \to L$

maps $\mathrm{Hom}_R(S,R)$ into S (see the definition of ϕ in appendix F,

section b), formula (2)). Hence

$$\vartheta_N(S/R) \;=\; \psi(\phi(\mathrm{Ann}(I))) \;\subset\; \psi(\mathrm{Hom}_S(\mathrm{Hom}_R(S,R),S)) \;=\; \vartheta_D(S/R).$$

b) If S/R is projective, then $\phi : \mathrm{Ann}_{S\otimes_R S}(I) \to \mathrm{Hom}_S(\mathrm{Hom}_R(S,R),S)$

is bijective by F.9, and we obtain equality.

<u>G.12. Corollary.</u>   If S/R has a presentation

$$S = R[X_1,\ldots,X_n]/(t_1,\ldots,t_n) = R[x_1,\ldots,x_n]$$

as a complete intersection, then

$$\vartheta_D(S,R) = \vartheta_N(S/R) = (\frac{\partial(t_1,\ldots,t_n)}{\partial(x_1,\ldots,x_n)}).$$

This follows from G.11 and G.3.

G.13. Proposition.   Let S/R and R/P be finite algebras where P is noetherian. Assume that any non-zerodivisor of P (of R) is also one of R (of S). With $K := Q(P)$, $L := Q(R)$, and $M := Q(S)$, suppose that M/L and L/K are étale. Then

a) $\mathcal{L}_{S/P} = \{z \in M \mid \sigma_{M/L}(sz) \in \mathcal{L}_{R/P}$ for all $s \in S\}$

b) If $\mathcal{L}_{R/P}$ is an invertible R-ideal (e.g. R/P is a Gorenstein algebra (G.10) or R is a Dedekind domain), then

$$\mathcal{L}_{S/P} = \mathcal{L}_{S/R}\cdot\mathcal{L}_{R/P} \text{ and } \vartheta_D(S/P) = \vartheta_D(S/R)\cdot\vartheta_D(R/P).$$

Proof.   a) For $z \in \mathcal{L}_{S/P}$, $s \in S$, and $r \in R$, we have $\sigma_{M/K}(rsz) = \sigma_{L/K}(r\sigma_{M/L}(sz)) \in P$, hence $\sigma_{M/L}(sz) \in \mathcal{L}_{R/P}$. Conversely, if $\sigma_{M/L}(sz) \in \mathcal{L}_{R/P}$ for some $z \in M$ and all $s \in S$, then $\sigma_{M/K}(sz) = \sigma_{L/K}(\sigma_{M/L}(sz)) \in P$, hence $z \in \mathcal{L}_{S/P}$.

b) We may assume P is local, since the complementary module $\mathcal{L}_{R/P}$ is clearly compatible with localization in P. Then $\mathcal{L}_{R/P} = c\cdot R$ with a unit $c \in L$. By a)

$$\mathcal{L}_{S/P} = \{z\in M \mid \sigma_{M/L}(s\cdot\frac{z}{c})\in R \text{ for all } s\in S\} = c\cdot\mathcal{L}_{S/R} = \mathcal{L}_{S/R}\cdot\mathcal{L}_{R/P}.$$

The transitive formula for the different follows by taking inverses.

G.14. Corollary.   Suppose $M = L$ and $f_{S/R}$ is the conductor from S to R. Then

a) $f_{S/R}\cdot\mathcal{L}_{R/P} \subset \mathcal{L}_{S/P} \subset \mathcal{L}_{R/P}.$

b) If $\mathcal{L}_{R/P}$ is an invertible R-ideal, then

$$\mathcal{L}_{S/P} = f_{S/R} \cdot \mathcal{L}_{R/P}.$$

c) (Dedekind's formula for conductor and different).

If $\mathcal{L}_{R/P}$ is an invertible R-ideal and $\mathcal{L}_{S/P}$ is an invertible S-ideal, then

$$f_{S/R} \cdot \vartheta_D(S/P) = S \cdot \vartheta_D(R/P).$$

Proof. a) results immediately from G.13a). As for b), observe that $\mathcal{L}_{S/R} = f_{S/R}$, since M = L. Then apply G.13b). To prove c) we may assume that $\mathcal{L}_{S/P} = S \cdot a$ and $\mathcal{L}_{R/P} = R \cdot b$ are prinicpal ideals generated by units a ∈ M, b ∈ L. By b) we have $S \cdot b^{-1} = f_{S/R} \cdot a^{-1}$, which is the formula of c).

Exercises

1) Under the assumptions of G.9a), suppose R is normal. Show that $\vartheta_D(S/R) \subset S$.

2) Under the assumptions of G.10, S/R is a Gorenstein algebra if and only if $\vartheta^\sigma(S/R)$ is an invertible ideal (Apply exercise 3) of appendix E).

3) Give an example of an algebra S/R for which $\vartheta_D(S/R)$ is defined and $\vartheta_N(S/R) \neq \vartheta_D(S/R)$.

BIBLIOGRAPHY

[EGA]   GROTHENDIECK,A.;DIEUDONNÉ;J.: Eléments de géomé-
          trie algébrique. Publ.Math.IHES 4(1960),8(1961),
          11(1961),17(1963),20(1964),24(1965),28(1966),
          32(1967)

[Ab]   ABHYANKAR,S.: Two Notes on Formal Power Series.
          Proc.AMS 7, 903-905 (1956)

[An]   ANGÉNIOL,B.: Familles de cycles algébriques-Schéma
          de Chow. Springer Lecture Notes in Math.896 (1981)

[And]   ANDRÉ,M.: Homologie des algèbres commutatives.
          Springer, Berlin-Heidelberg-New York (1974)

[Ang]   ANGERMÜLLER,G.: On Some Conditions for a Polynomial
          Map with Constant Jacobian to be Invertible .
          Arch.Math. 40, 415-420 (1983)

[Av]   AVRAMOV,L.: Local Flat Extensions and Complete
          Intersection Defects. Math.Ann. 228, 27-37 (1977)

$[B_1]$ BOURBAKI,N.: Algèbre. Hermann. Paris

$[B_2]$ BOURBAKI,N.: Algèbre commutative. Hermann. Paris

[BCW]   BASS,H.;CONNELL,E.;WRIGHT,D.: The Jacobian Con-
          jecture: Reduction of Degree and Formal Expansion
          of the Inverse. Bull.AMS 7, 287-330 (1982)

[Ba]   BASSEIN,R.: On Smoothable Curve Singularities:
          Local Methods. Math.Ann. 230, 273-277 (1977)

[Be]   BERGER,R.: Differentialmoduln eindimensionaler
          lokaler Ringe. Math.Z. 81, 326-354 (1963)

[BG]   BUCHWEITZ,R.;GREUEL,G.M.: The Milnor Number and
          Deformations of Complexe Curve Singularities.
          Invent.Math. 52, 241-281 (1980)

[BK]   BERGER,R.;KUNZ,E.: Über die Struktur der Diffe-
          rentialmoduln von diskreten Bewertungsringen.
          Math.Z. 78, 97-115 (1962)

[BKKN]   BERGER,R.;KIEHL,R.;KUNZ,E.;NASTOLD,H.J.: Diffe-
          rentialrechnung in der analytischen Geometrie.
          Springer Lect.Notes in Math. 38 (1967)

$[BR_1]$ BREZULEANU,A.;RADU,N.: Excellent Rings and Good
          Separation of the Module of Differentials. Rev.
          Roum.Math.Pures et Appl. 23, 1455-1470 (1978)

[BR_2] BREZULEANU,A.;RADU,N.: Lectii de Algebra III.
Algebra Locala. (Roumainian) Universitatea din
Bucuresti (1982)

[Bru] BRUNS,W.: Zur Reflexivität analytischer Differentialmoduln. J.reine ang. Math. 227, 63-73 (1975)

[BV] BRUNS,W.;VETTER,U.: Zur Längenberechnung der
Torison äußerer Potenzen. Manuscr.Math. 14,
337-348 (1975).

[C] CARTIER,P.: Questions de rationalité de diviseurs
en géométrie algébrique. Bull.Soc.Math.France 86,
177-251 (1958)

[Ch] CHEVALLEY,C.: Introduction to the Theory of Algebraic Functions of One Variable. Amer.Math.Soc.
Surveys. New York (1951)

[DG] DIEUDONNÉ,J.;GROTHENDIECK,A.: Critères différen-
tiels de régularité pour les localisés des
algèbres analytiques. J.Alg. 5, 305-324 (1967)

[Ev] EVANS,G.: A Generalization of Zariski's Main
Theorem. Proc.AMS 26, 45-48 (1970)

[Fe] Ferrand,D.: Suite régulières et intersection
complète. C.R.Acad.Sci.Paris 264, 427-428 (1967)

[Fi] FITTING,H.: Die Determinantenideale eines Moduls.
Jber.Deutsch.Math.-Verein.46, 195-228 (1936)

[Fl] FLENNER,H.: Die Sätze von Bertini für lokale
Ringe. Math.Ann. 229, 253-294 (1977)

[G] GANONG,R.: Plane Frobenius Sandwiches.
Proc.AMS 84, 474-478 (1982)

[Gu] GULLIKSEN,T.H.: A Homological Characterization
of Local Complete Intersections. Compositio math.
23, 251-255 (1971)

[Ha] HARPER,L.: On Differentiably Simple Algebras.
Transa. AMS 100, 63-72 (1960)

[He_1] HERZOG,J.: Ein Cohen-Macaulay-Kriterium mit An-
wendungen auf den Konormalenmodul und den Diffe-
rentialmodul. Math.Z. 163, 149-162 (1978)

[He_2] HERZOG,J.: Deformationen von Cohen-Macaulay-Alge-
bren. J.reine ang.Math. 318, 83-105 (1980)

[He$_3$]   HERZOG,J.: Homological Properties of the Module
         of Differentials. Atas da 6^a Escola de Álgebra
         14, Soc. Brasiliera de Mat., 33-64 (1981)

[HK]   HERZOG,J.;KUNZ,E. (Ed.): Der kanonische Modul
         eines Cohen-Macaulay-Rings. Springer Lecture Notes
         in Math. 238 (1971)

[HW]   HERZOG,J.;WALDI,R.: Differentials of Linked Curve
         Singularities. Arch.Math. 42, 335-343 (1984)

[Ho]   HOCHSTER,M.: The Zariski-Lipman Conjecture in the
         Graded Case. J.Alg. 47, 411-424 (1977)

[I]   IVERSEN,B.: Generic Local Structure of the
         Morphisms in Commutative Algebra. Springer Lect.
         Notes in Math. 310 (1973).

[Kä]   KÄHLER,E.: Algebra und Differentialrechnung.
         Bericht über die Mathematikertagung in Berlin 1953

[Ki]   KIEHL,R.: Ausgezeichnete Ringe in der nichtarchi-
         medischen analytischen Geometrie. J.reine ang.
         Math. 234, 89-98 (1969)

[KK]   KIEHL,R.;KUNZ,E.: Vollständige Durchschnitte und
         p-Basen. Arch.Math. 16, 348-362 (1965)

[KN$_1$]   KIMURA,T.;NIITSUMA,H.: On Kunz's Conjecture.
         J.Math.Soc.Japan 34, 371-378 (1982)

[KN$_2$]   KIMURA,T.;NIITSUMA,H.: Differential Basis and
         p-Basis in a Regular Local Ring. Proc. AMS 92,
         335-338 (1984)

[Ko]   KOCH,J.: Über die Torsion des Differentialmoduls
         von Kurvensingularitäten. Regensburger Math.
         Schriften 5 (1983)

[Ku]   KUNZ,E.: Introduction to Commutative Algebra and
         Algebraic Geometry. Birkhäuser, Boston (1985)

[L]   LANG,S.: Introduction to Algebraic and Abelian
         Functions. Addison-Wesley, Reading Mass. (1972)

[Le]   LECH,Chr.: Inequalities Related to Certain Couples
         of Local Rings. Acta Math. 112, 69-89 (1964)

[Lin]   LINDEL,H.: On a Question of Bass, Quillen and
         Suslin Concerning Projective Modules over Poly-
         nomial Rings. Invent.Math. 65, 319-323 (1981)

[Lip$_1$]   LIPMAN,J.: Free Derivation Modules on Algebraic
Varieties. Amer.J.Math.87, 874-898 (1965)

[Lip$_2$]   LIPMAN,J.: On the Jacobian Ideal of the Module
of Differentials. Proc. AMS 21, 422-426 (1969)

[Lip$_3$]   LIPMAN,J.: Dualizing Sheaves, Differentials and
Residues on Algebraic Varieties. Astérisque 117
(1984)

[Lip$_4$]   LIPMAN,J.: Residues and Traces of Differential
Forms via Hochschild Homology.
Preprint (1985)

[M$_1$]   MATSUMURA,H.: Commutative Algebra
(sec.ed.) Benjamin, Reading Mass. (1980)

[M$_2$]   MATSUMURA,H.: A Proof of a Conjecture of E.Kunz.
Preprint (1983)

[MK]   MOHAN KUMAR,N.: Complete Intersections.
J.Math.Kyoto 17, 533-538 (1977)

[N]   NASTOLD,H.J.: Zum Dualitätssatz in inseparablen
Funktionenkörpern der Dimension 1. Math.Z. 76,
75-84 (1961)

[No]   NOETHER,E.: Idealdifferentiation und Differente.
J.reine ang.Math. 188, 1-21 (1950)

[Pe]   PESKINE,Chr.: Une généralization du "main theorem"
de Zariski. Bull.Sci.Math. 90, 116-127 (1966)

[Pi]   PINKHAM,H.: Deformations of Algebraic Varieties
with G$_m$-action. Astérisque 20 (1974)

[Pl$_1$]   PLATTE,E.: Ein elementarer Beweis des Zariski-
Lipman-Problems für graduierte analytische Algebren. Arch.Math. 31, 143-145 (1978)

[Pl$_2$]   PLATTE,E.: Differentielle Eigenschaften der Invarianten regulärer Algebren. J.Alg.62,1-12 (1980)

[Pl$_3$]   PLATTE,E.: Zur endlichen homologischen Dimension
von Differentialmoduln. Manuscr.math. 32, 295-302
(1980)

[Ray]   RAYNAUD,M.: Anneaux locaux henséliens. Springer
Lect.Notes in Math.169 (1970)

[RS]   RUDAKOV,A.N.;SHAFAREVITCH,I.R.: Inseparable
Morphisms of Algebraic Surfaces. Math.USSR
Isvestija 10, 1205-1237 (1976)

[S]   SERRE,J.P.: Groupes algébriques et corps de
           classes. Hermann, Paris (1959)

[Sch]   SCHEJA,G.: Differentialmoduln lokaler analytischer
           Algebren. Schriftenreihe Math.Inst.Univ. Fribourg
           (Suisse) Nr. 2 (1969/7o)

[Se$_1$]   SEIDENBERG,A.: Derivations and Integral Closure.
           Pac.J.Math. 16, 167-173 (1966)

[Se$_2$]   SEIDENBERG,A.: Differential Ideals in Rings of
           Finitely Generated Type. Am.J.Math.89, 22-42 (1967)

[SS$_1$]   SCHEJA,G.;STORCH,U.: Lokale Verzweigungstheorie.
           Schriftenreihe Math.Inst.Univ.Fribourg (Suisse)
           Nr. 5 (1974)

[SS$_2$]   SCHEJA,G.;STORCH,U.: Differentielle Eigenschaften
           der Lokalisierungen analytischer Algebren.
           Math.Ann. 197, 137-170 (1970)

[SS$_3$]   SCHEJA,G.;STORCH,U.: Über Spurfunktionen bei voll-
           ständigen Durchschnitten. J.reine angew.Math.
           278/279, 174-190 (1975)

[SS$_4$]   SCHEJA,G.;STORCH,U.: Residuen bei vollständigen
           Durchschnitten. Math.Nachr. 91, 157-170 (1979)

[T$_1$]   TATE,J.: Genus Change in Inseparable Extensions
           of Function Fields. Proc.AMS 3, 400-406 (1952)

[T$_2$]   TATE,J.: Residues of Differentials on Curves.
           Ann.Scient.Ec.Norm.Sup. 1, 149-159 (1968)

[U]   ULRICH,B.: Torsion des Differentialmoduls und Ko-
           tangentenmodul von Kurvensingularitäten.
           Arch.Math. 36, 510-523 (1981)

[Va$_1$]   VASCONCELOS,W.: Ideals Generated by R-Sequences.
           J.Alg. 6, 309-316 (1967)

[Va$_2$]   VASCONCELOS,W.: On the Homology of I/I^2. Comm.in
           Alg. 6, 1801-1809 (1978)

[Ve]   VETTER,U.: Äußere Potenzen von Differentialmoduln
           reduzierter vollständiger Durchschnitte. Manuscr.
           Math. 2, 67-75 (1970)

[vdW]   VAN DER WAERDEN: Algebra I. Springer,Berlin-
           Heidelberg-New York

[Wa]   WALDI,R.: Zur Konstruktion von Weierstraßpunkten
        mit vorgegebener Halbgruppe. Manuscr.Math. 30,
        257-278 (1980)
[Wi]   WIEBE,H.: Über homologische Invarianten lokaler
        Ringe. Math.Ann. 179, 257-274 (1969)
$[Y_1]$ YUAN,S.: Differentiably Simple Rings of Prime
        Characteristic. Duke Math.J. 31, 623-630 (1964)
$[Y_2]$ YUAN,S.: Inseparable Galois Theory of Exponent
        One. Transa.AMS 149, 163-170 (1970)

SYMBOL INDEX

The numbers indicate on which page the symbol is defined.

SUBJECT INDEX